Biotic and Abiotic Stress in Plants

Biotic and Abiotic Stress in Plants

Edited by

Peyton Turner

Biotic and Abiotic Stress in Plants
Edited by Peyton Turner
ISBN: 978-1-63549-051-0 (Hardback)

Published by Larsen and Keller Education,
5 Penn Plaza,
19th Floor,
New York, NY 10001, USA

Cataloging-in-Publication Data

Biotic and abiotic stress in plants / edited by Peyton Turner.
p. cm.
Includes bibliographical references and index.
ISBN 978-1-63549-051-0
1. Plants--Effect of stress on. 2. Plants--Effect of stress on--Genetic aspects. 3. Plant physiology.
4. Plant genomes. I. Turner, Peyton.
QK754 .B56 2017
581.7--dc23

Printed and bound in the United States of America.

For more information regarding Larsen and Keller Education and its products, please visit the publisher's website www.larsen-keller.com

Table of Contents

Preface

This book provides comprehensive insights into the field of plant biology. It talks in detail about the biotic and abiotic stress in plants. It outlines the processes and applications of the subject in detail. Abotic stress refers to negative impact of non-living things on plants and animals whereas, biotic stress refers to the living organisms which cause problems for other living beings like weeds or harmful insects. The topics covered in this extensive book deal with the core subjects of biotic and abiotic stress in plants. For all those who are interested in this subject, the textbook can prove to be an essential guide. It attempts to assist those with a goal of dealing into the field of biotic and abiotic stress.

A detailed account of the significant topics covered in this book is provided below:

Chapter 1- Living organisms, such as viruses, parasites and fungi, usually damage plants. The stress caused by this damage is known as biotic stress. Biotic stress, unlike abiotic stress is not caused by non-living factors such as wind, sunlight and drought. This section will provide an integrated understanding of biotic stress.

Chapter 2- The various biotic stressors are plant viruses, viroids, insects, herbivores and parasitic plants. The virus that affects plants is known as plant virus whereas the infection that occurs in the plants is referred to as viroid. Likewise this text also explains to the reader, stressors such as insects, herbivores and parasitic plants. The chapter strategically encompasses and incorporates the major components and key concepts of biotic stress, providing a complete understanding.

Chapter 3- Oomycetes are absorptive organisms that reproduce sexually and asexually. These cause devastating diseases in plants, such as late blight of potatoes and the sudden death of oak trees. Oomycete has numerous types; some of these are phytophthora, Phytophthora infestans, Phytophthora cinnamon and Phytophthora capsici. The major categories of oomycete are dealt with great detail in the chapter.

Chapter 4- Fungus is among the most widely distributed organisms on Earth. They are in large numbers and can be found in every part of our world. The types of fungus discussed in this section are ascomycota, Sclerotinia sclerotiorum, Magnaporthe grisea, sclerotium and armillaria. This text is an overview of the subject matter incorporating all the major aspects of fungus.

Chapter 5- Bacteria can be found in a number of shapes and sizes. They are usually a few micrometers in length. It is the oldest habitant of this world and is very important for the process of recycling nutrients. Some of the bacteria discussed in the content are beet vascular necrosis, phytoplasma, Rhodococcus fascians, Agrobacterium tumefaciens, xanthomonas etc. The chapter provides the reader with an in-depth understanding on bacteria.

Chapter 6- Abiotic stress is the negative stress caused by non-living factors. Abiotic stress causes the most damage to the growth and the productivity of crops across the globe. In agriculture, abiotic stress causes stress by natural environment factors such as high winds, droughts and floods. This chapter elucidates all the factors related to abiotic stress.

Chapter 7- The basic abiotic stressors are wildfire and drought. Wildfire is the fire that occurs either in forests or in rural areas whereas droughts are the shortage of water supply in a particular area. Droughts can last for months or even years, and it certainly has substantial impact on the ecosystem. The text discusses the major stressors of abiotic stress in critical manner providing key analysis to the subject matter.

Chapter 8- The study of the diseases found in plants is known as plant pathology. The diseases discussed in this section are blackleg (potatoes), soybean rust, clubroot, citrus canker and cherry X disease. The topics elaborated in this chapter will help in gaining a better perspective about the effects of biotic and abiotic stress.

Chapter 9- Precautions need to be taken to prevent plants from biotic as well as abiotic stress. Some of these conservation methods are pesticide, insecticide, herbicide, fungicide and bactericide. Insecticides are used to kill insects, whereas pesticides are meant for attracting and then destroying any pests. This chapter discusses in detail the conservation methods of biotic and abiotic stress.

It gives me an immense pleasure to thank our entire team for their efforts. Finally in the end, I would like to thank my family and colleagues who have been a great source

Editor

1

Introduction to Biotic Stress

Living organisms, such as viruses, parasites and fungi, usually damage plants. The stress caused by this damage is known as biotic stress. Biotic stress, unlike abiotic stress is not caused by non-living factors such as wind, sunlight and drought. This section will provide an integrated understanding of biotic stress.

Biotic Stress

Biotic stress is stress that occurs as a result of damage done to plants by other living organisms, such as bacteria, viruses, fungi, parasites, beneficial and harmful insects, weeds, and cultivated or native plants. Not to be confused with abiotic stress, which is the negative impact of non-living factors on the organisms in a specific environment such as sunlight, wind, salinity, over watering and drought. The types of biotic stresses imposed on a plant depend on both geography and climate and on the host plant and its ability to resist particular stresses. Although there are many kinds of biotic stress, the majority of plant diseases are caused by fungi. Biotic stress remains a broadly defined term and those who study it face many challenges, such as the greater difficulty in controlling biotic stresses in an experimental context compared to abiotic stress.

The damage caused by these various living and nonliving agents can appear very similar.Even with close observation, accurate diagnosis can be difficult. For example, browning of leaves on an oak tree caused by drought stress may appear similar to leaf browning caused by oak wilt, a serious vascular disease, or the browning cause by anthracnose, a fairly minor leaf disease.

Agriculture

It is a major focus of agricultural research, due to the vast economic losses caused by biotic stress to cash crops. The relationship between biotic stress and plant yield affects economic decisions as well as practical development. The impact of biotic injury on crop yield impacts population dynamics, plant-stressor coevolution, and ecosystem nutrient cycling.

Biotic stress also impacts horticultural plant health and natural habitats ecology. It also has dramatic changes in the host recipient.Plants are exposed to many stress factors, such as drought, high salinity or pathogens, which reduce the yield of the cultivated plants or affect the quality of the harvested products. *Arabidopsis thaliana* is often used as a model plant to study the responses of plants to different sources of stress.

In History

Biotic stresses have had huge repercussions for humanity; an example of this is the potato blight, an oomycete which caused widespread famine in England, Ireland and Belgium in the 1840s. Another example is grape phylloxera coming from North America in the 19th century, which led to the Great French Wine Blight.

Today

Losses to pests and disease in crop plants continue to pose a significant threat to agriculture and food security. During the latter half of the 20th century, agriculture became increasingly reliant on synthetic chemical pesticides to provide control of pests and diseases, especially within the intensive farming systems common in the developed world. However, in the 21st century, this reliance on chemical control is becoming unsustainable. Pesticides tend to have a limited lifespan due to the emergence of resistance in the target pests, and are increasingly recognised in many cases to have negative impacts on biodiversity, and on the health of agricultural workers and even consumers.

Tomorrow

Due to the implications of climate change, it is suspected that plants will have increased susceptibility to pathogens. Additionally, elevated threat of abiotic stresses (i.e. drought and heat) are likely to contribute to plant pathogen susceptibility.

Effect on Plant Growth

Photosynthesis

Many biotic stresses affect photosynthesis, as chewing insects reduce leaf area and virus infections reduce the rate of photosynthesis per leaf area. Vascular- wilt fungi compromise the water transport and photosynthesis by inducing stomata closure.

Response to Stress

Plants have co-evolved with their parasites for several hundred million years. This co-evolutionary process has resulted in the selection of a wide range of plant defences against microbial pathogens and herbivorous pests which act to minimise frequency and impact of attack. These defences include both physical and chemical adaptations, which may either be expressed constitutively, or in many cases, are activated only in response to attack. For example, utilization of high metal ion concentrations derived from the soil allow plants to reduce the harmful effects of biotic stressors (pathogens, herbivores etc.); meanwhile preventing the infliction of severe metal toxicity by way of safeguarding metal ion distribution throughout the plant with protective physiological pathways. Such induced resistance provides a mechanism whereby the costs of defence are avoided until defense is beneficial to the plant. At the same time, successful pests and pathogens have evolved mechanisms to overcome both constitutive and induced resistance in their particular host species. In order to fully understand and manipulate plant biotic stress resistance, we require a detailed knowledge of these interactions at a wide range of scales, from the molecular to the community level.

Cross Tolerance with Abiotic Stress

- Evidence shows that a plant undergoing multiple stresses, both abiotic and biotic (usually pathogen or herbivore attack), can produce a positive effect on plant performance, by reducing their susceptibility to biotic stress compared to how they respond to individual stresses. The interaction leads to a crosstalk between their respective hormone signalling pathways which will either induce or antagonize another restructuring genes machinery to increase tolerance of defense reactions.

- Reactive oxygen species (ROS) are key signalling molecules produced in response to biotic and abiotic stress cross tolerance. ROS are produced in response to biotic stresses during the oxidative burst.
- Dual stress imposed by ozone (O_3) and pathogen affects tolerance of crop and leads to altered host pathogen interaction (Fuhrer, 2003). Alteration in pathogenesis potential of pest due to O_3 exposure is of ecological and economical importance.
- Tolerance to both biotic and abiotic stresses has been achieved. In maize, breeding programmes have led to plants which are tolerant to drought and have additional resistance to the parasitic weed *Striga hermonthica.*

Remote Sensing

The Agricultural Research Service (ARS) and various government agencies and private institutions have provided a great deal of fundamental information relating spectral reflectance and thermal emittance properties of soils and crops to their agronomic and biophysical characteristics. This knowledge has facilitated the development and use of various remote sensing methods for non-destructive monitoring of plant growth and development and for the detection of many environmental stresses that limit plant productivity. Coupled with rapid advances in computing and position locating technologies, remote sensing from ground-, air-, and space-based platforms is now capable of providing detailed spatial and temporal information on plant response to their local environment that is needed for site specific agricultural management approaches. This is very important in today's society because with increasing pressure on global food productivity due to population increase, result in a demand for stress-tolerant crop varieties that has never been greater.

Biotic Component

Biotic components are the living things that shape an ecosystem.

Biotic components usually include:

- Producers, i.e. autotrophs: e.g. plants, they convert the energy [from photosynthesis (the transfer of sunlight, water, and carbon dioxide into energy), or other sources such as hydrothermal vents] into food.
- Consumers, i.e. heterotrophs: e.g. animals, they depend upon producers (occasionally other consumers) for food.
- Decomposers, i.e. detritivores: e.g. fungi and bacteria, they break down chemicals from producers and consumers (usually dead) into simpler form which can be reused.

A biotic factor is any living component that affects the population of another organism, or the environment. This includes animals that consume the organism, and the living food that the organism consumes. Biotic factors also include human influence, pathogens and disease outbreaks. Each biotic factor needs energy to do work and food for proper growth.

All species are influenced by biotic factors in one way or another. For example, If the number of predators will increase, the whole food web will be affected (the population number of organisms that are lower in the food web will decrease). Similarly, when organisms have more food to eat, they will grow quicker and will be more likely to reproduce, so the population size will obviously increase. Pathogens and disease outbreaks, however, are most likely to cause a decrease in population size. Humans make the most sudden changes in an environment (e.g. building cities and factories, disposing of waste into the water). These changes are most likely to cause a decrease in the population of any species, due to the sudden appearance of pollutants.

Biotic components are contrasted to abiotic components, which are non-living components that influence population size and the environment. Examples of abiotic factors are: temperature, light intensity, moisture and water levels, air currents, carbon dioxide levels and the pH of water and soil.

The factors mentioned above may either cause an increase or a decrease in population size, depending on the organism. For example, rainfall may encourage the growth of new plants, but too much of it may cause flooding, which may drastically decrease the population size.

Different Types of Biotic Components

Autotroph

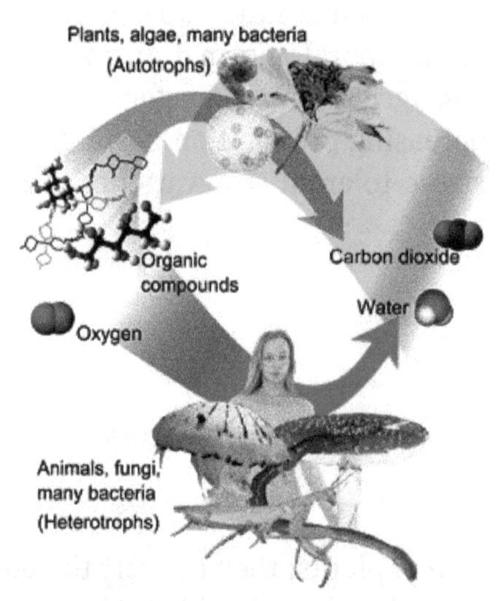

Overview of cycle between autotrophs and heterotrophs. Photosynthesis is the main means by which plants, algae and many bacteria produce organic compounds and oxygen from carbon dioxide and water (green arrow).

An autotroph ("self-feeding", from the Greek autos "self" and trophe "nourishing") or producer, is an organism that produces complex organic compounds (such as carbohydrates, fats, and proteins) from simple substances present in its surroundings, generally using energy from light (photosynthesis) or inorganic chemical reactions (chemosynthesis). They are the producers in a food chain, such as plants on land or algae in water, in contrast to heterotrophs as consumers of autotrophs. They do not need a living source of energy or organic carbon. Autotrophs can reduce carbon dioxide to make organic compounds for biosynthesis and also create a store of chemical energy. Most autotrophs use water as the reducing agent, but some can use other hydrogen compounds such as hydrogen sulfide. Phototrophs (green plants and algae), a type of autotroph, convert electromagnetic energy from sunlight into chemical energy in the form of reduced carbon.

Autotrophs can be photoautotrophs or chemoautotrophs. Phototrophs use light as an energy source, while chemotrophs utilize electron donors as a source of energy, whether from organic or inorganic sources; however in the case of autotrophs, these electron donors come from inorganic chemical sources. Such chemotrophs are lithotrophs. Lithotrophs use inorganic compounds, such as hydrogen sulfide, elemental sulfur, ammonium and ferrous iron, as reducing agents for biosynthesis and chemical energy storage. Photoautotrophs and lithoautotrophs use a portion of the ATP produced during photosynthesis or the oxidation of inorganic compounds to reduce NADP to NADPH to form organic compounds.

History

The term was coined by Albert Bernhard Frank in 1892.

Variants

Some organisms rely on organic compounds as a source of carbon, but are able to use light or inorganic compounds as a source of energy. Such organisms are not defined as autotrophic, but rather as heterotrophic. An organism that obtains carbon from organic compounds but obtains energy from light is called a *photoheterotroph,* while an organism that obtains carbon from organic compounds but obtains energy from the oxidation of inorganic compounds is termed a *chemoheterotroph, chemolithoheterotroph,* or *lithoheterotroph.*

Evidence suggests that some fungi may also obtain energy from radiation. Such radiotrophic fungi were found growing inside a reactor of the Chernobyl nuclear power plant.

Flowchart to determine if a species is autotroph, heterotroph, or a subtype

Ecology

Green fronds of a maidenhair fern, a photoautotroph

Autotrophs are fundamental to the food chains of all ecosystems in the world. They take energy from the environment in the form of sunlight or inorganic chemicals and use it to create energy-rich molecules such as carbohydrates. This mechanism is called primary production. Other organisms, called heterotrophs, take in autotrophs as food to carry out functions necessary for their life. Thus, heterotrophs — all animals, almost all fungi, as well as most bacteria and protozoa — depend on autotrophs, or primary producers, for the energy and raw materials they need. Heterotrophs obtain energy by breaking down organic molecules (carbohydrates, fats, and proteins) obtained in food. Carnivorous organisms rely on autotrophs indirectly, as the nutrients obtained from their heterotroph prey come from autotrophs they have consumed.

Most ecosystems are supported by the autotrophic primary production of plants that capture photons initially released by the sun. The process of photosynthesis splits a water molecule (H_2O), releasing oxygen (O_2) into the atmosphere, and reducing carbon dioxide (CO_2) to release the hydrogen atoms that fuel the metabolic process of primary production. Plants convert and store the energy of the photon into the chemical bonds of simple sugars during photosynthesis. These plant sugars are polymerized for storage as long-chain carbohydrates, including other sugars, starch, and cellulose; glucose is also used to make fats and proteins. When autotrophs are eaten by heterotrophs, i.e., consumers such as animals, the carbohydrates, fats, and proteins contained in them become energy sources for the heterotrophs. Proteins can be made using nitrates, sulfates, and phosphates in the soil.

Heterotroph

A heterotroph is an organism that cannot fix carbon and uses organic carbon for growth. Heterotrophs can be further divided based on how they obtain energy; if the heterotroph uses light for energy, then it is considered a photoheterotroph, while if the heterotroph uses chemical energy, it is considered a chemoheterotroph.

Heterotrophs contrast with autotrophs, such as plants and algae, which can use energy from sunlight (photoautotrophs) or inorganic compounds (lithoautotrophs) to produce organic compounds such as carbohydrates, fats, and proteins from inorganic carbon dioxide. These reduced carbon compounds can be used as an energy source by the autotroph and provide the energy in food consumed by heterotrophs. Ninety-five percent or more of all types of living organisms are heterotrophic, including all animals and fungi and most bacteria and protists.

History

The term was coined by Albert Bernhard Frank in 1892.

Types

Organotrophs exploit reduced carbon compounds as energy sources, like carbohydrates, fats, and proteins from plants and animals. Photoorganoheterotrophs such as Rhodospirillaceae and purple non-sulfur bacteria synthesize organic compounds by utilization of sunlight coupled with oxidation of inorganic substances, including hydrogen sulfide, elemental sulfur, thiosulfate, and molecular hydrogen. They use organic compounds to build structures. They do not fix carbon dioxide and apparently do not have the Calvin cycle. Chemolithoheterotrophs can be distinguished from mixotrophs (or facultative chemolithotroph), which can utilize either carbon dioxide or organic carbon as the carbon source.

Heterotrophs, by consuming reduced carbon compounds, are able to use all the energy that they obtain from food for growth and reproduction, unlike autotrophs, which must use some of their energy for carbon fixation. Both heterotrophs and autotrophs alike are usually dependent on the metabolic activities of other organisms for nutrients other than carbon, including nitrogen, phosphorus, and sulfur, and can die from lack of food that supplies these nutrients. This applies not only to animals and fungi but also to bacteria.

Flowchart

- Autotroph
    - Chemoautotroph
    - Photoautotroph
- Heterotroph
    - Chemoheterotroph
    - Photoheterotroph

Ecology

Most heterotrophs are chemoorganoheterotrophs (or simply organotrophs) who utilize organic compounds both as a carbon source and an energy source. The term "heterotroph" very often refers to chemoorganoheterotrophs. Heterotrophs function as consumers in food chains: they obtain organic carbon by eating autotrophs or other heterotrophs. They break down complex organic compounds (e.g., carbohydrates, fats, and proteins) produced by autotrophs into simpler compounds (e.g., carbohydrates into glucose, fats into fatty acids and glycerol, and proteins into amino acids). They release energy by oxidizing carbon and hydrogen atoms present in carbohydrates, lipids, and proteins to carbon dioxide and water, respectively.

Most opisthokonts and prokaryotes are heterotrophic; in particular, all animals and fungi are heterotrophs. Some animals, such as corals, form symbiotic relationships with autotrophs and obtain organic carbon in this way. Furthermore, some parasitic plants have also turned fully or partially heterotrophic, while carnivorous plants consume animals to augment their nitrogen supply while remaining autotrophic.

Animals are heterotrophs by ingestion, fungi are heterotrophs by absorption.

Decomposer

The fungi on this tree are decomposers

Decomposers are organisms that break down dead or decaying organisms, and in doing so, they carry out the natural process of decomposition. Like herbivores and predators, decomposers are heterotrophic, meaning that they use organic substrates to get their energy, carbon and nutrients for growth and development. While the terms decomposer and detritivore are often interchangeably used, however, detritivores must digest dead matter via internal processes while decomposers can break down cells of other organisms using biochemical reactions without need for internal digestion. Thus, invertebrates such as earthworms, woodlice, and sea cucumbers are detritivores, not decomposers, in the technical sense, since they must ingest nutrients and are unable to absorb them externally.

Fungi

The primary decomposers of litter in many ecosystems are fungi. Unlike bacteria, which are unicellular organisms, most saprotrophic fungi grow as a branching network of hyphae. While bacteria are restricted to growing and feeding on the exposed surfaces of organic matter, fungi can use their hyphae to penetrate larger pieces of organic matter. Additionally, only wood-decay fungi have evolved the enzymes necessary to decompose lignin, a chemically complex substance found in wood. These two factors make fungi the primary decomposers in forests, where litter has high concentrations of lignin and often occurs in large pieces. Fungi decompose organic matter by releasing enzymes to break down the decaying material, after which they absorb the nutrients in the decaying material. Hyphae used to break down matter and absorb nutrients are also used in reproduction. When two compatible fungi's hyphae grow close to each other, they will then fuse together for reproduction and form another fungus.

References

- Mauseth, James D. (2008). Botany: An Introduction to Plant Biology (4 ed.). Jones & Bartlett Publishers. p. 252. ISBN 978-0-7637-5345-0.
- Beckett, Brian S. (1981). Illustrated Human and Social Biology. Oxford University Press. p. 38. ISBN 978-0-19-914065-7.
- Mauseth, James D. (2008). Botany: an introduction to plant biology (4th ed.). Jones & Bartlett Publishers. p. 252. ISBN 978-0-7637-5345-0.
- Libes, Susan M. (2009). Introduction to marine biogeochemistry (2nd ed.). Academic Press. p. 192. ISBN 978-0-12-088530-5.
- Dworkin, Martin (2006). The prokaryotes: ecophysiology and biochemistry (3rd ed.). Springer. p. 988. ISBN 978-0-387-25492-0.
- Blanchette, Robert (September 1991). "Delignification by Wood-Decay Fungi". Annual Review of Phytopathology. 29: 281–403. doi:10.1146/annurev.py.29.090191.002121. Retrieved 20 April 2015.
- Waggoner, Ben; Speer, Brian. "Fungi: Life History and Ecology". Introduction to the Fungi. Retrieved 24 January 2014.

2

Various Biotic Stressors

The various biotic stressors are plant viruses, viroids, insects, herbivores and parasitic plants. The virus that affects plants is known as plant virus whereas the infection that occurs in the plants is referred to as viroid. Likewise this text also explains to the reader, stressors such as insects, herbivores and parasitic plants. The chapter strategically encompasses and incorporates the major components and key concepts of biotic stress, providing a complete understanding.

Plant Virus

Pepper mild mottle virus

Plant viruses are viruses that affect plants. Like all other viruses, plant viruses are obligate intracellular parasites that do not have the molecular machinery to replicate without a host. Plant viruses are pathogenic to higher plants. While this article does not intend to list all plant viruses, it discusses some important viruses as well as their uses in plant molecular biology.

Overview

Although plant viruses are not nearly as well understood as the animal counterparts, one plant virus has become iconic. The first virus to be discovered was *Tobacco mosaic virus* (TMV). This and other viruses cause an estimated US$60 billion loss in crop yields worldwide each year. Plant viruses are grouped into 73 genera and 49 families. However, these figures relate only to cultivated plants that represent only a tiny fraction of the total number of plant species. Viruses in wild plants have been poorly studied, but those studies that exist almost overwhelming show that such interactions between wild plants and their viruses do not appear to cause disease in the host plants.

To transmit from one plant to another and from one plant cell to another, plant viruses must use strategies that are usually different from animal viruses. Plants do not move, and so plant-to-plant transmission usually involves vectors (such as insects). Plant cells are surrounded by solid cell walls, therefore transport through plasmodesmata is the preferred path for virions to move between plant cells. Plants probably have specialized mechanisms for transporting mRNAs through plasmodesmata, and these mechanisms are thought to be used by RNA viruses to spread from one cell to another.

Plant defenses against viral infection include, among other measures, the use of siRNA in response to dsRNA. Most plant viruses encode a protein to suppress this response. Plants also reduce transport through plasmodesmata in response to injury.

History

The discovery of plant viruses causing disease is often accredited to A. Mayer (1886) working in the Netherlands demonstrated that the sap of mosaic obtained from tobacco leaves developed mosaic symptom when injected in healthy plants. However the infection of the sap was destroyed when it was boiled. He thought that the causal agent was the bacteria. However, after larger inoculation with a large number of bacteria, he failed to develop a mosaic symptom.

In 1898, Martinus Beijerinck, who was a Professor of Microbiology at the Technical University the Netherlands, put forth his concepts that viruses were small and determined that the "mosaic disease" remained infectious when passed through a Chamberland filter-candle. This was in contrast to bacteria microorganisms, which were retained by the filter. Beijerinck referred to the infectious filtrate as a "contagium vivum fluidum", thus the coinage of the modern term "virus".

After the initial discovery of the 'viral concept' there was need to classify any other known viral diseases based on the mode of transmission even though microscopic observation proved fruitless. In 1939 Holmes published a classification list of 129 plant viruses. This was expanded and in 1999 there were 977 officially recognized, and some provisional, plant virus species.

The purification (crystallization) of TMV was first performed by Wendell Stanley, who published his findings in 1935, although he did not determine that the RNA was the infectious material. However, he received the Nobel Prize in Chemistry in 1946. In the 1950s a discovery by two labs simultaneously proved that the purified RNA of the TMV was infectious which reinforced the argument. The RNA carries genetic information to code for the production of new infectious particles.

More recently virus research has been focused on understanding the genetics and molecular biology of plant virus genomes, with a particular interest in determining how the virus can replicate, move and infect plants. Understanding the virus genetics and protein functions has been used to explore the potential for commercial use by biotechnology companies. In particular, viral-derived sequences have been used to provide an understanding of novel forms of resistance. The recent boom in technology allowing humans to manipulate plant viruses may provide new strategies for production of value-added proteins in plants.

Structure

Viruses are extremely small and can only be observed under an electron microscope. The structure of a virus is given by its coat of proteins, which surround the viral genome. Assembly of viral particles takes place spontaneously.

Over 50% of known plant viruses are rod-shaped (flexuous or rigid). The length of the particle is normally dependent on the genome but it is usually between 300–500 nm with a diameter of 15–20 nm. Protein subunits can be placed around the circumference of a circle to form a disc. In the presence of the viral genome, the discs are stacked, then a tube is created with room for the nucleic acid genome in the middle.

The second most common structure amongst plant viruses are isometric particles. They are 25–50 nm in diameter. In cases when there is only a single coat protein, the basic structure consists of 60 T subunits, where T is an integer. Some viruses may have 2 coat proteins that associate to form an icosahedral shaped particle.

There are three genera of *Geminiviridae* that possess geminate particles which are like two isometric particles stuck together.

A very small number of plant viruses have, in addition to their coat proteins, a lipid envelope. This is derived from the plant cell membrane as the virus particle buds off from the cell.

Transmission of Plant Viruses

Through Sap

Viruses can be spread by direct transfer of sap by contact of a wounded plant with a healthy one. Such contact may occur during agricultural practices, as by damage caused by tools or hands, or naturally, as by an animal feeding on the plant. Generally TMV, potato viruses and cucumber mosaic viruses are transmitted via sap.

Insects

Plant viruses need to be transmitted by a vector, most often insects such as leafhoppers. One class of viruses, the Rhabdoviridae, has been proposed to actually be insect viruses that have evolved to replicate in plants. The chosen insect vector of a plant virus will often be the determining factor in that virus's host range: it can only infect plants that the insect vector feeds upon. This was shown in part when the old world white fly made it to the United States, where it transferred many plant viruses into new hosts. Depending on the way they are transmitted, plant viruses are classified as non-persistent, semi-persistent and persistent. In non-persistent transmission, viruses become attached to the distal tip of the stylet of the insect and on the next plant it feeds on, it inoculates it with the virus. Semi-persistent viral transmission involves the virus entering the foregut of the insect. Those viruses that manage to pass through the gut into the haemolymph and then to the salivary glands are known as persistent. There are two sub-classes of persistent viruses: propagative and circulative. Propagative viruses are able to replicate in both the plant and the insect (and may have originally been insect viruses), whereas circulative can not. Circulative viruses are protected inside aphids by the chaperone protein symbionin, produced by bacterial symbionts. Many plant viruses encode within their genome polypeptides with domains essential for transmission by insects. In non-persistent and semi-persistent viruses, these domains are in the coat protein and another protein known as the helper component. A bridging hypothesis has been proposed to explain how these proteins aid in insect-mediated viral transmission. The helper component will bind to the specific domain of the coat protein, and then the insect mouthparts — creating a bridge. In persistent propagative viruses, such as tomato spotted wilt virus (TSWV), there is often a lipid coat surrounding the proteins that is not seen in other classes of plant viruses. In the case of TSWV, 2 viral proteins are expressed in this lipid envelope. It has been proposed that the viruses bind via these proteins and are then taken into the insect cell by receptor-mediated endocytosis.

Nematodes

Soil-borne nematodes also have been shown to transmit viruses. They acquire and transmit them by feeding on infected roots. Viruses can be transmitted both non-persistently and persistently, but there is no evidence of viruses being able to replicate in nematodes. The virions attach to the stylet (feeding organ) or to the gut when they feed on an infected plant and can then detach during later feeding to infect other plants. Examples of viruses that can be transmitted by nematodes include tobacco ringspot virus and tobacco rattle virus.

Plasmodiophorids

A number of virus genera are transmitted, both persistently and non-persistently, by soil borne zoosporic protozoa. These protozoa are not phytopathogenic themselves, but parasitic. Transmission of the virus takes place when they become associated with the plant roots. Examples include *Polymyxa graminis*, which has been shown to transmit plant viral diseases in cereal crops and *Polymyxa betae* which transmits Beet necrotic yellow vein virus. Plasmodiophorids also create wounds in the plant's root through which other viruses can enter.

Seed and Pollen Borne Viruses

Plant virus transmission from generation to generation occurs in about 20% of plant viruses. When viruses are transmitted by seeds, the seed is infected in the generative cells and the virus is maintained in the germ cells and sometimes, but less often, in the seed coat. When the growth and development of plants is delayed because of situations like unfavourable weather, there is an increase in the amount of virus infections in seeds. There does not seem to be a correlation between the location of the seed on the plant and its chances of being infected. Little is known about the mechanisms involved in the transmission of plant viruses via seeds, although it is known that it is environmentally influenced and that seed transmission occurs because of a direct invasion of the embryo via the ovule or by an indirect route with an attack on the embryo mediated by infected gametes. These processes can occur concurrently or separately depending on the host plant. It is unknown how the virus is able to directly invade and cross the embryo and boundary between the parental and progeny generations in the ovule. Many plants species can be infected through seeds including but not limited to the families Leguminosae, Solanaceae, Compositae, Rosaceae, Cucurbitaceae, Gramineae. Bean common mosaic virus is transmitted through seeds.

Direct Plant-to-human Transmission

Researchers from the University of the Mediterranean in Marseille, France have found tenuous evidence that suggest a virus common to peppers, the Pepper Mild Mottle Virus (PMMoV) may have moved on to infect humans. This is a very rare and highly unlikely event as, to enter a cell and replicate, a virus must "bind to a receptor on its surface, and a plant virus would be highly unlikely to recognize a receptor on a human cell. One possibility is that the virus does not infect human cells directly. Instead, the naked viral RNA may alter the function of the cells through a mechanism similar to RNA interference, in which the presence of certain RNA sequences can turn genes on and off," according to Virologist Robert Garry from the Tulane University in New Orleans, Louisiana.

Translation of Plant Viral Proteins

75% of plant viruses have genomes that consist of single stranded RNA (ssRNA). 65% of plant viruses have +ssRNA, meaning that they are in the same sense orientation as messenger RNA but 10% have -ssRNA, meaning they must be converted to +ssRNA before they can be translated. 5% are double stranded RNA and so can be immediately translated as +ssRNA viruses. 3% require a reverse transcriptase enzyme to convert between RNA and DNA. 17% of plant viruses are ssDNA and very few are dsDNA, in contrast a quarter of animal viruses are dsDNA and three quarters of bacteriophage are dsDNA. Viruses use the plant ribosomes to produce the 4-10 proteins encoded by their genome. However, since many of the proteins are encoded on a single strand (that is, they are polycistronic) this will mean that the ribosome will either only produce one protein, as it will terminate translation at the first stop codon, or that a polyprotein will be produced. Plant viruses have had to evolve special techniques to allow the production of viral proteins by plant cells.

5' Cap

For translation to occur, eukaryotic mRNAs require a 5' Cap structure. This means that viruses must also have one. This normally consists of 7MeGpppN where N is normally adenine or guanine. The viruses encode a protein, normally a replicase, with a methyltransferase activity to allow this.

Some viruses are cap-snatchers. During this process, a 7mG-capped host mRNA is recruited by the viral transcriptase complex and subsequently cleaved by a virally encoded endonuclease. The resulting capped leader RNA is used to prime transcription on the viral genome.

However some plant viruses do not use cap, yet translate efficiently due to cap-independent translation enhancers present in 5' and 3' untranslated regions of viral mRNA.

Readthrough

Some viruses (e.g. tobacco mosaic virus (TMV)) have RNA sequences that contain a "leaky" stop codon. In TMV 95% of the time the host ribosome will terminate the synthesis of the polypeptide at this codon but the rest of the time it continues past it. This means that 5% of the proteins produced are larger than and different from the others normally produced, which is a form of translational regulation. In TMV, this extra sequence of polypeptide is an RNA polymerase that replicates its genome.

Production of Sub-genomic RNAs

Some viruses use the production of subgenomic RNAs to ensure the translation of all proteins within their genomes. In this process the first protein encoded on the genome, and this the first to be translated, is a replicase. This protein will act on the rest of the genome producing negative strand sub-genomic RNAs then act upon these to form positive strand sub-genomic RNAs that are essentially mRNAs ready for translation.

Segmented Genomes

Some viral families, such as the *Bromoviridae* instead opt to have multipartite genomes, genomes split between multiple viral particles. For infection to occur, the plant must be infected with all particles across the genome. For instance *Brome mosaic virus* has a genome split between 3 viral particles, and all 3 particles with the different RNAs are required for infection to take place.

Polyprotein Processing

This strategy is adopted by viral genera such as the Potyviridae and Tymoviridae. The ribosome translates a single protein from the viral genome. Within the polyprotein is an enzyme (or enzymes) with proteinase function that is able to cleave the polyprotein into the various single proteins or just cleave away the protease, which can then cleave other polypeptides producing the mature proteins.

Well Understood Plant Viruses

Tobacco mosaic virus (TMV) and Cauliflower mosaic virus (CaMV) are frequently used in plant molecular biology. Of special interest is the CaMV 35S promoter, which is a very strong promoter most frequently used in plant transformations.

Viroid

Viroids are among the smallest infectious pathogens known, larger only than prions, which are misfolded proteins. Viroids consist solely of short strands of circular, single-stranded RNA without protein coats. They are mostly plant pathogens, some of which are of economic importance. Viroid genomes are extremely small in size, ranging from 246 to 467 nucleobases. In comparison, the genome of the smallest known viruses capable of causing an infection by themselves are around 2,000 nucleobases in size. The human pathogen hepatitis D virus is a defective RNA virus similar to viroids.

Viroids, the first known representatives of a new domain of "sub-viral pathogens", were discovered, initially characterized, and named by Theodor Otto Diener, plant pathologist at the U.S Department of Agriculture's Research Center in Beltsville, Maryland, in 1971. The first viroid to be identified was *Potato spindle tuber viroid* (PSTVd). Some 33 species have been identified.

Viroids do not code for any protein. Viroid's replication mechanism uses RNA polymerase II, a host cell enzyme normally associated with synthesis of messenger RNA from DNA, which instead catalyzes "rolling circle" synthesis of new RNA using the viroid's RNA as a template. Some viroids are ribozymes, having catalytic properties which allow self-cleavage and ligation of unit-size genomes from larger replication intermediates.

With Diener's 1989 hypothesis that viroids may represent "living relics" from the widely assumed, ancient, and non-cellular RNA world—extant before the evolution of DNA or proteins—viroids have assumed significance beyond plant pathology to evolutionary science, by representing the most plausible RNAs capable of performing crucial steps in abiogenesis, the evolution of life from inanimate matter.

Taxonomy

- Family Pospiviroidae
    - Genus *Pospiviroid*; type species: *Potato spindle tuber viroid* ; 356–361 nucleotides(nt)
    - Genus *Pospiviroid*; type species: *Citrus exocortis* ; 368–467 nt
    - Genus *Hostuviroid*; type species: *Hop stunt viroid* ; 294–303 nt
    - Genus *Cocadviroid*; type species: *Coconut cadang-cadang viroid*; 246–247 nt
    - Genus *Apscaviroid*; type species: *Apple scar skin viroid* ; 329–334 nt
    - Genus *Coleviroid*; type species: *Coleus blumei viroid 1* ; 248–251 nt

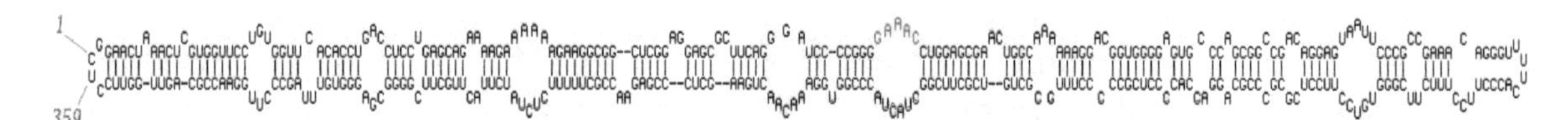

Putative secondary structure of the PSTVd viroid

- Family Avsunviroidae
    - Genus *Avsunviroid*; type species: *Avocado sunblotch viroid* ; 246–251 nt
    - Genus *Pelamoviroid*; type species: *Peach latent mosaic viroid* ;335–351 nt
    - Genus *Elaviroid*; type species: *Eggplant latent viroid* ; 332–335 nt

Transmission

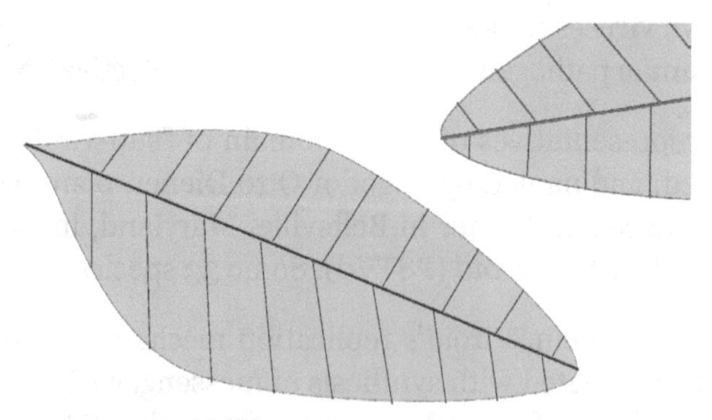

The reproduction mechanism of a typical viroid. Leaf contact transmits the viroid. The viroid enters the cell via its plasmodesmata. RNA polymerase II catalyzes rolling-circle synthesis of new viroids.

Viroid infections are transmitted by cross contamination following mechanical damage to plants as a result of horticultural or agricultural practices. Some are transmitted by aphids and they can also be transferred from plant to plant by leaf contact.

Replication

Viroids replicate in the nucleus (*Pospiviroidae*) or chloroplasts (*Avsunviroidae*) of plant cells in three steps through an RNA-based mechanism. They require RNA polymerase II, a host cell enzyme normally associated with synthesis of messenger RNA from DNA, which instead catalyzes "rolling circle" synthesis of new RNA using the viroid as template Some viroids are ribozymes, having catalytic properties which allow self-cleavage and ligation of unit-size genomes from larger replication intermediates.

RNA Silencing

There has long been uncertainty over how viroids induce symptoms in plants without encoding any protein products within their sequences. Evidence suggests that RNA silencing is involved in the process. First, changes to the viroid genome can dramatically alter its virulence. This reflects the fact that any siRNAs produced would have less complementary base pairing with target messenger RNA. Secondly, siRNAs corresponding to sequences from viroid genomes have been isolated from infected plants. Finally, transgenic expression of the noninfectious hpRNA of potato spindle tuber viroid develops all the corresponding viroid-like symptoms. This indicates that when viroids replicate via a double stranded intermediate RNA, they are targeted by a dicer enzyme and cleaved into siRNAs that are then loaded onto the RNA-induced silencing complex. The viroid siRNAs contain sequences capable of complementary base pairing with the plant's own messenger RNAs, and induction of degradation or inhibition of translation causes the classic viroid symptoms.

Living Relics of the RNA World

Diener's 1989 hypothesis proposed that unique properties of viroids make them more plausible macromolecules than introns, or other RNAs considered in the past as possible "living relics" of a hypothetical, pre-cellular RNA world. If so, viroids have assumed significance beyond plant virology for evolutionary science, because their properties make them more plausible candidates than other RNAs to perform crucial steps in the evolution of life from inanimate matter (abiogenesis). These properties are:

1. viroids' small size, imposed by error-prone replication
2. their high guanine and cytosine content, which increases stability and replication fidelity
3. their circular structure, which assures complete replication without genomic tags
4. existence of structural periodicity, which permits modular assembly into enlarged genomes
5. their lack of protein-coding ability, consistent with a ribosome-free habitat
6. replication mediated in some by ribozymes—the fingerprint of the RNA world

Diener's hypothesis was mostly forgotten until 2014, when it was resurrected in a review article by Flores et al., in which the authors summarized Diener's evidence supporting his hypothesis. In the same year, *New York Times* science writer Carl Zimmer published a popularized piece that mistakenly credited Flores et al. with the hypothesis' original conception.

The presence, in extant cells, of RNAs with molecular properties predicted for RNAs of the RNA World constitutes another powerful argument supporting the RNA World hypothesis.

History

In the 1920s, symptoms of a previously unknown potato disease were noticed in New York and New Jersey fields. Because tubers on affected plants become elongated and misshaped, they named it the potato spindle tuber disease.

The symptoms appeared on plants onto which pieces from affected plants had been budded—indicating that the disease was caused by a transmissible pathogenic agent. However, a fungus or bacterium could not be found consistently associated with symptom-bearing plants, and therefore, it was assumed the disease was caused by a virus. Despite numerous attempts over the years to isolate and purify the assumed virus, using increasingly sophisticated methods, these were unsuccessful when applied to extracts from potato spindle tuber disease-afflicted plants.

In 1971 Theodor O. Diener showed that the agent was not a virus, but a totally unexpected novel type of pathogen, one-80th the size of typical viruses, for which he proposed the term "viroid". Parallel to agriculture-directed studies, more basic scientific research elucidated many of viroids' physical, chemical, and macromolecular properties. Viroids were shown to consist of short stretches (a few hundred nucleobases) of single-stranded RNA and, unlike viruses, did not have a protein coat. Compared with other infectious plant pathogens, viroids are extremely small in size, ranging from 246 to 467 nucleobases; they thus consist of fewer than 10,000 atoms. In comparison, the genomes of the smallest known viruses capable of causing an infection by themselves are around 2,000 nucleobases long.

In 1976, Sänger et al. presented evidence that potato spindle tuber viroid is a "single-stranded, covalently closed, circular RNA molecule, existing as a highly base-paired rod-like structure"—believed to be the first such molecule described. Circular RNA, unlike linear RNA, forms a covalently closed continuous loop, in which the 3' and 5' ends present in linear RNA molecules have been joined together. Sänger et al. also provided evidence for the true circularity of viroids by finding that the RNA could not be phosphorylated at the 5' terminus. Then, in other tests, they failed to find even one free 3' end, which ruled out the possibility of the molecule having two 3' ends. Viroids thus are true circular RNAs.

The single-strandedness and circularity of viroids was confirmed by electron microscopy, and Gross et al. determined the complete nucleotide sequence of potato spindle tuber viroid in 1978. PSTVd was the first pathogen of a eukaryotic organism for which the complete molecular structure has been established. Over thirty plant diseases have since been identified as viroid-, not virus-caused, as had been assumed.

Insect

Insects are a class of invertebrates within the arthropod phylum that have a chitinous exoskeleton, a three-part body (head, thorax and abdomen), three pairs of jointed legs, compound eyes and one pair of antennae. They are the most diverse group of animals on the planet, including more than a million described species and representing more than half of all known living organisms. The number of extant species is estimated at between six and ten million, and potentially represent over

90% of the differing animal life forms on Earth. Insects may be found in nearly all environments, although only a small number of species reside in the oceans, a habitat dominated by another arthropod group, crustaceans.

The life cycles of insects vary but most hatch from eggs. Insect growth is constrained by the inelastic exoskeleton and development involves a series of molts. The immature stages can differ from the adults in structure, habit and habitat, and can include a passive pupal stage in those groups that undergo 4-stage metamorphosis. Insects that undergo 3-stage metamorphosis lack a pupal stage and adults develop through a series of nymphal stages. The higher level relationship of the Hexapoda is unclear. Fossilized insects of enormous size have been found from the Paleozoic Era, including giant dragonflies with wingspans of 55 to 70 cm (22–28 in). The most diverse insect groups appear to have coevolved with flowering plants.

Adult insects typically move about by walking, flying or sometimes swimming. As it allows for rapid yet stable movement, many insects adopt a tripedal gait in which they walk with their legs touching the ground in alternating triangles. Insects are the only invertebrates to have evolved flight. Many insects spend at least part of their lives under water, with larval adaptations that include gills, and some adult insects are aquatic and have adaptations for swim-ming. Some species, such as water striders, are capable of walking on the surface of water. Insects are mostly solitary, but some, such as certain bees, ants and termites, are social and live in large, well-organized colonies. Some insects, such as earwigs, show maternal care, guarding their eggs and young. Insects can communicate with each other in a variety of ways. Male moths can sense the pheromones of female moths over great distances. Other species communicate with sounds: crickets stridulate, or rub their wings together, to attract a mate and repel other males. Lampyridae in the beetle order communicate with light.

Humans regard certain insects as pests, and attempt to control them using insecticides and a host of other techniques. Some insects damage crops by feeding on sap, leaves or fruits. A few parasitic species are pathogenic. Some insects perform complex ecological roles; blow-flies, for example, help consume carrion but also spread diseases. Insect pollinators are essential to the life-cycle of many flowering plant species on which most organisms, including humans, are at least partly dependent; without them, the terrestrial portion of the biosphere (including humans) would be devastated. Many other insects are considered ecologically beneficial as predators and a few provide direct economic benefit. Silkworms and bees have been used extensively by humans for the production of silk and honey, respectively. In some cultures, people eat the larvae or adults of certain insects.

Phylogeny and Evolution

The evolutionary relationship of insects to other animal groups remains unclear.

Although traditionally grouped with millipedes and centipedes—possibly on the basis of convergent adaptations to terrestrialisation—evidence has emerged favoring closer evolutionary ties with crustaceans. In the Pancrustacea theory, insects, together with Entognatha, Remipedia, and Cephalocarida, make up a natural clade labeled Miracrustacea.

A report in November 2014 unambiguously places the insects in one clade, with the crustaceans and myriapods, as the nearest sister clades. This study resolved insect phylogeny of all extant insect orders, and provides "a robust phylogenetic backbone tree and reliable time estimates of insect evolution."

Other terrestrial arthropods, such as centipedes, millipedes, scorpions, and spiders, are sometimes confused with insects since their body plans can appear similar, sharing (as do all arthropods) a jointed exoskeleton. However, upon closer examination, their features differ significantly; most noticeably, they do not have the six-legged characteristic of adult insects.

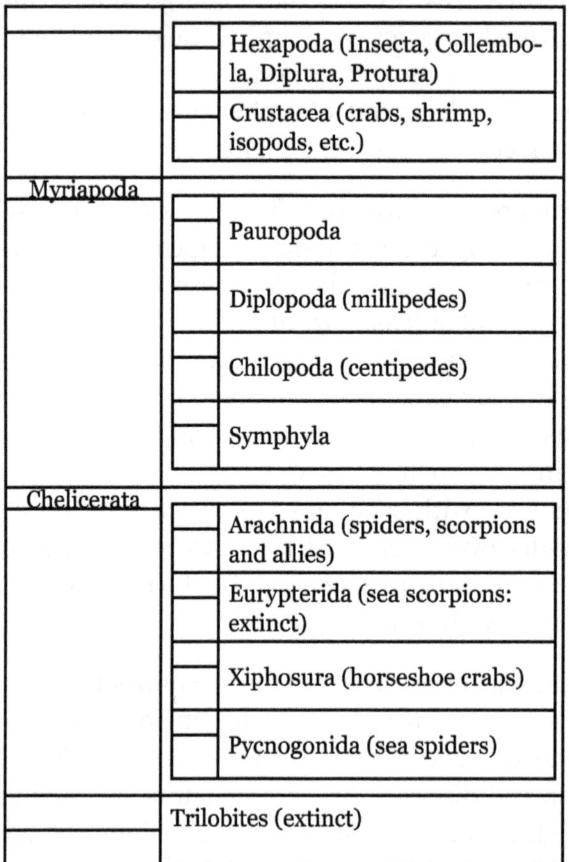

	Hexapoda (Insecta, Collembola, Diplura, Protura)
	Crustacea (crabs, shrimp, isopods, etc.)
Myriapoda	Pauropoda
	Diplopoda (millipedes)
	Chilopoda (centipedes)
	Symphyla
Chelicerata	Arachnida (spiders, scorpions and allies)
	Eurypterida (sea scorpions: extinct)
	Xiphosura (horseshoe crabs)
	Pycnogonida (sea spiders)
	Trilobites (extinct)

A phylogenetic tree of the arthropods and related groups

The higher-level phylogeny of the arthropods continues to be a matter of debate and research. In 2008, researchers at Tufts University uncovered what they believe is the world's oldest known full-body impression of a primitive flying insect, a 300 million-year-old specimen from the Carboniferous period. The oldest definitive insect fossil is the Devonian *Rhyniognatha hirsti*, from the 396-million-year-old Rhynie chert. It may have superficially resembled a modern-day silverfish insect. This species already possessed dicondylic mandibles (two articulations in the mandible), a feature associated with winged insects, suggesting that wings may already have evolved at this time. Thus, the first insects probably appeared earlier, in the Silurian period.

Four super radiations of insects have occurred: beetles (evolved about 300 million years ago), flies (evolved about 250 million years ago), and moths and wasps (evolved about 150 million years ago). These four groups account for the majority of described species. The flies and moths along with the fleas evolved from the Mecoptera.

The origins of insect flight remain obscure, since the earliest winged insects currently known appear to have been capable fliers. Some extinct insects had an additional pair of winglets attaching to the first segment of the thorax, for a total of three pairs. As of 2009, no evidence suggests the insects were a particularly successful group of animals before they evolved to have wings.

Late Carboniferous and Early Permian insect orders include both extant groups, their stem groups, and a number of Paleozoic groups, now extinct. During this era, some giant dragonfly-like forms reached wingspans of 55 to 70 cm (22 to 28 in), making them far larger than any living insect. This gigantism may have been due to higher atmospheric oxygen levels that allowed increased respiratory efficiency relative to today. The lack of flying vertebrates could have been another factor. Most extinct orders of insects developed during the Permian period that began around 270 million years ago. Many of the early groups became extinct during the Permian-Triassic extinction event, the largest mass extinction in the history of the Earth, around 252 million years ago.

The remarkably successful Hymenoptera appeared as long as 146 million years ago in the Cretaceous period, but achieved their wide diversity more recently in the Cenozoic era, which began 66 million years ago. A number of highly successful insect groups evolved in conjunction with flowering plants, a powerful illustration of coevolution.

Many modern insect genera developed during the Cenozoic. Insects from this period on are often found preserved in amber, often in perfect condition. The body plan, or morphology, of such specimens is thus easily compared with modern species. The study of fossilized insects is called paleoentomology.

Evolutionary Relationships

Insects are prey for a variety of organisms, including terrestrial vertebrates. The earliest vertebrates on land existed 400 million years ago and were large amphibious piscivores. Through gradual evolutionary change, insectivory was the next diet type to evolve.

Insects were among the earliest terrestrial herbivores and acted as major selection agents on plants. Plants evolved chemical defenses against this herbivory and the insects, in turn, evolved mechanisms to deal with plant toxins. Many insects make use of these toxins to protect themselves from their predators. Such insects often advertise their toxicity using warning colors. This successful evolutionary pattern has also been used by mimics. Over time, this has led to complex groups of coevolved species. Conversely, some interactions between plants and insects, like pollination, are beneficial to both organisms. Coevolution has led to the development of very specific mutualisms in such systems.

Taxonomy

Traditional morphology-based or appearance-based systematics have usually given the Hexapoda the rank of superclass, and identified four groups within it: insects (Ectognatha), springtails (Collembola), Protura, and Diplura, the latter three being grouped together as the Entognatha on the basis of internalized mouth parts. Supraordinal relationships have undergone numerous changes with the advent of methods based on evolutionary history and genetic data. A recent theory is that the Hexapoda are polyphyletic (where the last common ancestor was not a member of the group), with the entognath classes having separate evolutionary histories from the Insecta. Many of the traditional appearance-based taxa have been shown to be paraphyletic, so rather than using ranks like subclass, superorder, and infraorder, it has proved better to use monophyletic groupings (in which the last common ancestor is a member of the group). The following represents the best-supported monophyletic groupings for the Insecta.

Insects can be divided into two groups historically treated as subclasses: wingless insects, known as Apterygota, and winged insects, known as Pterygota. The Apterygota consist of the primitively wingless order of the silverfish (Thysanura). Archaeognatha make up the Monocondylia based on the shape of their mandibles, while Thysanura and Pterygota are grouped together as Dicondylia. The Thysanura themselves possibly are not monophyletic, with the family Lepidotrichidae being a sister group to the Dicondylia (Pterygota and the remaining Thysanura).

Paleoptera and Neoptera are the winged orders of insects differentiated by the presence of hardened body parts called sclerites, and in the Neoptera, muscles that allow their wings to fold flatly over the abdomen. Neoptera can further be divided into incomplete metamorphosis-based (Polyneoptera and Paraneoptera) and complete metamorphosis-based groups. It has proved difficult to clarify the relationships between the orders in Polyneoptera because of constant new findings calling for revision of the taxa. For example, the Paraneoptera have turned out to be more closely related to the Endopterygota than to the rest of the Exopterygota. The recent molecular finding that the traditional louse orders Mallophaga and Anoplura are derived from within Psocoptera has led to the new taxon Psocodea. Phasmatodea and Embiidina have been suggested to form the Eukinolabia. Mantodea, Blattodea, and Isoptera are thought to form a monophyletic group termed Dictyoptera.

The Exopterygota likely are paraphyletic in regard to the Endopterygota. Matters that have incurred controversy include Strepsiptera and Diptera grouped together as Halteria based on a reduction of one of the wing pairs – a position not well-supported in the entomological community. The Neuropterida are often lumped or split on the whims of the taxonomist. Fleas are now thought to be closely related to boreid mecopterans. Many questions remain in the basal relationships amongst endopterygote orders, particularly the Hymenoptera.

The study of the classification or taxonomy of any insect is called systematic entomology. If one works with a more specific order or even a family, the term may also be made specific to that order or family, for example systematic dipterology.

Diversity

Though the true dimensions of species diversity remain uncertain, estimates range from 2.6–7.8 million species with a mean of 5.5 million. This probably represents less than 20% of all species on Earth and with only about 20,000 new species of all organisms being described each year, most species likely will remain undescribed for many years unless species descriptions increase in rate. About 850,000–1,000,000 of all described species are insects. Of the 24 orders of insects, four dominate in terms of numbers of described species, with at least 3 million species included in Coleoptera, Diptera, Hymenoptera and Lepidoptera. A recent study estimated the number of beetles at 0.9–2.1 million with a mean of 1.5 million.

Comparison of the estimated number of species in the four most speciose insect orders			
	Described species	**Average description rate (species per year)**	**Publication effort**
Coleoptera	300,000–400,000	2308	0.01
Lepidoptera	110,000–120,000	642	0.03
Diptera	90,000–150,000	1048	0.04
Hymenoptera	100,000–125,000	1196	0.02

Morphology and Physiology

External

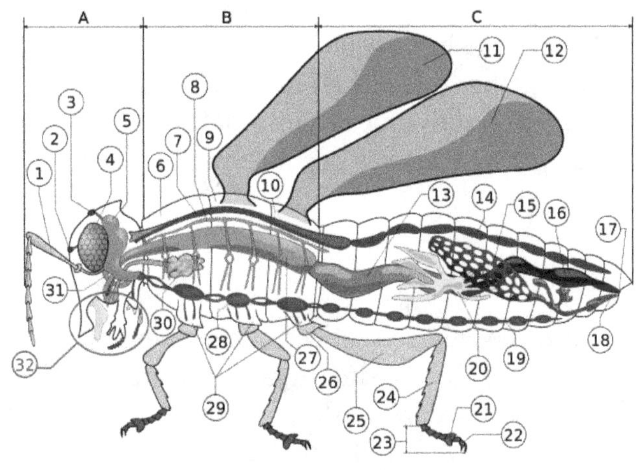

Insect morphology
A- Head **B**- Thorax **C**- Abdomen
1. antenna
2. ocelli (lower)
3. ocelli (upper)
4. compound eye
5. brain (cerebral ganglia)
6. prothorax
7. dorsal blood vessel
8. tracheal tubes (trunk with spiracle)
9. mesothorax
10. metathorax
11. forewing
12. hindwing
13. mid-gut (stomach)
14. dorsal tube (Heart)
15. ovary
16. hind-gut (intestine, rectum & anus)
17. anus
18. oviduct
19. nerve chord (abdominal ganglia)
20. Malpighian tubes
21. tarsal pads
22. claws
23. tarsus
24. tibia
25. femur

26. trochanter
27. fore-gut (crop, gizzard)
28. thoracic ganglion
29. coxa
30. salivary gland
31. subesophageal ganglion
32. mouthparts

Insects have segmented bodies supported by exoskeletons, the hard outer covering made mostly of chitin. The segments of the body are organized into three distinctive but interconnected units, or tagmata: a head, a thorax and an abdomen. The head supports a pair of sensory antennae, a pair of compound eyes, and, if present, one to three simple eyes (or ocelli) and three sets of variously modified appendages that form the mouthparts. The thorax has six segmented legs—one pair each for the prothorax, mesothorax and the metathorax segments making up the thorax—and, none, two or four wings. The abdomen consists of eleven segments, though in a few species of insects, these segments may be fused together or reduced in size. The abdomen also contains most of the digestive, respiratory, excretory and reproductive internal structures. Considerable variation and many adaptations in the body parts of insects occur, especially wings, legs, antenna and mouthparts.

Segmentation

The head is enclosed in a hard, heavily sclerotized, unsegmented, exoskeletal head capsule, or epicranium, which contains most of the sensing organs, including the antennae, ocellus or eyes, and the mouthparts. Of all the insect orders, Orthoptera displays the most features found in other insects, including the sutures and sclerites. Here, the vertex, or the apex (dorsal region), is situated between the compound eyes for insects with a hypognathous and opisthognathous head. In prognathous insects, the vertex is not found between the compound eyes, but rather, where the ocelli are normally. This is because the primary axis of the head is rotated 90° to become parallel to the primary axis of the body. In some species, this region is modified and assumes a different name. The thorax is a tagma composed of three sections, the prothorax, mesothorax and the metathorax. The anterior segment, closest to the head, is the prothorax, with the major features being the first pair of legs and the pronotum. The middle segment is the mesothorax, with the major features being the second pair of legs and the anterior wings. The third and most posterior segment, abutting the abdomen, is the metathorax, which features the third pair of legs and the posterior wings. Each segment is dilineated by an intersegmental suture. Each segment has four basic regions. The dorsal surface is called the tergum (or *notum*) to distinguish it from the abdominal terga. The two lateral regions are called the pleura (singular: pleuron) and the ventral aspect is called the sternum. In turn, the notum of the prothorax is called the pronotum, the notum for the mesothorax is called the mesonotum and the notum for the metathorax is called the metanotum. Continuing with this logic, the mesopleura and metapleura, as well as the mesosternum and metasternum, are used.

The abdomen is the largest tagma of the insect, which typically consists of 11–12 segments and is less strongly sclerotized than the head or thorax. Each segment of the abdomen is represented by a sclerotized tergum and sternum. Terga are separated from each other and from the adjacent sterna or pleura by membranes. Spiracles are located in the pleural area. Variation of this ground plan includes the fusion of terga or terga and sterna to form continuous dorsal or ventral shields or a conical tube. Some insects bear a sclerite in the pleural area called a laterotergite.

Ventral sclerites are sometimes called laterosternites. During the embryonic stage of many insects and the postembryonic stage of primitive insects, 11 abdominal segments are present. In modern insects there is a tendency toward reduction in the number of the abdominal segments, but the primitive number of 11 is maintained during embryogenesis. Variation in abdominal segment number is considerable. If the Apterygota are considered to be indicative of the ground plan for pterygotes, confusion reigns: adult Protura have 12 segments, Collembola have 6. The orthopteran family Acrididae has 11 segments, and a fossil specimen of Zoraptera has a 10-segmented abdomen.

Exoskeleton

The insect outer skeleton, the cuticle, is made up of two layers: the epicuticle, which is a thin and waxy water resistant outer layer and contains no chitin, and a lower layer called the procuticle. The procuticle is chitinous and much thicker than the epicuticle and has two layers: an outer layer known as the exocuticle and an inner layer known as the endocuticle. The tough and flexible endocuticle is built from numerous layers of fibrous chitin and proteins, criss-crossing each other in a sandwich pattern, while the exocuticle is rigid and hardened. The exocuticle is greatly reduced in many soft-bodied insects (e.g., caterpillars), especially during their larval stages.

Insects are the only invertebrates to have developed active flight capability, and this has played an important role in their success. Their muscles are able to contract multiple times for each single nerve impulse, allowing the wings to beat faster than would ordinarily be possible. Having their muscles attached to their exoskeletons is more efficient and allows more muscle connections; crustaceans also use the same method, though all spiders use hydraulic pressure to extend their legs, a system inherited from their pre-arthropod ancestors. Unlike insects, though, most aquatic crustaceans are biomineralized with calcium carbonate extracted from the water.

Internal

Nervous System

The nervous system of an insect can be divided into a brain and a ventral nerve cord. The head capsule is made up of six fused segments, each with either a pair of ganglia, or a cluster of nerve cells outside of the brain. The first three pairs of ganglia are fused into the brain, while the three following pairs are fused into a structure of three pairs of ganglia under the insect's esophagus, called the subesophageal ganglion.

The thoracic segments have one ganglion on each side, which are connected into a pair, one pair per segment. This arrangement is also seen in the abdomen but only in the first eight segments. Many species of insects have reduced numbers of ganglia due to fusion or reduction. Some cockroaches have just six ganglia in the abdomen, whereas the wasp *Vespa crabro* has only two in the thorax and three in the abdomen. Some insects, like the house fly *Musca domestica*, have all the body ganglia fused into a single large thoracic ganglion.

At least a few insects have nociceptors, cells that detect and transmit signals responsible for the sensation of pain. This was discovered in 2003 by studying the variation in reactions of larvae of the common fruitfly Drosophila to the touch of a heated probe and an unheated one. The larvae reacted to the touch of the heated probe with a stereotypical rolling behavior that was not exhibited when the larvae were touched by the unheated probe. Although nociception has been demonstrated in insects, there is no consensus that insects feel pain consciously

Insects are capable of learning.

Digestive System

An insect uses its digestive system to extract nutrients and other substances from the food it consumes. Most of this food is ingested in the form of macromolecules and other complex substances like proteins, polysaccharides, fats and nucleic acids. These macromolecules must be broken down by catabolic reactions into smaller molecules like amino acids and simple sugars before being used by cells of the body for energy, growth, or reproduction. This break-down process is known as digestion.

The main structure of an insect's digestive system is a long enclosed tube called the alimentary canal, which runs lengthwise through the body. The alimentary canal directs food unidirectionally from the mouth to the anus. It has three sections, each of which performs a different process of digestion. In addition to the alimentary canal, insects also have paired salivary glands and salivary reservoirs. These structures usually reside in the thorax, adjacent to the foregut.

The salivary glands (element 30 in numbered diagram) in an insect's mouth produce saliva. The salivary ducts lead from the glands to the reservoirs and then forward through the head to an opening called the salivarium, located behind the hypopharynx. By moving its mouthparts (element 32 in numbered diagram) the insect can mix its food with saliva. The mixture of saliva and food then travels through the salivary tubes into the mouth, where it begins to break down. Some insects, like flies, have extra-oral digestion. Insects using extra-oral digestion expel digestive enzymes onto their food to break it down. This strategy allows insects to extract a significant proportion of the available nutrients from the food source. The gut is where almost all of insects' digestion takes place. It can be divided into the foregut, midgut and hindgut.

Foregut

The first section of the alimentary canal is the foregut (element 27 in numbered diagram), or stomodaeum. The foregut is lined with a cuticular lining made of chitin and proteins as protection from tough food. The foregut includes the buccal cavity (mouth), pharynx, esophagus and crop and proventriculus (any part may be highly modified) which both store food and signify when to continue passing onward to the midgut.

Digestion starts in buccal cavity (mouth) as partially chewed food is broken down by saliva from the salivary glands. As the salivary glands produce fluid and carbohydrate-digesting enzymes (mostly amylases), strong muscles in the pharynx pump fluid into the buccal cavity, lubricating the food like the salivarium does, and helping blood feeders, and xylem and phloem feeders.

From there, the pharynx passes food to the esophagus, which could be just a simple tube passing it on to the crop and proventriculus, and then onward to the midgut, as in most insects. Alternately, the foregut may expand into a very enlarged crop and proventriculus, or the crop could just be a diverticulum, or fluid-filled structure, as in some Diptera species.

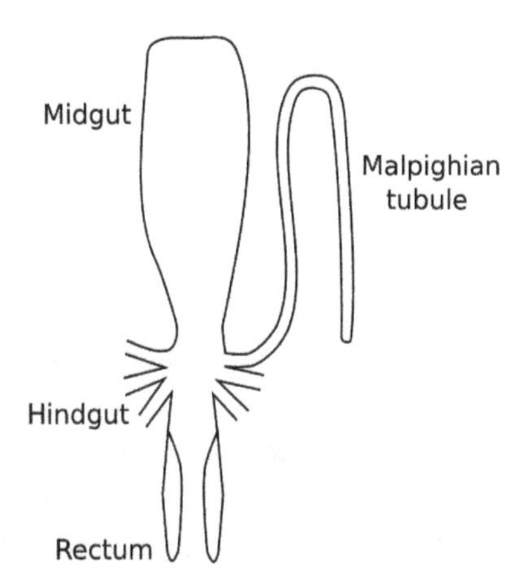

Stylized diagram of insect digestive tract showing malpighian tubule, from an insect of the order Orthoptera

Bumblebee defecating. Note the contraction of the abdomen to provide internal pressure

Midgut

Once food leaves the crop, it passes to the midgut (element 13 in numbered diagram), also known as the mesenteron, where the majority of digestion takes place. Microscopic projections from the midgut wall, called microvilli, increase the surface area of the wall and allow more nutrients to be absorbed; they tend to be close to the origin of the midgut. In some insects, the role of the microvilli and where they are located may vary. For example, specialized microvilli producing digestive enzymes may more likely be near the end of the midgut, and absorption near the origin or beginning of the midgut.

Hindgut

In the hindgut (element 16 in numbered diagram), or proctodaeum, undigested food particles are joined by uric acid to form fecal pellets. The rectum absorbs 90% of the water in these fecal pellets, and the dry pellet is then eliminated through the anus (element 17), completing the process of digestion. The uric acid is formed using hemolymph waste products diffused from the Malpighian tubules (element 20). It is then emptied directly into the alimentary canal, at the junction between the midgut and hindgut. The number of Malpighian tubules possessed by a given insect varies between species, ranging from only two tubules in some insects to over 100 tubules in others.

Reproductive System

The reproductive system of female insects consist of a pair of ovaries, accessory glands, one or more spermathecae, and ducts connecting these parts. The ovaries are made up of a number of egg tubes, called ovarioles, which vary in size and number by species. The number of eggs that the insect is able to make vary by the number of ovarioles with the rate that eggs can be develop being also influenced by ovariole design. Female insects are able make eggs, receive and store sperm, manipulate sperm from different males, and lay eggs. Accessory glands or glandular parts of the oviducts produce a variety of substances for sperm maintenance, transport and fertilization, as well as for protection of eggs. They can produce glue and protective substances for coating eggs or tough coverings for a batch of eggs called oothecae. Spermathecae are tubes or sacs in which sperm can be stored between the time of mating and the time an egg is fertilized.

For males, the reproductive system is the testis, suspended in the body cavity by tracheae and the fat body. Most male insects have a pair of testes, inside of which are sperm tubes or follicles that are enclosed within a membranous sac. The follicles connect to the vas deferens by the vas efferens, and the two tubular vasa deferentia connect to a median ejaculatory duct that leads to the outside. A portion of the vas deferens is often enlarged to form the seminal vesicle, which stores the sperm before they are discharged into the female. The seminal vesicles have glandular linings that secrete nutrients for nourishment and maintenance of the sperm. The ejaculatory duct is derived from an invagination of the epidermal cells during development and, as a result, has a cuticular lining. The terminal portion of the ejaculatory duct may be sclerotized to form the intromittent organ, the aedeagus. The remainder of the male reproductive system is derived from embryonic mesoderm, except for the germ cells, or spermatogonia, which descend from the primordial pole cells very early during embryogenesis.

Respiratory System

The tube-like heart (green) of the mosquito *Anopheles gambiae* extends horizontally across the body, interlinked with the diamond-shaped wing muscles (also green) and surrounded by pericardial cells (red). Blue depicts cell nuclei.

Insect respiration is accomplished without lungs. Instead, the insect respiratory system uses a system of internal tubes and sacs through which gases either diffuse or are actively pumped, delivering oxygen directly to tissues that need it via their trachea (element 8 in numbered diagram). Since oxygen is delivered directly, the circulatory system is not used to carry oxygen, and is therefore greatly reduced. The insect circulatory system has no veins or arteries, and instead consists of little more than a single, perforated dorsal tube which pulses peristaltically. Toward the thorax, the dorsal tube (element 14) divides into chambers and acts like the insect's heart. The opposite end of the dorsal tube is like the aorta of the insect circulating the hemolymph, arthropods' fluid analog of blood, inside the body cavity. Air is taken in through openings on the sides of the abdomen called spiracles.

The respiratory system is an important factor that limits the size of insects. As insects get bigger, this type of oxygen transport gets less efficient and thus the heaviest insect currently weighs less than 100 g. However, with increased atmospheric oxygen levels, as happened in the late Paleozoic, larger insects were possible, such as dragonflies with wingspans of more than two feet.

There are many different patterns of gas exchange demonstrated by different groups of insects. Gas exchange patterns in insects can range from continuous and diffusive ventilation, to discontinuous gas exchange. During continuous gas exchange, oxygen is taken in and carbon dioxide is released in a continuous cycle. In discontinuous gas exchange, however, the insect takes in oxygen while it is active and small amounts of carbon dioxide are released when the insect is at rest. Diffusive ventilation is simply a form of continuous gas exchange that occurs by diffusion rather than physically taking in the oxygen. Some species of insect that are submerged also have adaptations to aid in respiration. As larvae, many insects have gills that can extract oxygen dissolved in water, while others need to rise to the water surface to replenish air supplies which may be held or trapped in special structures.

Circulatory System

The insect circulatory system utilizes hemolymph, a tissue analogous to blood that circulates in the interior of the insect body, while remaining in direct contact with the animal's tissues. It is composed of plasma in which hemocytes are suspended. In addition to hemocytes, the plasma also contains many chemicals. It is also the major tissue type of the open circulatory system of arthropods, characteristic of spiders, crustaceans and insects.

Reproduction and Development

A pair of Simosyrphus grandicornis hoverflies mating in flight.

The majority of insects hatch from eggs. The fertilization and development takes place inside the egg, enclosed by a shell (chorion) that consists of maternal tissue. In contrast to eggs of other arthropods, most insect eggs are drought resistant. This is because inside the chorion two additional membranes develop from embryonic tissue, the amnion and the serosa. This serosa secretes a cuticle rich in chitin that protects the embryo against desiccation. In Schizophora however the serosa does not develop, but these flies lay their eggs in damp places, such as rotting matter. Some species of insects, like the cockroach *Blaptica dubia*, as well as juvenile aphids and tsetse flies, are ovoviviparous. The eggs of ovoviviparous animals develop entirely inside the female, and then hatch immediately upon being laid. Some other species, such as those in the genus of cockroaches known as *Diploptera*, are viviparous, and thus gestate inside the mother and are born alive. Some insects, like parasitic wasps, show polyembryony, where a single fertilized egg divides into many and in some cases thousands of separate embryos. Insects may be *univoltine*, *bivoltine* or *multivoltine*, i.e. they may have one, two or many broods (generations) in a year.

A pair of grasshoppers mating.

Other developmental and reproductive variations include haplodiploidy, polymorphism, paedomorphosis or peramorphosis, sexual dimorphism, parthenogenesis and more rarely hermaphroditism. In haplodiploidy, which is a type of sex-determination system, the offspring's sex is determined by the number of sets of chromosomes an individual receives. This system is typical in bees and wasps. Polymorphism is where a species may have different *morphs* or *forms*, as in the oblong winged katydid, which has four different varieties: green, pink and yellow or tan. Some insects may retain phenotypes that are normally only seen in juveniles; this is called paedomorphosis. In peramorphosis, an opposite sort of phenomenon, insects take on previously unseen traits after they have matured into adults. Many insects display sexual dimorphism, in which males and females have notably different appearances, such as the moth *Orgyia recens* as an exemplar of sexual dimorphism in insects.

Some insects use parthenogenesis, a process in which the female can reproduce and give birth without having the eggs fertilized by a male. Many aphids undergo a form of parthenogenesis, called cyclical parthenogenesis, in which they alternate between one or many generations of asexual and sexual reproduction. In summer, aphids are generally female and parthenogenetic; in the autumn, males may be produced for sexual reproduction. Other insects produced by parthenogenesis are bees, wasps and ants, in which they spawn males. However, overall, most individuals are female, which are produced by fertilization. The males are haploid and the females are diploid. More rarely, some insects display hermaphroditism, in which a given individual has both male and female reproductive organs.

The different forms of the male (top) and female (bottom) tussock moth *Orgyia recens* is an example of sexual dimorphism in insects.

Insect life-histories show adaptations to withstand cold and dry conditions. Some temperate region insects are capable of activity during winter, while some others migrate to a warmer climate or go into a state of torpor. Still other insects have evolved mechanisms of diapause that allow eggs or pupae to survive these conditions.

Metamorphosis

Metamorphosis in insects is the biological process of development all insects must undergo. There are two forms of metamorphosis: incomplete metamorphosis and complete metamorphosis.

Incomplete Metamorphosis

Hemimetabolous insects, those with incomplete metamorphosis, change gradually by undergoing a series of molts. An insect molts when it outgrows its exoskeleton, which does not stretch and would otherwise restrict the insect's growth. The molting process begins as the insect's epidermis secretes a new epicuticle inside the old one. After this new epicuticle is secreted, the epidermis releases a mixture of enzymes that digests the endocuticle and thus detaches the old cuticle. When this stage is complete, the insect makes its body swell by taking in a large quantity of water or air, which makes the old cuticle split along predefined weaknesses where the old exocuticle was thinnest.

Immature insects that go through incomplete metamorphosis are called nymphs or in the case of dragonflies and damselflies, also naiads. Nymphs are similar in form to the adult except for the presence of wings, which are not developed until adulthood. With each molt, nymphs grow larger and become more similar in appearance to adult insects.

This Southern Hawker dragonfly molts its exoskeleton several times during its life as a nymph; shown is the final molt to become a winged adult (eclosion).

Complete Metamorphosis

Holometabolism, or complete metamorphosis, is where the insect changes in four stages, an egg or embryo, a larva, a pupa and the adult or imago. In these species, an egg hatches to produce a larva, which is generally worm-like in form. This worm-like form can be one of several varieties: eruciform (caterpillar-like), scarabaeiform (grub-like), campodeiform (elongated, flattened and active), elateriform (wireworm-like) or vermiform (maggot-like). The larva grows and eventually becomes a pupa, a stage marked by reduced movement and often sealed within a cocoon. There are three types of pupae: obtect, exarate or coarctate. Obtect pupae are compact, with the legs and other appendages enclosed. Exarate pupae have their legs and other appendages free and extended. Coarctate pupae develop inside the larval skin. Insects undergo considerable change in form during the pupal stage, and emerge as adults. Butterflies are a well-known example of insects that undergo complete metamorphosis, although most insects use this life cycle. Some insects have evolved this system to hypermetamorphosis.

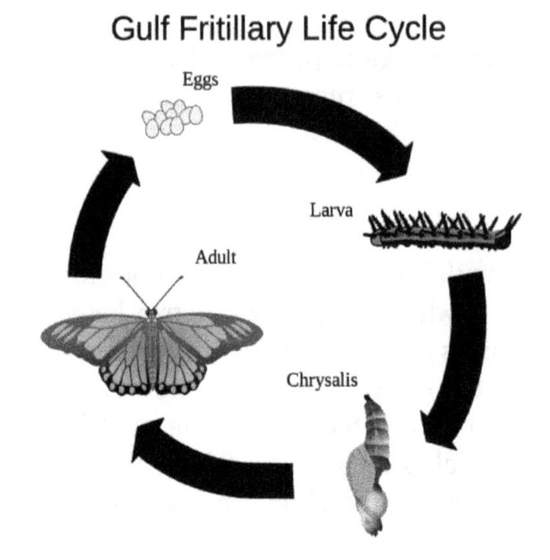

Gulf Fritillary life cycle, an example of holometabolism.

Some of the oldest and most successful insect groups, such Endopterygota, use a system of complete metamorphosis. Complete metamorphosis is unique to a group of certain insect orders including Diptera, Lepidoptera and Hymenoptera. This form of development is exclusive and not seen in any other arthropods.

Senses and Communication

Many insects possess very sensitive and, or specialized organs of perception. Some insects such as bees can perceive ultraviolet wavelengths, or detect polarized light, while the antennae of male moths can detect the pheromones of female moths over distances of many kilometers. The yellow paper wasp (*Polistes versicolor*) is known for its wagging movements as a form of communication within the colony; it can waggle with a frequency of 10.6±2.1 Hz (n=190). These wagging movements can signal the arrival of new material into the nest and aggression between workers can be used to stimulate others to increase foraging expeditions. There is a pronounced tendency for there to be a trade-off between visual acuity and chemical or tactile acuity, such that most insects with well-developed eyes have reduced or simple antennae, and vice versa. There are a variety of different mechanisms by which insects perceive sound, while the patterns are not universal, insects can generally hear sound if they can produce it. Different insect species can have varying hearing, though most insects can hear only a narrow range of frequencies related to the frequency of the sounds they can produce. Mosquitoes have been found to hear up to 2 kHz, and some grasshoppers can hear up to 50 kHz. Certain predatory and parasitic insects can detect the characteristic sounds made by their prey or hosts, respectively. For instance, some nocturnal moths can perceive the ultrasonic emissions of bats, which helps them avoid predation. Insects that feed on blood have special sensory structures that can detect infrared emissions, and use them to home in on their hosts.

Some insects display a rudimentary sense of numbers, such as the solitary wasps that prey upon a single species. The mother wasp lays her eggs in individual cells and provides each egg with a number of live caterpillars on which the young feed when hatched. Some species of wasp always provide five, others twelve, and others as high as twenty-four caterpillars per cell. The number of caterpillars is different among species, but always the same for each sex of larva. The male solitary wasp in the genus *Eumenes* is smaller than the female, so the mother of one species supplies him with only five caterpillars; the larger female receives ten caterpillars in her cell.

Light Production and Vision

Insects have compound eyes and two antennae.

A few insects, such as members of the families Poduridae and Onychiuridae (Collembola), Mycetophilidae (Diptera) and the beetle families Lampyridae, Phengodidae, Elateridae and Staphylinidae are bioluminescent. The most familiar group are the fireflies, beetles of the family Lampyridae. Some species are able to control this light generation to produce flashes. The function varies with some species using them to attract mates, while others use them to lure prey. Cave dwelling larvae of *Arachnocampa* (Mycetophilidae, Fungus gnats) glow to lure small flying insects into sticky strands of silk. Some fireflies of the genus *Photuris* mimic the flashing of female *Photinus* species to attract males of that species, which are then captured and devoured. The colors of emitted light vary from dull blue (*Orfelia fultoni*, Mycetophilidae) to the familiar greens and the rare reds (*Phrixothrix tiemanni*, Phengodidae).

Most insects, except some species of cave crickets, are able to perceive light and dark. Many species have acute vision capable of detecting minute movements. The eyes may include simple eyes or ocelli as well as compound eyes of varying sizes. Many species are able to detect light in the infrared, ultraviolet and the visible light wavelengths. Color vision has been demonstrated in many species and phylogenetic analysis suggests that UV-green-blue trichromacy existed from at least the Devonian period between 416 and 359 million years ago.

Sound Production and Hearing

Insects were the earliest organisms to produce and sense sounds. Insects make sounds mostly by mechanical action of appendages. In grasshoppers and crickets, this is achieved by stridulation. Cicadas make the loudest sounds among the insects by producing and amplifying sounds with special modifications to their body and musculature. The African cicada *Brevisana brevis* has been measured at 106.7 decibels at a distance of 50 cm (20 in). Some insects, such as the *Helicoverpa zea* moths, hawk moths and Hedylid butterflies, can hear ultrasound and take evasive action when they sense that they have been detected by bats. Some moths produce ultrasonic clicks that were once thought to have a role in jamming bat echolocation. The ultrasonic clicks were subsequently found to be produced mostly by unpalatable moths to warn bats, just as warning colorations are used against predators that hunt by sight. Some otherwise palatable moths have evolved to mimic these calls. More recently, the claim that some moths can jam bat sonar has been revisited. Ultrasonic recording and high-speed infrared videography of bat-moth interactions suggest the palatable tiger moth really does defend against attacking big brown bats using ultrasonic clicks that jam bat sonar.

Very low sounds are also produced in various species of Coleoptera, Hymenoptera, Lepidoptera, Mantodea and Neuroptera. These low sounds are simply the sounds made by the insect's movement. Through microscopic stridulatory structures located on the insect's muscles and joints, the normal sounds of the insect moving are amplified and can be used to warn or communicate with other insects. Most sound-making insects also have tympanal organs that can perceive airborne sounds. Some species in Hemiptera, such as the corixids (water boatmen), are known to communicate via underwater sounds. Most insects are also able to sense vibrations transmitted through surfaces.

Communication using surface-borne vibrational signals is more widespread among insects because of size constraints in producing air-borne sounds. Insects cannot effectively produce low-frequency sounds, and high-frequency sounds tend to disperse more in a dense environment (such as foliage), so insects living in such environments communicate primarily using substrate-borne vibrations. The mechanisms of production of vibrational signals are just as diverse as those for producing sound in insects.

Some species use vibrations for communicating within members of the same species, such as to attract mates as in the songs of the shield bug *Nezara viridula*. Vibrations can also be used to communicate between entirely different species; lycaenid (gossamer-winged butterfly) caterpillars which are myrmecophilous (living in a mutualistic association with ants) communicate with ants in this way. The Madagascar hissing cockroach has the ability to press air through its spiracles to make a hissing noise as a sign of aggression; the Death's-head Hawkmoth makes a squeaking noise by forcing air out of their pharynx when agitated, which may also reduce aggressive worker honey bee behavior when the two are in close proximity.

Chemical Communication

Chemical communications in animals rely on a variety of aspects including taste and smell. Chemoreception is the physiological response of a sense organ (i.e. taste or smell) to a chemical stimulus where the chemicals act as signals to regulate the state or activity of a cell. A semiochemical is a message-carrying chemical that is meant to attract, repel, and convey information. Types of semiochemicals include pheromones and kairomones. One example is the butterfly *Phengaris arion* which uses chemical signals as a form of mimicry to aid in predation.

In addition to the use of sound for communication, a wide range of insects have evolved chemical means for communication. These chemicals, termed semiochemicals, are often derived from plant metabolites include those meant to attract, repel and provide other kinds of information. Pheromones, a type of semiochemical, are used for attracting mates of the opposite sex, for aggregating conspecific individuals of both sexes, for deterring other individuals from approaching, to mark a trail, and to trigger aggression in nearby individuals. Allomonea benefit their producer by the effect they have upon the receiver. Kairomones benefit their receiver instead of their producer. Synomones benefit the producer and the receiver. While some chemicals are targeted at individuals of the same species, others are used for communication across species. The use of scents is especially well known to have developed in social insects.

Social Behavior

A cathedral mound created by termites (Isoptera).

Social insects, such as termites, ants and many bees and wasps, are the most familiar species of eusocial animal. They live together in large well-organized colonies that may be so tightly integrated and genetically similar that the colonies of some species are sometimes considered superorganisms. It is sometimes argued that the various species of honey bee are the only invertebrates (and indeed one of the few non-human groups) to have evolved a system of abstract symbolic communication where a behavior is used to *represent* and convey specific information about something in the environment. In this communication system, called dance language, the angle at which a bee dances represents a direction relative to the sun, and the length of the dance represents the distance to be flown. Though perhaps not as advanced as honey bees, bumblebees also potentially have some social communication behaviors. *Bombus terrestris*, for example, exhibit a faster learning curve for visiting unfamiliar, yet rewarding flowers, when they can see a conspecific foraging on the same species.

Only insects which live in nests or colonies demonstrate any true capacity for fine-scale spatial orientation or homing. This can allow an insect to return unerringly to a single hole a few millimeters in diameter among thousands of apparently identical holes clustered together, after a trip of up to several kilometers' distance. In a phenomenon known as philopatry, insects that hibernate have shown the ability to recall a specific location up to a year after last viewing the area of interest. A few insects seasonally migrate large distances between different geographic regions (e.g., the overwintering areas of the Monarch butterfly).

Care of Young

The eusocial insects build nest, guard eggs, and provide food for offspring full-time. Most insects, however, lead short lives as adults, and rarely interact with one another except to mate or compete for mates. A small number exhibit some form of parental care, where they will at least guard their eggs, and sometimes continue guarding their offspring until adulthood, and possibly even feeding them. Another simple form of parental care is to construct a nest (a burrow or an actual construction, either of which may be simple or complex), store provisions in it, and lay an egg upon those provisions. The adult does not contact the growing offspring, but it nonetheless does provide food. This sort of care is typical for most species of bees and various types of wasps.

Locomotion

Flight

White-lined sphinx moth feeding in flight

Insects are the only group of invertebrates to have developed flight. The evolution of insect wings has been a subject of debate. Some entomologists suggest that the wings are from paranotal lobes, or extensions from the insect's exoskeleton called the nota, called the *paranotal theory*. Other theories are based on a pleural origin. These theories include suggestions that wings originated from modified gills, spiracular flaps or as from an appendage of the epicoxa. The *epicoxal theory* suggests the insect wings are modified epicoxal exites, a modified appendage at the base of the legs or coxa. In the Carboniferous age, some of the *Meganeura* dragonflies had as much as a 50 cm (20 in) wide wingspan. The appearance of gigantic insects has been found to be consistent with high atmospheric oxygen. The respiratory system of insects constrains their size, however the high oxygen in the atmosphere allowed larger sizes. The largest flying insects today are much smaller and include several moth species such as the Atlas moth and the White Witch (*Thysania agrippina*).

Insect flight has been a topic of great interest in aerodynamics due partly to the inability of steady-state theories to explain the lift generated by the tiny wings of insects. But insect wings are in motion, with flapping and vibrations, resulting in churning and eddies, and the misconception that physics says "bumblebees can't fly" persisted throughout most of the twentieth century.

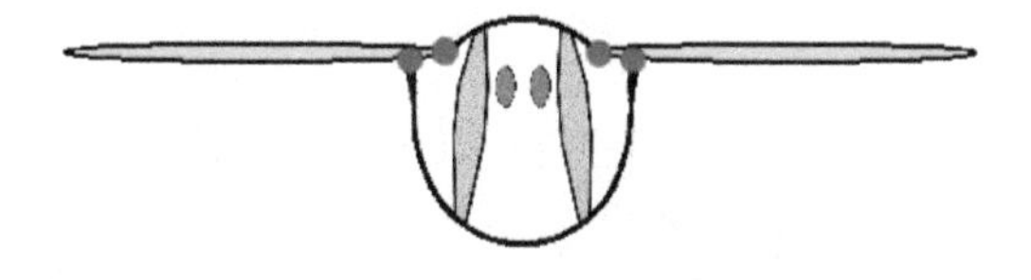

a b c d

Basic motion of the insect wing in insect with an indirect flight mechanism scheme of dorsoventral cut through a thorax segment with
a wings
b joints
c dorsoventral muscles
d longitudinal muscles.

Unlike birds, many small insects are swept along by the prevailing winds although many of the larger insects are known to make migrations. Aphids are known to be transported long distances by low-level jet streams. As such, fine line patterns associated with converging winds within weather radar imagery, like the WSR-88D radar network, often represent large groups of insects.

Walking

Many adult insects use six legs for walking and have adopted a tripedal gait. The tripedal gait allows for rapid walking while always having a stable stance and has been studied extensively in cockroaches. The legs are used in alternate triangles touching the ground. For the first step, the middle right leg and the front and rear left legs are in contact with the ground and move the insect forward, while the front and rear right leg and the middle left leg are lifted and moved forward to a new position. When they touch the ground to form a new stable triangle the other legs can be lifted and brought forward in turn and so on. The purest form of the tripedal gait is seen in insects moving at high speeds. However, this type of locomotion is not rigid and insects can adapt a variety of gaits. For example, when moving slowly, turning, or avoiding obstacles, four or more feet may be touching the ground. Insects can also adapt their gait to cope with the loss of one or more limbs.

Cockroaches are among the fastest insect runners and, at full speed, adopt a bipedal run to reach a high velocity in proportion to their body size. As cockroaches move very quickly, they need to be video recorded at several hundred frames per second to reveal their gait. More sedate locomotion is seen in the stick insects or walking sticks (Phasmatodea). A few insects have evolved to walk on the surface of the water, especially members of the Gerridae family, commonly known as water striders. A few species of ocean-skaters in the genus *Halobates* even live on the surface of open oceans, a habitat that has few insect species.

Use in Robotics

Insect walking is of particular interest as an alternative form of locomotion in robots. The study of insects and bipeds has a significant impact on possible robotic methods of transport. This may allow new robots to be designed that can traverse terrain that robots with wheels may be unable to handle.

Swimming

The backswimmer Notonecta glauca underwater, showing its paddle-like hindleg adaptation

A large number of insects live either part or the whole of their lives underwater. In many of the more primitive orders of insect, the immature stages are spent in an aquatic environment. Some groups of insects, like certain water beetles, have aquatic adults as well.

Many of these species have adaptations to help in under-water locomotion. Water beetles and water bugs have legs adapted into paddle-like structures. Dragonfly naiads use jet propulsion, forcibly expelling water out of their rectal chamber. Some species like the water striders are capable of walking on the surface of water. They can do this because their claws are not at the tips of the legs as in most insects, but recessed in a special groove further up the leg; this prevents the claws from piercing the water's surface film. Other insects such as the Rove beetle *Stenus* are known to emit pygidial gland secretions that reduce surface tension making it possible for them to move on the surface of water by Marangoni propulsion (also known by the German term *Entspannungsschwimmen*).

Ecology

Insect ecology is the scientific study of how insects, individually or as a community, interact with the surrounding environment or ecosystem. Insects play one of the most important roles in their ecosystems, which includes many roles, such as soil turning and aeration, dung burial, pest control, pollination and wildlife nutrition. An example is the beetles, which are scavengers that feed on dead animals and fallen trees and thereby recycle biological materials into forms found useful by other organisms. These insects, and others, are responsible for much of the process by which topsoil is created.

Defense and Predation

Insects are mostly soft bodied, fragile and almost defenseless compared to other, larger lifeforms. The immature stages are small, move slowly or are immobile, and so all stages are exposed to predation and parasitism. Insects then have a variety of defense strategies to avoid being attacked by predators or parasitoids. These include camouflage, mimicry, toxicity and active defense.

Perhaps one of the most well-known examples of mimicry, the viceroy butterfly (top) appears very similar to the noxious-tasting monarch butterfly (bottom).

Camouflage is an important defense strategy, which involves the use of coloration or shape to blend into the surrounding environment. This sort of protective coloration is common and widespread among beetle families, especially those that feed on wood or vegetation, such as many of the leaf beetles (family Chrysomelidae) or weevils. In some of these species, sculpturing or various colored scales or hairs cause the beetle to resemble bird dung or other inedible objects. Many of those that live in sandy environments blend in with the coloration of the substrate. Most phasmids are known for effectively replicating the forms of sticks and leaves, and the bodies of some species (such as *O. macklotti* and *Palophus centaurus*) are covered in mossy or lichenous outgrowths that supplement their disguise. Some species have the ability to change color as their surroundings shift (*B. scabrinota, T. californica*). In a further behavioral adaptation to supplement crypsis, a number of species have been noted to perform a rocking motion where the body is swayed from side to side that is thought to reflect the movement of leaves or twigs swaying in the breeze. Another method by which stick insects avoid predation and resemble twigs is by feigning death (catalepsy), where the insect enters a motionless state that can be maintained for a long period. The nocturnal feeding habits of adults also aids Phasmatodea in remaining concealed from predators.

Another defense that often uses color or shape to deceive potential enemies is mimicry. A number of longhorn beetles (family Cerambycidae) bear a striking resemblance to wasps, which helps them avoid predation even though the beetles are in fact harmless. Batesian and Müllerian mimicry complexes are commonly found in Lepidoptera. Genetic polymorphism and natural selection give rise to otherwise edible species (the mimic) gaining a survival advantage by resembling inedible species (the model). Such a mimicry complex is referred to as *Batesian* and is most commonly known by the mimicry by the limenitidine Viceroy butterfly of the inedible danaine Monarch. Later research has discovered that the Viceroy is, in fact more toxic than the Monarch and this resemblance should be considered as a case of Müllerian mimicry. In Müllerian mimicry, inedible species, usually within a taxonomic order, find it advantageous to resemble each other so as to reduce the sampling rate by predators who need to learn about the insects' inedibility. Taxa from the toxic genus *Heliconius* form one of the most well known Müllerian complexes.

Chemical defense is another important defense found amongst species of Coleoptera and Lepidoptera, usually being advertised by bright colors, such as the Monarch butterfly. They obtain their toxicity by sequestering the chemicals from the plants they eat into their own tissues. Some Lepidoptera manufacture their own toxins. Predators that eat poisonous butterflies and moths may become sick and vomit violently, learning not to eat those types of species; this is actually the basis of Müllerian mimicry. A predator who has previously eaten a poisonous lepidopteran may avoid other species with similar markings in the future, thus saving many other species as well. Some ground beetles of the Carabidae family can spray chemicals from their abdomen with great accuracy, to repel predators.

Pollination

European honey bee carrying pollen in a pollen basket back to the hive

Pollination is the process by which pollen is transferred in the reproduction of plants, thereby enabling fertilisation and sexual reproduction. Most flowering plants require an animal to do the transportation. While other animals are included as pollinators, the majority of pollination is done by insects. Because insects usually receive benefit for the pollination in the form of energy rich nectar it is a grand example of mutualism. The various flower traits (and combinations thereof) that differentially attract one type of pollinator or another are known as pollination syndromes. These arose through complex plant-animal adaptations. Pollinators find flowers through bright colorations, including ultraviolet, and attractant pheromones. The study of pollination by insects is known as *anthecology*.

Parasitism

Many insects are parasites of other insects such as the parasitoid wasps. These insects are known as entomophagous parasites. They can be beneficial due to their devastation of pests that can destroy crops and other resources. Many insects have a parasitic relationship with humans such as the mosquito. These insects are known to spread diseases such as malaria and yellow fever and because of such, mosquitoes indirectly cause more deaths of humans than any other animal.

Relationship to Humans

As Pests

Many insects are considered pests by humans. Insects commonly regarded as pests include those that are parasitic (*e.g.* lice, bed bugs), transmit diseases (mosquitoes, flies), damage structures (termites), or destroy agricultural goods (locusts, weevils). Many entomologists are involved in various forms of pest control, as in research for companies to produce insecticides, but increasingly rely on methods of biological pest control, or biocontrol. Biocontrol uses one organism to reduce the population density of another organism — the pest — and is considered a key element of integrated pest management.

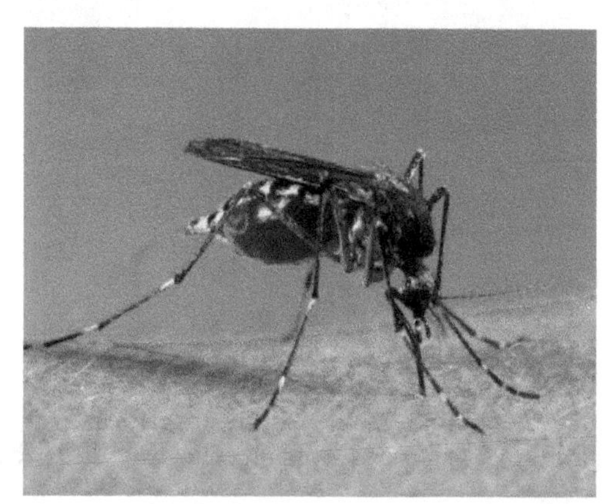

Aedes aegypti, a parasite, is the vector of dengue fever and yellow fever

Despite the large amount of effort focused at controlling insects, human attempts to kill pests with insecticides can backfire. If used carelessly, the poison can kill all kinds of organisms in the area, including insects' natural predators, such as birds, mice and other insectivores. The effects of DDT's use exemplifies how some insecticides can threaten wildlife beyond intended populations of pest insects.

In Beneficial Roles

Although pest insects attract the most attention, many insects are beneficial to the environment and to humans. Some insects, like wasps, bees, butterflies and ants, pollinate flowering plants. Pollination is a mutualistic relationship between plants and insects. As insects gather nectar from different plants of the same species, they also spread pollen from plants on which they have previously fed. This greatly increases plants' ability to cross-pollinate, which maintains and possibly even improves their evolutionary fitness. This ultimately affects humans since ensuring healthy crops is critical to agriculture. As well as pollination ants help with seed distribution of plants. This helps to spread the plants which increases plant diversity. This leads to an overall better environment. A serious environmental problem is the decline of populations of pollinator insects, and a number of species of insects are now cultured primarily for pollination management in order to have sufficient pollinators in the field, orchard or greenhouse at bloom time. Another solution, as shown in Delaware, has been to raise native plants to help support native pollinators like *L. vierecki*. Insects also produce useful substances such as honey, wax, lacquer and silk. Honey bees have been cultured by humans for thousands of years for honey, although contracting for crop pollination is becoming more significant for beekeepers. The silkworm has greatly affected human history, as silk-driven trade established relationships between China and the rest of the world.

Insectivorous insects, or insects which feed on other insects, are beneficial to humans because they eat insects that could cause damage to agriculture and human structures. For example, aphids feed on crops and cause problems for farmers, but ladybugs feed on aphids, and can be used as a means to get significantly reduce pest aphid populations. While birds are perhaps more visible predators of insects, insects themselves account for the vast majority of insect consumption. Ants also help control animal populations by consuming small vertebrates. Without predators to keep them in check, insects can undergo almost unstoppable population explosions.

Because they help flowering plants to cross-pollinate, some insects are critical to agriculture. This European honey bee is gathering nectar while pollen collects on its body.

Insects are also used in medicine, for example fly larvae (maggots) were formerly used to treat wounds to prevent or stop gangrene, as they would only consume dead flesh. This treatment is finding modern usage in some hospitals. Recently insects have also gained attention as potential sources of drugs and other medicinal substances. Adult insects, such as crickets and insect larvae of various kinds, are also commonly used as fishing bait.

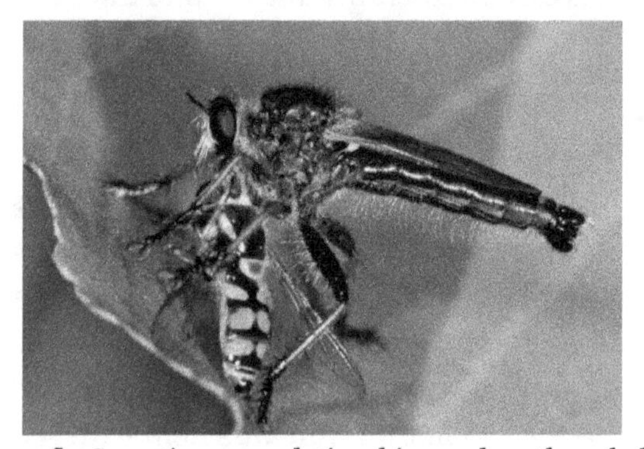

A robberfly with its prey, a hoverfly. Insectivorous relationships such as these help control insect populations.

In Research

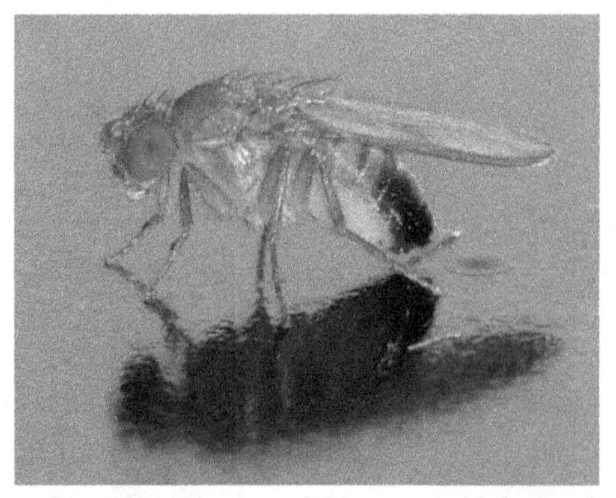

The common fruitfly Drosophila melanogaster is one of the most widely used organisms in biological research.

Insects play important roles in biological research. For example, because of its small size, short generation time and high fecundity, the common fruit fly *Drosophila melanogaster* is a model organism for studies in the genetics of higher eukaryotes. *D. melanogaster* has been an essential part of studies into principles like genetic linkage, interactions between genes, chromosomal genetics, development, behavior and evolution. Because genetic systems are well conserved among eukaryotes, understanding basic cellular processes like DNA replication or transcription in fruit flies can help to understand those processes in other eukaryotes, including humans. The genome of *D. melanogaster* was sequenced in 2000, reflecting the organism's important role in biological research. It was found that 70% of the fly genome is similar to the human genome, supporting the evolution theory.

As Food

In some cultures, insects, especially deep-fried cicadas, are considered to be delicacies, whereas in other places they form part of the normal diet. Insects have a high protein content for their mass, and some authors suggest their potential as a major source of protein in human nutrition. In most first-world countries, however, entomophagy (the eating of insects), is taboo. Since it is impossible to entirely eliminate pest insects from the human food chain, insects are inadvertently present in many foods, especially grains. Food safety laws in many countries do not prohibit insect parts in food, but rather limit their quantity. According to cultural materialist anthropologist Marvin Harris, the eating of insects is taboo in cultures that have other protein sources such as fish or livestock.

Due to the abundance of insects and a worldwide concern of food shortages, the Food and Agriculture Organisation of the United Nations considers that the world may have to, in the future, regard the prospects of eating insects as a food staple. Insects are noted for their nutrients, having a high content of protein, minerals and fats and are eaten by one-third of the global population.

In Culture

Scarab beetles held religious and cultural symbolism in Old Egypt, Greece and some shamanistic Old World cultures. The ancient Chinese regarded cicadas as symbols of rebirth or immortality. In Mesopotamian literature, the epic poem of Gilgamesh has allusions to Odonata which signify the impossibility of immortality. Amongst the Aborigines of Australia of the Arrernte language groups, honey ants and witchety grubs served as personal clan totems. In the case of the 'San' bush-men of the Kalahari, it is the praying mantis which holds much cultural significance including creation and zen-like patience in waiting.

Herbivore

A herbivore is an animal anatomically and physiologically adapted to eating plant material, for example foliage, for the main component of its diet. As a result of their plant diet, herbivorous animals typically have mouthparts adapted to rasping or grinding. Horses and other herbivores have wide flat teeth that are adapted to grinding grass, tree bark, and other tough plant material.

A deer and two fawns feeding on foliage

A large percentage of herbivores have mutualistic gut flora that help them digest plant matter, which is more difficult to digest than animal prey. This gut flora is made up of cellulose-digesting protozoans or bacteria living in the herbivores' intestines.

A caterpillar feeding on a leaf

Etymology

Herbivore is the anglicized form of a modern Latin coinage, *herbivora,* cited in Charles Lyell's 1830 *Principles of Geology*. Richard Owen employed the anglicized term in an 1854 work on fossil teeth and skeletons. *Herbivora* is derived from the Latin *herba* meaning a small plant or herb, and *vora,* from *vorare,* to eat or devour.

Definition and Related Terms

Herbivory is a form of consumption in which an organism principally eats autotrophs such as plants, algae and photosynthesizing bacteria. More generally, organisms that feed on autotrophs in general are known as primary consumers. *Herbivory* usually refers to animals eating plants; fungi, bacteria and protists that feed on living plants are usually termed plant pathogens (plant diseases), and microbes that feed on dead plants are saprotrophs. Flowering plants that obtain nutrition from other living plants are usually termed parasitic plants. There is however no single exclusive and definitive ecological classification of consumption patterns; each textbook has its own variations on the theme.

Evolution of Herbivory

Our understanding of herbivory in geological time comes from three sources: fossilized plants, which may preserve evidence of defence (such as spines), or herbivory-related damage; the observation of plant debris in fossilised animal faeces; and the construction of herbivore mouthparts.

A fossil *Viburnum lesquereuxii* leaf with evidence of insect herbivory; Dakota Sandstone (Cretaceous) of Ellsworth County, Kansas. Scale bar is 10 mm.

Although herbivory was long thought to be a Mesozoic phenomenon, fossils have shown that within less than 20 million years after the first land plants evolved, plants were being consumed by arthropods. Insects fed on the spores of early Devonian plants, and the Rhynie chert also provides evidence that organisms fed on plants using a "pierce and suck" technique.

During the next 75 million years plants evolved a range of more complex organs, such as roots and seeds. There is no evidence of any organism being fed upon until the middle-late Mississippian, 330.9 million years ago. There was a gap of 50 to 100 million years between the time each organ evolved and the time organisms evolved to feed upon them; this may be due to the low levels of oxygen during this period, which may have suppressed evolution. Further than their arthropod status, the identity of these early herbivores is uncertain. Hole feeding and skeletonisation are recorded in the early Permian, with surface fluid feeding evolving by the end of that period.

Herbivory among four-limbed terrestrial vertebrates, the tetrapods developed in the Late Carboniferous (307 - 299 million years ago). Early tetrapods were large amphibious piscivores. While amphibians continued to feed on fish and insects, some reptiles began exploring two new food types, tetrapods (carnivory) and plants (herbivory). The entire dinosaur order ornithischia was composed with herbivores dinosaurs. Carnivory was a natural transition from insectivory for medium and large tetrapods, requiring minimal adaptation. In contrast, a complex set of adaptations was necessary for feeding on highly fibrous plant materials.

Arthropods evolved herbivory in four phases, changing their approach to it in response to changing plant communities.

Tetrapod herbivores made their first appearance in the fossil record of their jaws near the Permio-Carboniferous boundary, approximately 300 million years ago. The earliest evidence of their herbivory has been attributed to dental occlusion, the process in which teeth from the upper jaw come in contact with teeth in the lower jaw is present. The evolution of dental occlusion led to a drastic increase in plant food processing and provides evidence about feeding strategies based on tooth wear patterns. Examination of phylogenetic frameworks of tooth and jaw morphologes has revealed that dental occlusion developed independently in several lineages tetrapod herbivores. This suggests that evolution and spread occurred simultaneously within various lineages.

Food Chain

Herbivores form an important link in the food chain; because they consume plants in order to digest the carbohydrates photosynthetically produced by a plant. Carnivores in turn consume herbivores for the same reason, while omnivores can obtain their nutrients from either plants or animals. Due to a herbivore's ability to survive solely on tough and fibrous plant matter, they are termed the primary consumers in the food cycle (chain). Herbivory, carnivory, and omnivory can be regarded as special cases of Consumer-Resource Systems.

Leaf miners feed on leaf tissue between the epidermal layers, leaving visible trails

Feeding Strategies

Two herbivore feeding strategies are grazing (e.g. cows) and browsing (e.g. moose). Although the exact definition of the feeding strategy may depend on the writer, most authors agree that to define a grazer at least 90% of the forage has to be grass, and for a browser at least 90% tree leaves and/or twigs. An intermediate feeding strategy is called "mixed-feeding". In their daily need to take up energy from forage, herbivores of different body mass may be selective in choosing their food. "Selective" means that herbivores may choose their forage source depending on, e.g., season or food availability, but also that they may choose high quality (and consequently highly nutritious) forage before lower quality. The latter especially is determined by the body mass of the herbivore, with small herbivores selecting for high quality forage, and with increasing body mass animals are less selective. Several theories attempt to explain and quantify the relationship between animals and their food, such as Kleiber's law, Holling's disk equation and the marginal value theorem.

Kleiber's law describes the relationship between an animal's size and its feeding strategy, saying that larger animals need to eat less food per unit weight than smaller animals. Kleiber's law states that the metabolic rate (q_0) of an animal is the mass of the animal (M) raised to the 3/4 power: $q_0=M^{3/4}$ Therefore, the mass of the animal increases at a faster rate than the metabolic rate.

Herbivores employ numerous types of feeding strategies. Many herbivores do not fall into one specific feeding strategy, but employ several strategies and eat a variety of plant parts.

Types of feeding strategies		
Feeding Strategy	**Diet**	**Example**

Algivores	Algae	krill, crabs, sea snail, sea urchin, parrotfish, surgeonfish, flamingo
Frugivores	Fruit	Ruffed lemurs
Folivores	Leaves	Koalas
Nectarivores	Nectar	Honey possum
Granivores	Seeds	Hawaiian honeycreepers
Palynivores	Pollen	Bees
Mucivores	Plant fluids, i.e. sap	Aphids
Xylophages	Wood	Termites

Optimal Foraging Theory is a model for predicting animal behavior while looking for food or other resource, such as shelter or water. This model assesses both individual movement, such as animal behavior while looking for food, and distribution within a habitat, such as dynamics at the population and community level. For example, the model would be used to look at the browsing behavior of a deer while looking for food, as well as that deer's specific location and movement within the forested habitat and its interaction with other deer while in that habitat.

This model has been criticized as circular and untestable. Critics have pointed out that its proponents use examples that fit the theory, but do not use the model when it does not fit the reality. Other critics point out that animals do not have the ability to assess and maximize their potential gains, therefore the optimal foraging theory is irrelevant and derived to explain trends that do not exist in nature.

Holling's disk equation models the efficiency at which predators consume prey. The model predicts that as the number of prey increases, the amount of time predators spend handling prey also increases and therefore the efficiency of the predator decreases. In 1959, S. Holling proposed an equation to model the rate of return for an optimal diet: Rate (R) = Energy gained in foraging (Ef)/ (time searching (Ts) + time handling (Th))

Where s = cost of search per unit time f = rate of encounter with items, h = handling time, e = energy gained per encounter

In effect, this would indicate that a herbivore in a dense forest would spend more time handling (eating) the vegetation because there was so much vegetation around than a herbivore in a sparse forest, who could easily browse through the forest vegetation. According to the Holling's disk equation, a herbivore in the sparse forest would be more efficient at eating than the herbivore in the dense forest

The marginal value theorem describes the balance between eating all the food in a patch for immediate energy, or moving to a new patch and leaving the plants in the first patch to regenerate for future use. The theory predicts that absent complicating factors, an animal should leave a resource patch when the rate of payoff (amount of food) falls below the average rate of payoff for the entire area. According to this theory, locus should move to a new patch of food when the patch they are currently feeding on requires more energy to obtain food than an average patch. Within this theory, two subsequent parameters emerge, the Giving Up Density (GUD) and the Giving Up Time (GUT). The Giving Up Density (GUD) quantifies the amount of food that remains in a patch when a forager moves to a new patch. The Giving Up Time (GUT) is used when an animal continuously assesses the patch quality.

Attacks and Counter-attacks

Herbivore Offense

The myriad defenses displayed by plants means that their herbivores need a variety of skills to overcome these defenses and obtain food. These allow herbivores to increase their feeding and use of a host plant. Herbivores have three primary strategies for dealing with plant defenses: choice, herbivore modification, and plant modification.

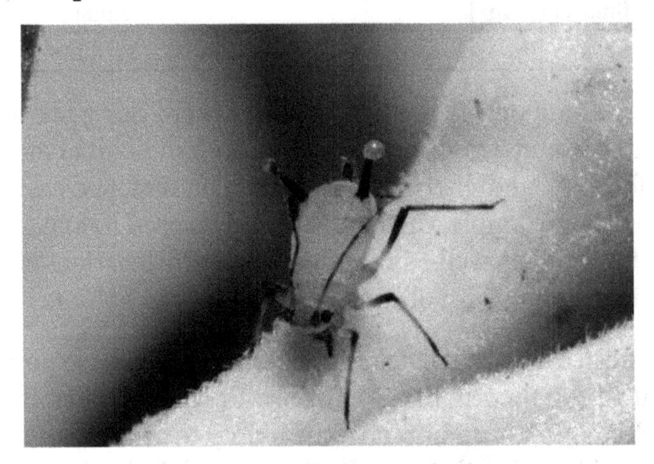

Aphids are fluid feeders on plant sap.

Feeding choice involves which plants a herbivore chooses to consume. It has been suggested that many herbivores feed on a variety of plants to balance their nutrient uptake and to avoid consuming too much of any one type of defensive chemical. This involves a tradeoff however, between foraging on many plant species to avoid toxins or specializing on one type of plant that can be detoxified.

Herbivore modification is when various adaptations to body or digestive systems of the herbivore allow them to overcome plant defenses. This might include detoxifying secondary metabolites, sequestering toxins unaltered, or avoiding toxins, such as through the production of large amounts of saliva to reduce effectiveness of defenses. Herbivores may also utilize symbionts to evade plant defences. For example, some aphids use bacteria in their gut to provide essential amino acids lacking in their sap diet.

Plant modification occurs when herbivores manipulate their plant prey to increase feeding. For example, some caterpillars roll leaves to reduce the effectiveness of plant defenses activated by sunlight.

Plant Defense

A plant defense is a trait that increases plant fitness when faced with herbivory. This is measured relative to another plant that lacks the defensive trait. Plant defenses increase survival and/or reproduction (fitness) of plants under pressure of predation from herbivores.

Defense can be divided into two main categories, tolerance and resistance. Tolerance is the ability of a plant to withstand damage without a reduction in fitness. This can occur by diverting herbivory to non-essential plant parts or by rapid regrowth and recovery from herbivory. Resistance refers to the ability of a plant to reduce the amount of damage it receives from a herbivore. This can occur via avoidance in space or time, physical defenses, or chemical defenses. Defenses can either be constitutive, always present in the plant, or induced, produced or translocated by the plant following damage or stress.

Physical, or mechanical, defenses are barriers or structures designed to deter herbivores or reduce intake rates, lowering overall herbivory. Thorns such as those found on roses or acacia trees are one example, as are the spines on a cactus. Smaller hairs known as trichomes may cover leaves or stems and are especially effective against invertebrate herbivores. In addition, some plants have waxes or resins that alter their texture, making them difficult to eat. Also the incorporation of silica into cell walls is analogous to that of the role of lignin in that it is a compression-resistant structural component of cell walls; so that plants with their cell walls impregnated with silica are thereby afforded a measure of protection against herbivory.

Chemical defenses are secondary metabolites produced by the plant that deter herbivory. There are a wide variety of these in nature and a single plant can have hundreds of different chemical defenses. Chemical defenses can be divided into two main groups, carbon-based defenses and nitrogen-based defenses.

1. Carbon-based defenses include terpenes and phenolics. Terpenes are derived from 5-carbon isoprene units and comprise essential oils, carotenoids, resins, and latex. They can have a number of functions that disrupt herbivores such as inhibiting adenosine triphosphate (ATP) formation, molting hormones, or the nervous system. Phenolics combine an aromatic carbon ring with a hydroxyl group. There are a number of different phenolics such as lignins, which are found in cell walls and are very indigestible except for specialized microorganisms; tannins, which have a bitter taste and bind to proteins making them indigestible; and furanocumerins, which produce free radicals disrupting DNA, protein, and lipids, and can cause skin irritation.

2. Nitrogen-based defenses are synthesized from amino acids and primarily come in the form of alkaloids and cyanogens. Alkaloids include commonly recognized substances such as caffeine, nicotine, and morphine. These compounds are often bitter and can inhibit DNA or RNA synthesis or block nervous system signal transmission. Cyanogens get their name from the cyanide stored within their tissues. This is released when the plant is damaged and inhibits cellular respiration and electron transport.

Plants have also changed features that enhance the probability of attracting natural enemies to herbivores. Some emit semiochemicals, odors that attract natural enemies, while others provide food and housing to maintain the natural enemies' presence, e.g. ants that reduce herbivory. A given plant species often has many types of defensive mechanisms, mechanical or chemical, constitutive or induced, which allow it to escape from herbivores.

Herbivore–plant Interactions per Predator–prey Theory

According to the theory of predator–prey interactions, the relationship between herbivores and plants is cyclic. When prey (plants) are numerous their predators (herbivores) increase in numbers, reducing the prey population, which in turn causes predator number to decline. The prey population eventually recovers, starting a new cycle. This suggests that the population of the herbivore fluctuates around the carrying capacity of the food source, in this case the plant.

Several factors play into these fluctuating populations and help stabilize predator–prey dynamics. For example, spatial heterogeneity is maintained, which means there will always be pockets of plants not found by herbivores. This stabilizing dynamic plays an especially important role for specialist herbivores that feed on one species of plant and prevents these specialists from wiping out their food source. Prey defenses also help stabilize predator–prey dynamics, and for more information on these relationships see the section on Plant Defenses. Eating a second prey type helps herbivores' populations stabilize. Alternating between two or more plant types provides population stability for the herbivore, while the populations of the plants oscillate. This plays an important role for generalist herbivores that eat a variety of plants. Keystone herbivores keep vegetation populations in check and allow for a greater diversity of both herbivores and plants. When an invasive herbivore or plant enters the system, the balance is thrown off and the diversity can collapse to a monotaxon system.

The back and forth relationship of plant defense and herbivore offense can be seen as a sort of "adaptation dance" in which one partner makes a move and the other counters it. This reciprocal change drives coevolution between many plants and herbivores, resulting in what has been referred to as a "coevolutionary arms race". The escape and radiation mechanisms for coevolution, presents the idea that adaptations in herbivores and their host plants, has been the driving force behind speciation.

While much of the interaction of herbivory and plant defense is negative, with one individual reducing the fitness of the other, some is actually beneficial. This beneficial herbivory takes the form of mutualisms in which both partners benefit in some way from the interaction. Seed dispersal by herbivores and pollination are two forms of mutualistic herbivory in which the herbivore receives a food resource and the plant is aided in reproduction.

Impacts

Herbivorous fish and marine animals are an indispensable part of the coral reef ecosystem. Since algae and seaweeds grow much faster than corals they can occupy spaces where corals could have settled. They can outgrow and thus outcompete corals on bare surfaces. In the absence of plant-eating fish, seaweeds deprive corals of sunlight. They can also physically damage corals with scrapes.

The impact of herbivory can be seen in areas ranging from economics to ecological, and both. For example, environmental degradation from white-tailed deer (Odocoileus virginianus) in the US alone has the potential to both change vegetative communities through over-browsing and cost forest restoration projects upwards of $750 million annually. Agricultural crop damage by the same species totals approximately $100 million every year. Insect crop damages also contribute largely to annual crop losses in the U.S. Herbivores affect economics through the revenue generated by hunting and ecotourism. For example, the hunting of herbivorous game species such as white-tailed deer, cottontail rabbits, antelope, and elk in the U.S. contributes greatly to the billion-dollar annually hunting industry. Ecotourism is a major source of revenue, particularly in Africa, where many large mammalian herbivores such as elephants, zebras, and giraffes help to bring in the equivalent of millions of US dollars to various nations annually.

Parasitic Plant

A parasitic plant is a plant that derives some or all of its nutritional requirements from another living plant. All parasitic plants have modified roots, named haustoria (singular: haustorium), which penetrate the host plants, connecting them to the conductive system – either the xylem, the phloem, or both. This provides them with the ability to extract water and nutrients from the host. The organisms which support the parasite are called hosts. Some parasitic plants are able to locate their host plants by detecting chemicals in the air or soil given off by host shoots or roots, respectively. About 4,100 species of parasitic plant in approximately 19 families of flowering plants are known.

Cuscuta, a stem holoparasite, on an acacia tree in Pakistan

Evolution of Parasitic Behavior

Parasitic behavior evolved in angiosperms roughly 12-13 times independently, a classic example of convergent evolution. Roughly 1% of all angiosperm species are parasitic, with a large degree of host dependence.

Seed Germination

Seed germination of parasitic plants occurs in a variety of ways. These means can either be chemical or mechanical and the means used by seeds often depends on whether or not the parasites are root parasites or stem parasites. Most parasitic plants need to germinate in close proximity to their host plants because their seeds are limited by the amount of resources available to survive without nutrients from their host plants. Resources are limited due in part to the fact that most parasitic plants are not able to use autotrophic nutrition to establish the early stages of seeding.

Root parasitic plant seeds tend to use chemical cues for germination. In order for germination to occur, seeds need to be fairly close to their host plant. For example, the seeds of the parasitic plant witchweed (*Striga asiatica*) need to be within 3 to 4 millimeters (mm) of its host in order to pick up chemical signals in the soil to signal germination. This range is important because *Striga asiatica* will only grow about 4 mm after germination. Chemical compound cues sensed by parasitic plant seeds are from host plant root exudates that are leached in close proximity from the host's root system into the surrounding soil. These chemical cues are a variety of compounds that are unstable and rapidly degraded in soil and are present within a radius of a few meters of the plant exuding them. Parasitic plants germinate and follow a concentration gradient of these compounds in the soil toward the host plants if close enough. These compounds are called strigolactones. Strigolactone stimulates ethylene biosynthesis in seeds causing them to germinate.

There are a variety of chemical germination stimulants. Strigol was the first of the germination stimulants to be isolated. It was isolated from a non-host cotton plant and has been found in true host plants such as corn and millets. The stimulants are usually plant specific, examples of other germination stimulants include sorgolactone from sorghum, orobanchol and alectrol from red clover, and 5-deoxystrigol from *Lotus japonicas*. Strigolactones are apocarotenoids that are produced via the carotenoid pathway of plants. Strigolactones and mycorrhizal fungi have a relationship in which strigolactone also cues the growth of mycorrhizal fungus.

Stem parasitic plants, unlike most root plants, germinate using the resources inside its endosperm and are able to survive for a small amount of time. An example, Dodder (*Cuscuta* spp.), is a parasitic plant whose seed falls to the ground and may remain dormant for up to five years before it is able to sense a host plants nearby. Using the resources in the seed endosperm, Dodder is able to germinate. Once germinated, the plant has 6 days to find and establish a connection with its host plant before its resources run out.

Dodder seeds germinate above ground and then the plant sends out stems in search of its host plant reaching up to 6 cm before it dies. It is believed that the plant uses two methods of finding a host. The stem is able to pick up its host plant's scent whereby it then is able to orient itself in the direction of its host. Scientists used volatiles from tomato plants (α-pinene, β-myrcene, and β-phellandrene) to test the reaction of C. *pentagona* and found that the stem will oriented itself in the direction of the odor. Some studies suggest that by using light reflecting from nearby plants dodders are able to select host with higher sugar because of the levels of chlorophyll in the leaves. Once Dodder finds its host, it wraps itself around the host plants stem. Using adventitious roots, Dodder taps into the host plant's stem and creates a haustorium, which is a special connection into the host plant vascular tissue. Dodder makes several of these connections with the host as it moves up the plant.

Host Range

Some parasitic plants are generalists and parasitize many different species, even several different species at once. Dodder (*Cassytha* spp., *Cuscuta* spp.) and red rattle (*Odontites vernus*) are generalist parasites. Other parasitic plants are specialists that parasitize a few or even just one species. Beech drops (*Epifagus virginiana*) is a root holoparasite only on American beech (*Fagus grandifolia*). *Rafflesia* is a holoparasite on the vine *Tetrastigma*. Plants such as *Pterospora* become parasites of mycorrhizal fungi. There is evidence that parasites also practice self-discrimination, species of *Tryphysaria* experience reduced haustorium development in the presence of other *Tryphysaria*. Although, the mechanism for self-discrimination in parasites is not yet known.

Aquatic Parasitic Plants

Parasitism also evolved within aquatic species of plants and algae. Parasitic marine plants are described as benthic, meaning that they are sedentary or attached to another structure. Plants and algae that grow on the host plant, using it as an attachment point are given the designation epiphytic (epilithic is the name given to plants/algae that use rocks or boulders for attachment), while not necessarily parasitic, some species occur in high correlation with a certain host species, suggesting that they rely on the host plant in some way or another. In contrast, endophytic plants and algae grow inside their host plant, these have a wide range of host dependence from obligate holoparasites to facultative hemiparasites.

Marine parasites occur as a higher proportion of marine flora in temperate rather than tropical waters. While no full explanation for this is available, many of the potential host plants such as kelp and other macroscopic brown algae are generally restricted to temperate areas. Roughly 75% of parasitic red algae infect hosts in the same taxonomic family as themselves, these are given the designation adelphoparasites. Other marine parasites, deemed endozoic, are parasites of marine invertebrates (molluscs, flatworms, sponges) and can be either holoparasitic or hemiparasitic, some retaining the ability to photosynthesize after infection.

Importance

Species within *Orobanchaceae* are some of the most economically destructive species on Earth. Species of *Striga* alone are estimated to cost billions of dollars a year in crop yield loss annually, infesting over 50 million hectares of cultivated land within Sub-Saharan Africa alone. *Striga* can infect both grasses and grains, including corn, rice and sorghum, undoubtedly some of the most important food crops. *Orobanche* also threatens a wide range of important crops, including peas, chickpeas, tomatoes, carrots, and varieties of the genus *Brassica* (e.g. cabbage, lettuce, and broccoli). Yield loss from *Orobanche* can reach 100% and has caused farmers in some regions of the world to abandon certain staple crops and begin importing others as an alternative. Much research has been devoted to the control of Orobanche and Striga species, which are even more devastating in developing areas of the world, though no method has been found to be entirely successful.

- Mistletoes cause economic damage to forest and ornamental trees.
- *Rafflesia arnoldii* produces the world's largest flowers at about one meter in diameter. It is a tourist attraction in its native habitat.
- Sandalwood trees (*Santalum* species) have many important cultural uses and their fragrant oils have high commercial value.
- Indian paintbrush (*Castilleja linariaefolia*) is the state flower of Wyoming.
- The Oak Mistletoe (*Phoradendron serotinum*) is the floral emblem of Oklahoma.
- A few other parasitic plants are occasionally cultivated for their attractive flowers, such as *Nutysia* and broomrape.
- Parasitic plants are important in research, especially on the loss of photosynthesis during evolution.

- A few dozen parasitic plants have occasionally been used as food by people.
- Western Australian Christmas tree (*Nuytsia floribunda*) sometimes damages underground cables. It mistakes the cables for host roots and tries to parasitize them using its sclerenchymatic guillotine.

Newly emergent snow plant (Sarcodes sanguinea), a fungus parasite

Plants Parasitic on Fungi

About 400 species of flowering plants, plus one gymnosperm (*Parasitaxus usta*), are parasitic on mycorrhizal fungi. This effectively gives these plants the ability to become associated with many of the other plants around them. They are termed myco-heterotrophs rather than parasitic plants. Some myco-heterotrophs are Indian pipe (*Monotropa uniflora*), snow plant (*Sarcodes sanguinea*), underground orchid (*Rhizanthella gardneri*), bird's nest orchid (*Neottia nidus-avis*), and sugarstick (*Allotropa virgata*). Within the taxonomic family *Ericaceae*, known for extensive mycorrhizal relationships, there are the Monotropoids. The Monotropoids include the genera *Monotropa*, *Monotropsis*, and *Pterospora* among others. Myco-heterotrophic behavior is commonly accompanied by the loss of chlorophyll.

References

- Alberts B, Johnson A, Lewis J, Raff M, Roberts K, Walter P (2002). Molecular Biology of the Cell. Garland Science. ISBN 0-8153-3218-1.
- Lewin, Benjamin.; Krebs, Jocelyn E.; Kilpatrick, Stephen T.; Goldstein, Elliott S.; Lewin, Benjamin. Genes IX. (2011). Lewin's genes. Sudbury, Mass.: Jones and Bartlett. p. 23. ISBN 9780763766320.
- Brian W. J. Mahy, Marc H. V. Van Regenmortel (ed.). Desk Encyclopedia of Plant and Fungal Virology. Academic Press. pp. 71–81. ISBN 978-0123751485.
- Pallas V, Martinez G, Gomez G (2012). "The interaction between plant viroid-induced symptoms and RNA silencing". Methods Mol. Biol. Methods in Molecular Biology. 894: 323–43. doi:10.1007/978-1-61779-882-5_22. ISBN 978-1-61779-881-8. PMID 22678590.
- Pommerville, Jeffrey C (2014). Fundamentals of Microbiology. Burlington, MA: Jones and Bartlett Learning. p. 482. ISBN 978-1-284-03968-9.
- Chapman, A. D. (2006). Numbers of living species in Australia and the World. Canberra: Australian Biological Resources Study. ISBN 978-0-642-56850-2.
- Gullan, P.J.; Cranston, P.S. (2005). The Insects: An Outline of Entomology (3 ed.). Oxford: Blackwell Publishing. ISBN 1-4051-1113-5.

- Kapoor, V.C. C. (1998). Principles and Practices of Animal Taxonomy. 1 (1 ed.). Science Publishers. p. 48. ISBN 1-57808-024-X.
- Resh, Vincent H.; Ring T. Carde (2009). Encyclopedia of Insects (2 ed.). U. S. A.: Academic Press. ISBN 0-12-374144-0.
- Barnes, R.S.K.; Calow, P.; Olive, P.; Golding, D.; and Spicer, J. (2001). "Invertebrates with Legs: the Arthropods and Similar Groups". The Invertebrates: A Synthesis. Blackwell Publishing. p. 168. ISBN 0-632-04761-5.
- Richard W. Merritt, Kenneth W. Cummins, and Martin B. Berg (editors) (2007). An Introduction to the Aquatic Insects of North America (4th ed.). Kendall Hunt Publishers. ISBN 978-0-7575-5049-2.
- Merritt, RW; KW Cummins & MB Berg (2007). An Introduction To The Aquatic Insects Of North America. Kendall Hunt Publishing Company. ISBN 0-7575-4128-3.
- Chapman, R.F. (1998). The Insects; Structure and Function (4th ed.). Cambridge, UK: Cambridge University Press. ISBN 0521578906.
- Ruppert, E.E.; Fox, R.S. & Barnes, R.D. (2004). Invertebrate Zoology (7 ed.). Brooks / Cole. pp. 523–524. ISBN 0-03-025982-7.
- Schowalter, Timothy Duane (2006). Insect ecology: an ecosystem approach (2(illustrated) ed.). Academic Press. p. 572. ISBN 978-0-12-088772-9.
- Evans, Arthur V.; Charles Bellamy (2000). An Inordinate Fondness for Beetles. University of California Press. ISBN 978-0-520-22323-3.
- Davidson, E. (2006). Big Fleas Have Little Fleas: How Discoveries of Invertebrate Diseases Are Advancing Modern Science. Tucson, Ariz.: University of Arizona Press. ISBN 0-8165-2544-7.
- Holldobler, Wilson (1994). Journey to the ants: a story of scientific exploration. Westminster college McGill Library: Cambridge, Mass.:Belknap Press of Haravard University Press, 1994. pp. 196–199. ISBN 0-674-48525-4.
- Smith, Deborah T (1991). Agriculture and the Environment: The 1991 Yearbook of Agriculture (1991 ed.). United States Government Printing. ISBN 0-16-034144-2.
- Davidson, RH; William F. Lyon (1979). Insect Pests of Farm, Garden, and Orchard. Wiley, John & Sons. p. 38. ISBN 0-471-86314-9.
- Pierce, BA (2006). Genetics: A Conceptual Approach (2nd ed.). New York: W.H. Freeman and Company. p. 87. ISBN 0-7167-8881-0.
- Michels, John (1880). John Michels, ed. Science. 1. American Association for the Advance of Science. 229 Broadway ave., N.Y.: American Association for the Advance of Science. pp. 2090pp. ISBN 1-930775-36-9.

3

Oomycete and its Types

Oomycetes are absorptive organisms that reproduce sexually and asexually. These cause devastating diseases in plants, such as late blight of potatoes and the sudden death of oak trees. Oomycete has numerous types; some of these are phytophthora, Phytophthora infestans, Phytophthora cinnamon and Phytophthora capsici. The major categories of oomycete are dealt with great detail in the chapter.

Oomycete

Oomycota or oomycetes form a distinct phylogenetic lineage of fungus-like eukaryotic microorganisms. They are filamentous, microscopic, absorptive organisms that reproduce both sexually and asexually. Oomycetes occupy both saprophytic and pathogenic lifestyles, and include some of the most notorious pathogens of plants, causing devastating diseases such as late blight of potato and sudden oak death. One oomycete, the mycoparasite Pythium oligandrum, is used for biocontrol, attacking plant pathogenic fungi. The oomycetes are also often referred to as water molds (or water moulds), although the water-preferring nature which led to that name is not true of most species, which are terrestrial pathogens. The Oomycota have a very sparse fossil record. A possible oomycete has been described from Cretaceous amber.

Morphology

The oomycetes rarely have septa, and if they do, they are scarce, appearing at the bases of sporangia, and sometimes in older parts of the filaments. Some are unicellular, but others are filamentous and branching.

Phylogenetic Relationships

This group was originally classified among the fungi (the name "oomycota" means "egg fungus") and later treated as protists, based on general morphology and lifestyle. A cladistic analysis based on modern discoveries about the biology of these organisms supports a relatively close relationship with some photosynthetic organisms, such as brown algae and diatoms. A common taxonomic classification based on these data, places the class Oomycota along with other classes such as Phaeophyceae (brown algae) within the phylum Heterokonta.

This relationship is supported by a number of observed differences in the characteristics of oomycetes and fungi. For instance, the cell walls of oomycetes are composed of cellulose rather than chitin and generally do not have septations. Also, in the vegetative state they have diploid nuclei, whereas fungi have haploid nuclei. Most oomycetes produce self-motile zoospores with two flagella. One flagellum has a "whiplash" morphology, and the other a branched "tinsel" morphology. The "tinsel" flagellum is unique to the Kingdom Heterokonta. Spores of the few fungal groups which retain flagella (such as the Chytridiomycetes) have only one whiplash flagellum. Oomycota and fungi have different metabolic pathways for synthesizing lysine and have a number of enzymes that differ. The ultrastructure is also different, with oomycota having tubular mitochondrial cristae and fungi having flattened cristae.

In spite of this evidence to the contrary, many species of oomycetes are still described or listed as types of fungi and may sometimes be referred to as pseudofungi, or lower fungi.

Classification

Previously the group was arranged into six orders.

- The Saprolegniales are the most widespread. Many break down decaying matter; others are parasites.
- The Leptomitales have wall thickenings that give their continuous cell body the appearance of septation. They bear chitin and often reproduce asexually.
- The Rhipidiales use rhizoids to attach their thallus to the bed of stagnant or polluted water bodies.
- The Albuginales are considered by some authors to be a family (Albuginaceae) within the Peronosporales, although it has been shown that they are phylogenetically distinct from this order.
- The Peronosporales too are mainly saprophytic or parasitic on plants, and have an aseptate, branching form. Many of the most damaging agricultural parasites belong to this order.
- The Lagenidiales are the most primitive; some are filamentous, others unicellular; they are generally parasitic.

However more recently this has been expanded considerably.

- Anisolpidiales Dick 2001
    - Anisolpidiaceae Karling 1943
- Lagenismatales Dick 2001
    - Lagenismataceae Dick 1995
- Salilagenidiales Dick 2001
    - Salilagenidiaceae Dick 1995
- Rozellopsidales Dick 2001
    - Rozellopsidaceae Dick 1995
    - Pseudosphaeritaceae Dick 1995

- Ectrogellales
  - Ectrogellaceae
- Haptoglossales
  - Haptoglossaceae
- Eurychasmales
  - Eurychasmataceae Petersen 1905
- Haliphthorales
  - Haliphthoraceae Vishniac 1958
- Olpidiopsidales
  - Sirolpidiaceae Cejp 1959
  - Pontismataceae Petersen 1909
  - Olpidiopsidaceae Cejp 1959
- Atkinsiellales
  - Atkinisellaceae
  - Crypticolaceae Dick 1995
- Saprolegniales
  - Achlyaceae
  - Verrucalvaceae Dick 1984
  - Saprolegniaceae Warm. 1884 [Leptolegniaceae]
- Leptomitales
  - Leptomitaceae Kuetz. 1843 [Apodachlyellaceae Dick 1986]
  - Leptolegniellaceae Dick 1971 [Ducellieriaceae Dick 1995]
- Rhipidiales
  - Rhipidiaceae Cejp 1959
- Albuginales
  - Albuginaceae Schroet. 1893
- Peronosporales [Pythiales; Sclerosporales; Lagenidiales]
  - Salisapiliaceae

- Pythiaceae Schroet. 1893 [Pythiogetonaceae; Lagenaceae Dick 1994; Lagenidiaceae; Peronophythoraceae; Myzocytiopsidaceae Dick 1995]
- Peronosporaceae Warm. 1884 [Sclerosporaceae Dick 1984]

Etymology

"Oomycota" means "egg fungi", referring to the large round oogonia, structures containing the female gametes, that are characteristic of the oomycetes.

The name "water mold" refers to their earlier classification as fungi and their preference for conditions of high humidity and running surface water, which is characteristic for the basal taxa of the oomycetes.

Biology

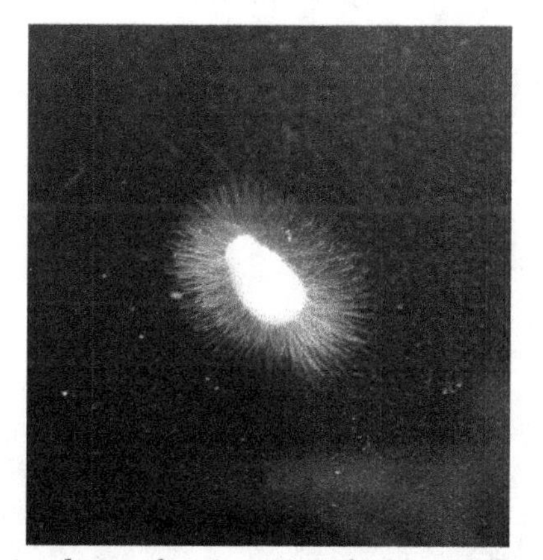

A culture of an oomycete from a stream

Reproduction

Most of the oomycetes produce two distinct types of spores. The main dispersive spores are asexual, self-motile spores called zoospores, which are capable of chemotaxis (movement toward or away from a chemical signal, such as those released by potential food sources) in surface water (including precipitation on plant surfaces). A few oomycetes produce aerial asexual spores that are distributed by wind. They also produce sexual spores, called oospores, that are translucent, double-walled, spherical structures used to survive adverse environmental conditions.

Pathogenicity

Many oomycetes species are economically important, aggressive plant pathogens. Some species can cause disease in fish, and at least one is a pathogen of mammals. The majority of the plant pathogenic species can be classified into four groups, although more exist.

- The *Phytophthora* group is a paraphyletic genus that causes diseases such as dieback, late blight in potatoes (the cause of the Irish Potato Famine of the 1840s that ravaged Ireland and other parts of Europe), sudden oak death, rhododendron root rot, and ink disease in the European chestnut

- The paraphyletic *Pythium* group is more prevalent than *Phytophthora* and individual species have larger host ranges, although usually causing less damage. *Pythium* damping off is a very common problem in greenhouses, where the organism kills newly emerged seedlings. Mycoparasitic members of this group (e.g. *P. oligandrum*) parasitize other oomycetes and fungi, and have been employed as biocontrol agents. One *Pythium* species, *Pythium insidiosum*, also causes Pythiosis in mammals.
- The third group are the downy mildews, which are easily identifiable by the appearance of white, brownish or olive "mildew" on the leaf undersides (although this group can be confused with the unrelated fungal powdery mildews).
- The fourth group are the white blister rusts, Albuginales, which cause white blister disease on a variety of flowering plants. White blister rusts sporulate beneath the epidermis of their hosts, causing spore-filled blisters on stems, leaves and the inflorescence. The Albuginales are currently divided into three genera, *Albugo* parasitic predominantly to Brassicales, *Pustula*, parasitic predominantly to Asterales, and *Wilsoniana*, predominantly parasitic to Caryophyllales. Like the downy mildews, the white blister rusts are obligate biotrophs, which means that they are unable to survive without the presence of a living host.

Phytophthora

Sudden oak death caused by Phytophthora ramorum

Phytophthora is a genus of plant-damaging Oomycetes (water molds), whose member species are capable of causing enormous economic losses on crops worldwide, as well as environmental damage in natural ecosystems. The cell wall of *Phytophthora* is made up of cellulose. The genus was first described by Heinrich Anton de Bary in 1875. Approximately 100 species have been described, although 100–500 undiscovered *Phytophthora* species are estimated to exist.

Pathogenicity

Phytophthora spp. are mostly pathogens of dicotyledons, and many are relatively host-specific parasites. Phytophthora cinnamomi, though, infects thousand of species ranging from club mosses, ferns, cycads, conifers, grasses, lilies, to members of many dicotyledonous families. Many species of *Phytophthora* are plant pathogens of considerable economic importance. *Phytophthora infestans* was the infective agent of the potato blight that caused the Great Irish Famine (1845–1849), and still remains the most destructive pathogen of solanaceous crops, including tomato and potato. The soya bean root and stem rot agent, *Phytophthora sojae*, has also caused longstanding problems for the agricultural industry. In general, plant diseases caused by this genus are difficult to control chemically, thus the growth of resistant cultivars is the main management strategy. Other important *Phytophthora* diseases are:

- *Phytophthora* taxon Agathis—causes collar-rot on New Zealand kauri (*Agathis australis*), New Zealand's most voluminous tree, an otherwise successful survivor of the Jurassic
- *Phytophthora cactorum*—causes rhododendron root rot affecting rhododendrons, azaleas and causes bleeding canker in hardwood trees
- *Phytophthora capsici*—infects Cucurbitaceae fruits, such as cucumbers and squash
- *Phytophthora cinnamomi*—causes cinnamon root rot affecting forest and fruit tress, and woody ornamentals including arborvitae, azalea, Chamaecyparis, dogwood, forsythia, Fraser fir, hemlock, Japanese holly, juniper, Pieris, rhododendron, Taxus, white pine, American chestnut and Australian Jarrah.
- *Phytophthora fragariae*—causes red root rot affecting strawberries
- *Phytophthora kernoviae*—pathogen of beech and rhododendron, also occurring on other trees and shrubs including oak, and holm oak. First seen in Cornwall, UK, in 2003.
- *Phytophthora lateralis*—causes cedar root disease in Port Orford cedar trees
- *Phytophthora megakarya*—one of the cocoa black pod disease species, is invasive and probably responsible for the greatest cocoa crop loss in Africa
- *Phytophthora nicotianae*—infects onions
- *Phytophthora palmivora*—causes fruit rot in coconuts and betel nuts
- *Phytophthora ramorum*—infects over 60 plant genera and over 100 host species; causes sudden oak death
- *Phytophthora quercina*—causes oak death
- *Phytophthora sojae*—causes soybean root rot

Research beginning in the 1990s has placed some of the responsibility for European forest dieback on the activity of imported Asian *Phytophthoras*.

Fungi Resemblance

Phytophthora is sometimes referred to as a fungus-like organism, but it is classified under a different kingdom altogether: Chromalveolata (formerly Stramenopila and previously Chromista). This is a good example of convergent evolution: *Phytophthora* is morphologically very similar to true fungi yet its evolutionary history is completely distinct. In contrast to fungi, chromalveolatas are more closely related to plants than to animals. Whereas fungal cell walls are made primarily of chitin, chromalveolata cell walls are constructed mostly of cellulose. Ploidy levels are different between these two groups; *Phytophthora* species have diploid (paired) chromosomes in the vegetative (growing, nonreproductive) stage of life, whereas fungi are almost always haploid in this stage. Biochemical pathways also differ, notably the highly conserved lysine synthesis path.

Biology

Phytophthora species may reproduce sexually or asexually. In many species, sexual structures have never been observed, or have only been observed in laboratory matings. In homothallic species, sexual structures occur in single culture. Heterothallic species have mating strains, designated as A1 and A2. When mated, antheridia introduce gametes into oogonia, either by the oogonium passing through the antheridium (amphigyny) or by the antheridium attaching to the proximal (lower) half of the oogonium (paragyny), and the union producing oospores. Like animals, but not like most true fungi, meiosis is gametic, and somatic nuclei are diploid. Asexual (mitotic) spore types are chlamydospores, and sporangia which produce zoospores. Chlamydospores are usually spherical and pigmented, and may have a thickened cell wall to aid in their role as a survival structure. Sporangia may be retained by the subtending hyphae (noncaducous) or be shed readily by wind or water tension (caducous) acting as dispersal structures. Also, sporangia may release zoospores, which have two unlike flagella which they use to swim towards a host plant.

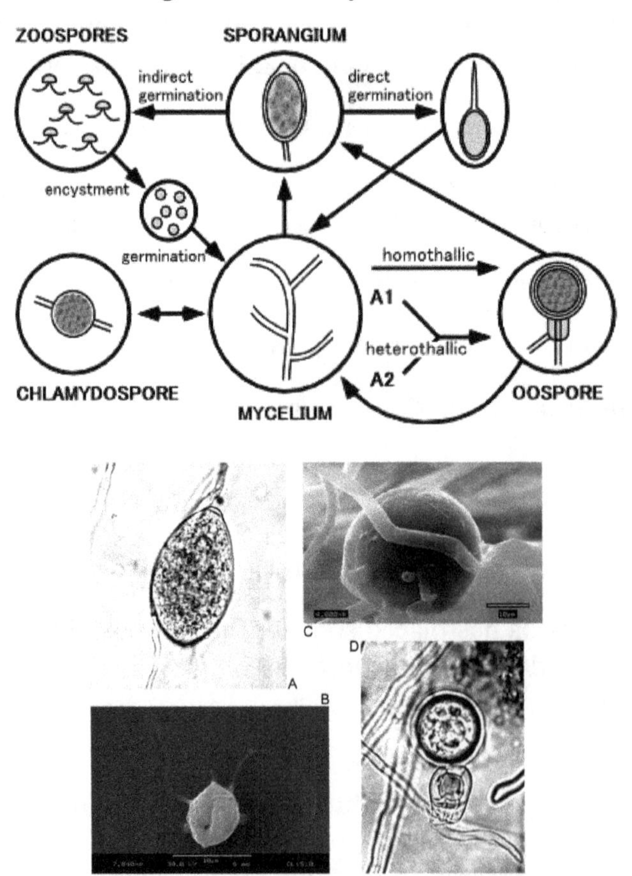

Phytophthora forms: A: Sporangia. B: Zoospore. C: Chlamydospore. D: Oospore

Phytophthora Infestans

Phytophthora infestans is an oomycete that causes the serious potato disease known as late blight or potato blight. (Early blight, caused by Alternaria solani, is also often called "potato blight".) Late blight was a major culprit in the 1840s European, the 1845 Irish and 1846 Highland potato famines. The organism can also infect tomatoes and some other members of the Solanaceae. At first, the spots are gray-green and water-soaked, but they soon enlarge and turn dark brown and firm, with a rough surface.

Biology

The asexual life cycle of *Phytophthora infestans* is characterized by alternating phases of hyphal growth, sporulation, sporangia germination (either through zoospore release or direct germination, i.e. germ tube emergence from the sporangium), and the re-establishment of hyphal growth. There is also a sexual cycle, which occurs when isolates of opposite mating type (A1 and A2) meet. Hormonal communication triggers the formation of the sexual spores, called oospores. The different types of spores play major roles in the dissemination and survival of *P. infestans*. Sporangia are spread by wind or water and enable the movement of *P. infestans* between different host plants. The zoospores released from sporangia are biflagellated and chemotactic, allowing further movement of *P. infestans* on water films found on leaves or soils. Both sporangia and zoospores are short-lived, in contrast to oospores which can persist in a viable form for many years.

Under ideal conditions, the life cycle can be completed on potato or tomato foliage in about five days. Sporangia develop on the leaves, spreading through the crop when temperatures are above 10 °C (50 °F) and humidity is over 75%-80% for 2 days or more. Rain can wash spores into the soil where they infect young tubers, and the spores can also travel long distances on the wind. The early stages of blight are easily missed. Symptoms include the appearance of dark blotches on leaf tips and plant stems. White mould will appear under the leaves in humid conditions and the whole plant may quickly collapse. Infected tubers develop grey or dark patches that are reddish brown beneath the skin, and quickly decay to a foul-smelling mush caused by the infestation of secondary soft bacterial rots. Seemingly healthy tubers may rot later when in store.

P. infestans survives poorly in nature apart from its plant hosts. Under most conditions, the hyphae and asexual sporangia can survive for only brief periods in plant debris or soil, and are generally killed off during frosts or very warm weather. The exceptions involve oospores, and hyphae present within tubers. The persistence of viable pathogen within tubers, such as those that are left in the ground after the previous year's harvest or left in cull piles is a major problem in disease management. In particular, volunteer plants sprouting from infected tubers are thought to be a major source of inoculum at the start of a growing season. This can have devastating effects by destroying entire crops.

Genetics

P. infestans is diploid, with about 11-13 chromosomes, and in 2009 scientists completed the sequencing of its genome. The genome was found to be considerably larger (240 Mbp) than that of most other *Phytophthora* species whose genomes have been sequenced; *Phytophthora sojae* has a

95 Mbp genome and *Phytophthora ramorum* had a 65 Mbp genome. About 18,000 genes were detected within the *P. infestans* genome. It also contained a diverse variety of transposons and many gene families encoding for effector proteins that are involved in causing pathogenicity. These proteins are split into two main groups depending on whether they are produced by the water mould in the symplast (inside plant cells) or in the apoplast (between plant cells). Proteins produced in the symplast included RXLR proteins, which contain an arginine-X-leucine-arginine (where X can be any amino acid) sequence at the amino terminus of the protein. Some RXLR proteins are avirulence proteins, meaning that they can be detected by the plant and lead to a hypersensitive response which restricts the growth of the pathogen. *P. infestans* was found to encode around 60% more of these proteins than most other *Phytophthora* species. Those found in the apoplast include hydrolytic enzymes such as proteases, lipases and glycosylases that act to degrade plant tissue, enzyme inhibitors to protect against host defence enzymes and necrotizing toxins. Overall the genome was found to have an extremely high repeat content (around 74%) and to have an unusual gene distribution in that some areas contain many genes whereas others contain very few.

Origin and Diversity of P. Infestans

Potatoes infected with late blight are shrunken on the outside, corky and rotted inside.

Historical model of a potato leaf with *Phytophthora infestans*, Botanical Museum Greifswald

The highlands of central Mexico are considered by many to be the center of origin of *P. infestans*, although others have proposed its origin to be in the Andes, which is also the origin of potatoes. A recent study evaluated these two alternate hypotheses and found conclusive support for central Mexico being the center of origin. Support for Mexico comes from multiple observations including the fact that populations are genetically most diverse in Mexico, late blight is observed in native tuber-bearing *Solanum* species, populations of the pathogen are in Hardy-Weinberg equilibrium, the two mating types occur in a 1:1 ratio, and detailed phylogeographic and evolutionary studies. Furthermore, the closest relatives of *P. infestans*, namely *P. mirabilis* and *P. ipomoeae* are endemic to central Mexico. On the other hand, the only close relative found in South America, namely *P. andina*, is a hybrid that does not share a single common ancestor with *P. infestans*. Finally, populations of *P. infestans* in South America lack genetic diversity and are clonal.

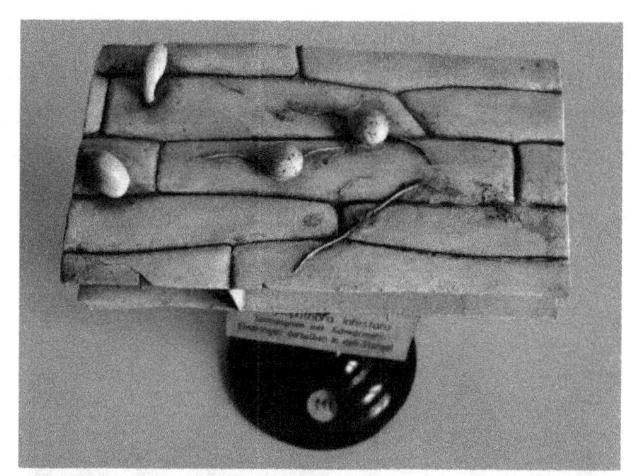

Historical model of Phytophthora infestans, Botanical Museum Greifswald

Migrations from Mexico to North America or Europe have occurred several times throughout history, probably linked to the movement of tubers. Until the 1970s, the A2 mating type was restricted to Mexico, but now in many regions of the world both A1 and A2 isolates can be found in the same region. The co-occurrence of the two mating types is significant due to the possibility of sexual recombination and formation of oospores, which can survive the winter. Only in Mexico and Scandinavia, however, is oospore formation thought to play a role in overwintering. In other parts of Europe, increasing genetic diversity has been observed as a consequence of sexual reproduction This is notable since different forms of *P. infestans* vary in their aggressiveness on potato or tomato, in sporulation rate, and sensitivity to fungicides. Variation in such traits also occurs in North America, however importation of new genotypes from Mexico appears to be the predominant cause of genetic diversity, as opposed to sexual recombination within potato or tomato fields. Many of the strains that appeared outside of Mexico since the 1980s have been more aggressive, leading to increased crop losses. Some of the differences between strains may be related to variation in the RXLR effectors that are present.

Disease Management

P. infestans is still a difficult disease to control. There are many chemical options in agriculture for the control of both damage to the foliage and infections of the tuber. A few of the most common foliar-applied fungicides are Ridomyl, a Gavel/SuperTin tank mix, and Previcur Flex. Orondis is a new product from Syngenta that shows promise but availability in 2016 will be limited. All of the aforementioned fungicides need to be tank mixed with a broad-spectrum fungicide such as mancozeb or chlorothalonil not just for resistance management but also because the potato plants will be attacked by other pathogens at the same time.

If adequate field scouting occurs and late blight is found soon after disease development, localized patches of potato plants can be killed with a dessicant (e.g. paraquat) through the use of a backpack sprayer. This management technique can be thought of as a field-scale hypersensitive response similar to what occurs in some plant-viral interactions whereby cells surrounding the initial point of infection are killed in order to prevent proliferation of the pathogen.

If infected tubers make it into the storage bin, there's a very high risk to the storage life of that bin. Once in storage, there isn't much that can be done besides emptying the parts of the bin that contain tubers infected with *Phytophthora infestans*. To increase the probability of successfully storing potatoes from a field where late blight was known to occur during the growing season, some products can be applied just prior to entering storage (e.g. Phostrol). The problem with products being sprayed on tubers just prior to storage is that you are applying these products in an aqueous solution and high moisture carries a high risk of tuber breakdown due to wide range of pathogens.

Around the world the disease causes around $6 billion of damage to crops each year.

Resistant Plants

Potatoes after exposure to Phytophthora infestans. The normal potatoes have blight but the cisgenic potatoes are healthy.

Breeding for resistance, particularly in potato, has had limited success in part due to difficulties in crossing cultivated potato to its wild relatives, which are the source of potential resistance genes. In addition, most resistance genes only work against a subset of *P. infestans* isolates, since effective plant disease resistance only results when the pathogen expresses a RXLR effector gene that matches the corresponding plant resistance (R) gene; effector-R gene interactions trigger a range of plant defenses, such as the production of compounds toxic to the pathogen.

Potato and tomato varieties vary in their susceptibility to blight. Most early varieties are very vulnerable; they should be planted early so that the crop matures before blight starts (usually in July in the Northern Hemisphere). Many old crop varieties, such as King Edward potato are also very susceptible but are grown because they are wanted commercially. Maincrop varieties which are very slow to develop blight include Cara, Stirling, Teena, Torridon, Remarka, and Romano. Some so-called resistant varieties can resist some strains of blight and not others, so their performance may vary depending on which are around. These crops have had polygenic resistance bred into them, and are known as "field resistant". New varieties such as Sarpo Mira and Sarpo Axona show great resistance to blight even in areas of heavy infestation. Defender is an American cultivar whose parentage includes Ranger Russet and Polish potatoes resistant to late blight. It is a long white-skinned cultivar with both foliar and tuber resistance to late blight. Defender was released in 2004.

Genetic engineering may also provide options for generating resistance cultivars. A resistance gene effective against most known strains of blight has been identified from a wild relative of the potato, *Solanum bulbocastanum*, and introduced by genetic engineering into cultivated varieties of potato. This is an example of cisgenic genetic engineering.

Reducing Inoculum

Blight can be controlled by limiting the source of inoculum. Only good-quality seed potatoes and tomatoes obtained from certified suppliers should be planted. Often discarded potatoes from the previous season and self-sown tubers can act as sources of inoculum.

Environmental Conditions

There are several environmental conditions that are conducive to *P. infestans*. An example of such took place in the United States during the 2009 growing season. As colder than average for the season and with greater than average rainfall, there was a major infestation of tomato plants, specifically in the eastern states. By using weather forecasting systems, such as BLITECAST, if the following conditions occur as the canopy of the crop closes, then the use of fungicides is recommended to prevent an epidemic.

- A Beaumont Period is a period of 48 consecutive hours, in at least 46 of which the hourly readings of temperature and relative humidity at a given place have not been less than 10 °C (50 °F) and 75%, respectively.
- A Smith Period is at least two consecutive days where min temperature is 10 °C (50 °F) or above and on each day at least 11 hours when the relative humidity is greater than 90%.

The Beaumont and Smith periods have traditionally been used by growers in the United Kingdom, with different criteria developed by growers in other regions. The Smith period has been the preferred system used in the UK since its introduction in the 1970s.

Based on these conditions and other factors, several tools have been developed to help growers manage the disease and plan fungicide applications. Often these are deployed as part of Decision Support Systems accessible through web sites or smart phones.

Use of Fungicides

Spraying in a potato field for prevention of potato blight in Nottinghamshire, England.

Fungicides for the control of potato blight are normally only used in a preventative manner, optionally in conjunction with disease forecasting. In susceptible varieties, sometimes fungicide applications may be needed weekly. An early spray is most effective. The choice of fungicide can depend on the nature of local strains of *P. infestans*. Metalaxyl is a fungicide that was marketed for use against *P. infestans*, but suffered serious resistance issues when used on its own. In some regions of the world during the 1980s and 1990s, most strains of *P. infestans* became resistant to metalaxyl, but in subsequent years many populations shifted back to sensitivity. To reduce the occurrence of resistance, it is strongly advised to use single-target fungicides such as metalaxyl along with carbamate compounds. A combination of other compounds are recommended for managing metalaxyl-resistant strains. These include mandipropamid, chlorothalonil, fluazinam, triphenytin, mancozeb and others. In the past, copper sulfate solution (called 'bluestone') was used to combat potato blight. Copper pesticides remain in use on organic crops, both in the form of copper hydroxide and copper sulfate. Given the dangers of copper toxicity, other organic control options that have been shown to be effective include horticultural oils, phosphorous acids, and rhamnolipid biosurfactants, while sprays containing "beneficial" microbes such as *Bacillus subtilis* or compounds that encourage the plant to produce defensive chemicals (such as knotweed extract) have not performed as well.

Control of Tuber Blight

Ridging is often used to reduce tuber contamination by blight. This normally involves piling soil or mulch around the stems of the potato blight meaning the pathogen has farther to travel to get to the tuber. Another approach is to destroy the canopy around five weeks before harvest, using a contact herbicide or sulfuric acid to burn off the foliage. Eliminating infected foliage reduces the likelihood of tuber infection.

Historical Impact

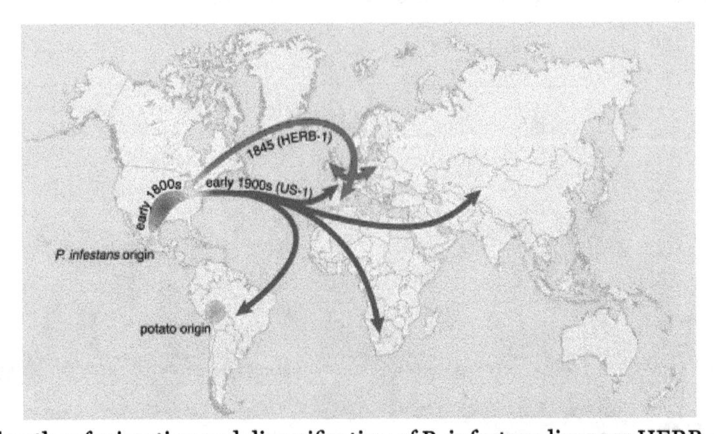

Suggested paths of migration and diversification of P. infestans lineages HERB-1 and US-1

The effects of *Phytophthora infestans* in Ireland in 1845–57 were one of the factors which caused over one million to starve to death and forced another two million to emigrate from affected countries. Most commonly referenced is the Great Irish Famine, during the late 1840s. The first recorded instances of the disease were in the United States, in Philadelphia and New York City in early 1843. Winds then spread the spores, and in 1845 it was found from Illinois to Nova Scotia, and from Virginia to Ontario. It crossed the Atlantic Ocean with a shipment of seed potatoes for

Belgian farmers in 1845. All of the potato-growing countries in Europe were affected, but the potato blight hit Ireland the hardest. Implicated in Ireland's fate was the island's disproportionate dependency on a single variety of potato, the Irish Lumper. The lack of genetic variability created a susceptible host population for the organism.

During the First World War, all of the copper in Germany was used for shell casings and electric wire and therefore none was available for making copper sulfate to spray potatoes. A major late blight outbreak on potato in Germany therefore went untreated, and the resulting scarcity of potatoes led to the deaths of 700,000 German civilians from starvation.

France, Canada, the United States, and the Soviet Union researched *P. infestans* as a biological weapon in the 1940s and 1950s. Potato blight was one of more than 17 agents that the United States researched as potential biological weapons before the nation suspended its biological weapons program. Whether a weapon based on the pathogen would be effective is questionable, due to the difficulties in delivering viable pathogen to an enemy's fields, and the role of uncontrollable environmental factors in spreading the disease.

Phytophthora Ramorum

Phytophthora ramorum is the oomycete plant pathogen known to cause the disease sudden oak death (SOD). The disease kills oak and other species of trees and has had devastating effects on the oak populations in California and Oregon, as well as being present in Europe. Symptoms include bleeding cankers on the tree's trunk and dieback of the foliage, in many cases eventually leading to the death of the tree.

P. ramorum also infects a great number of other plant species, significantly woody ornamentals such as *Rhododendron*, *Viburnum*, and *Pieris*, causing foliar symptoms known as ramorum dieback or ramorum blight. Such plants can act as a source of inoculum for new infections, with the pathogen-producing spores that can be transmitted by rainsplash and rainwater.

P. ramorum was first reported in 1995, and the origins of the pathogen are still unclear, but most evidence suggests it was repeatedly introduced as an exotic species. Very few control mechanisms exist for the disease, and they rely upon early detection and proper disposal of infected plant material.

Presence

The disease is known to exist in California's coastal region between Big Sur (in Monterey County) and southern Humboldt County. It is confirmed to exist in all coastal counties in this range, as well as in all immediately inland counties from Santa Clara County north to Lake County. It has not been found east of the California Coast Ranges, however. It was reported in Curry County, Oregon (just north of the California border), in 2001. Sonoma County has been hit hardest, having more than twice the area of new mortality of any other county in California.

About the same time, a similar disease in continental Europe and the UK was also identified as *Phytophthora ramorum*.

Hosts and Symptoms

In North America

It was first discovered in California in 1995 when large numbers of tanoaks (*Notholithocarpus densiflorus*) died mysteriously, and was described as a new species of *Phytophthora* in 2000. It has subsequently been found in many other areas, including Britain, Germany, and some other U.S. states, either accidentally introduced in nursery stock, or already present undetected.

In tanoaks, the disease may be recognized by wilting new shoots, older leaves becoming pale green, and after a period of two to three weeks, foliage turning brown while clinging to the branches. Dark brown sap may stain the lower trunk's bark. Bark may split and exude gum, with visible discoloration. After the tree dies back, suckers try to sprout the next year, but their tips soon bend and die. Ambrosia beetles (*Monarthrum scutellare*) will most likely infest a dying tree during midsummer, producing piles of fine white dust near tiny holes. Later, bark beetles (*Pseudopityophthorus pubipennis*) produce fine, red boring dust. Small black domes, the fruiting bodies of the *Hypoxylon* fungus, may also be present on the bark. Leaf death may occur more than a year after the initial infection and months after the tree has been girdled by beetles.

A hillside in Big Sur, California, devastated by sudden oak death

In coast live oaks and Californian black oaks, the first symptom is a burgundy-red to tar-black thick sap bleeding from the bark surface. These are often referred to as bleeding cankers.

In addition to oaks, many other forest species may be hosts for the disease; in fact, it was observed in the USA that nearly all woody plants in some Californian forests were susceptible to *P. ramorum*. including rhododendron, madrone (*Arbutus menziesii*), evergreen huckleberry (*Vaccinium ovatum*), California bay laurel (*Umbellularia californica*), buckeye (*Aesculus californica*), bigleaf maple (*Acer macrophyllum*), toyon (*Heteromeles arbutifolia*), manzanita (*Arctostaphylos spp.*), coast redwood (*Sequoia sempervirens*), Douglas fir (*Pseudotsuga menziesii*), coffeeberry (*Rhamnus californica*), honeysuckle (*Lonicera hispidula*), and Shreve oak (*Quercus parvula*). *P. ramorum* more commonly causes a less severe disease known as ramorum dieback/leaf blight on these hosts. Characteristic symptoms are dark spots on foliage and in some hosts the dieback of the stems and twigs. The disease is capable of killing some hosts, such as rhododendron, but most survive. Disease progression on these species is not well documented. Redwoods exhibit needle discoloration and cankers on small branches, with purple lesions on sprouts that may lead to sprout mortality.

In Europe

Leaf death caused by P. ramorum

In Europe, *Ramorum* blight was first observed on rhododendron and viburnum in the early 1990s, where it was initially found mainly on container-grown plants in nurseries. The principal symptoms were leaf and twig blight. By 2007, it had spread throughout nurseries and retail centers in 16 European countries, and had been detected in gardens, parks, and woodlands in at least eight countries. It has not caused significant harm to European oak species.

In 2009, the pathogen was found to be infecting and killing large numbers of Japanese larch trees (*Larix kaempferi*) in the United Kingdom at sites in the English counties of Somerset, Devon, and Cornwall. It was the first time in the world that *Phytophthora ramorum* had been found infecting this species. Since then, it has also been found extensively in larch plantations in Wales and in south-west Scotland, leading to the deliberate felling of large areas. The UK Forestry Commission noted that eradication of the disease would not be possible, and instead adopted a strategy of containing the disease to reduce its spread. Symptoms of the disease on larch trees include dieback of the tree's crown and branches, and a distinctive yellowing or ginger colour beneath the bark. In August 2010, disease was found in Japanese larch trees, in Counties Waterford and Tipperary in Ireland.

The closely related *Phytophthora kernoviae* causes similar symptoms to *P. ramorum*, but infects the European beech (*Fagus sylvatica*).

Transmission

P. ramorum produces both resting spores (chlamydospores) and zoospores, which have flagella enabling swimming. *P. ramorum* is spread by air; one of the major mechanisms of dispersal is rainwater splashing spores onto other susceptible plants, and into watercourses to be carried for greater distances. Chlamydospores can withstand harsh conditions and are able to overwinter. The pathogen will take advantage of wounding, but it is not necessary for infection to occur.

As mentioned above, *P. ramorum* does not kill every plant that can be used as a host, and these plants are most important in the epidemiology of the disease as they act as sources of inoculum. In California, California bay laurel (*Umbellularia californica*) seems to be the main source of inoculum. Green waste, such as leaf litter and tree stumps, are also capable of supporting *P. ramorum* as a saprotroph and acting as a source of inoculum. Because *P. ramorum* is able to infect many ornamental plants, it can be transmitted by ornamental plant movement.

Cannabis cultivation and associated traffic and movement of supplies and soil amendments in Northern California watersheds correlate with areas of introduction of *P. ramorum*. Hikers, mountain bikers, equestrians, and other people engaged in various outdoor activities may also unwittingly move the pathogen into areas where it was not previously present. Those travelling in an area known to be infested with SOD can help prevent the spread of the disease by cleaning their (and their animals') feet, tires, tools, camping equipment, etc. before returning home or entering another uninfected area, especially if they have been in muddy soil. Additionally, the movement of firewood could introduce sudden oak death to otherwise uninfected areas. Both homeowners and travelers are advised to buy and burn local firewood.

The Two Mating Types

P. ramorum is heterothallic and has two mating types, A1 and A2, required for sexual reproduction. Interestingly, the European population is predominantly A1 while both mating types A1 and A2 are found in North America. Genetics of the two isolates indicate that they are reproductively isolated. On average, the A1 mating type is more virulent than the A2 mating type, but more variation occurs in the pathogenicity of A2 isolates. It is currently not clear whether this pathogen can reproduce sexually in nature and genetic work has suggested that the lineages of the two mating types might be isolated reproductively or geographically given the evolutionary divergence observed.

Possible Origins

P. ramorum is a relatively new disease, and several debates have occurred about where it may have originated or how it evolved.

Introduction as an Exotic Species

Evidence suggests *P. ramorum* may be an introduced species, and these introductions occurred separately for the European and North American populations, hence why only one mating type exists on each continent – this is called a founder effect. The differences between the two populations are thus caused by adaptation to separate climates. Evidence includes little genetic variability, as *P. ramorum* has not had time to diversify since being introduced. Existing variability may be explained by multiple introductions with a few individuals adapting best to their respective environments. The behavior of the pathogen in California is also indicative of being introduced; it is assumed that such a high mortality rate of trees would have been noticed sooner if *P. ramorum* were native.

Where *P. ramorum* originated remains unclear, but most researchers feel Asia is the most likely, since many of the hosts of *P. ramorum* originated there. Since certain climates are best suited to *P. ramorum*, the most likely sources are the southern Himalayas, Tibet, or Yunnan province.

Hybridization Events

Species of *Phytophthora* have been shown to have evolved by the interspecific hybridization of two different species from the genus. When a species is introduced into a new environment, it causes episodic selection. The invading species is exposed to other resident taxa, and hybridization may occur to produce a new species. If these hybrids are successful, they may outcompete their parent species. Thus, *P, ramorum* is possibly a hybrid between two species. Its unique morphology does support this. Also, three sequences studied to establish the phylogeny of *Phytophthora*: ITS, cox II and nad 5, were identical, supporting *P. ramorum* having recently evolved.

A Native Organism

P. ramorum may be native to the United States. Infection rates could have previously been at a low level, but changes in the environment caused a change to the population structure. Alternatively, the symptoms of *P. ramorum* may have been mistaken for that of other pathogens. When SOD first appeared in the United States, many other pathogens and conditions were blamed before *P. ramorum* was found to be the causal agent. With many of the most seriously affected plants being in the forest, the likelihood of seeing diseased trees is also low.

Ecological Impacts

In relation to human ecology, the loss of tanoak as the pathogen spreads to culturally sensitive indigenous lands represents a loss of tanoak acorn as one of the most important traditional and ceremonial foods still used in Northern California such as among Yurok people, Hupa, Miwok, and Karuk peoples. Similar impact applies to the decline of other native plant species that are traditional food sources in tanoak and oak regimens infected by the pathogen.

In forest ecology, the pathogen contributes to loss of environmental services provided by diversity of plant species and interdependent wildlife.

The mortality caused by this emerging disease is expected to cause many indirect effects. Several predictions of long-term impacts have been discussed in the scientific literature. While such predictions are necessarily speculative, indirect impacts occurring on shorter time scales have been documented in a few cases. For instance, one study demonstrated that redwood trees (*Sequoia sempervirens*) grew faster after neighboring tanoaks were killed by sudden oak death. Other studies have combined current observations and reconstruction/projection techniques to document short-term impacts while also inferring future conditions. One study used this approach to investigate the effects of SOD on the structural characteristics of redwood forests.

Additional long-term impacts of SOD may be inferred from regeneration patterns in areas that have experienced severe mortality. These patterns may indicate which tree species will replace tanoak in diseased areas. Such transitions will be of particular importance in forest types that were relatively poor in tree species diversity before the introduction of SOD, e.g., redwood forest. As of 2011, the only study to comprehensively examine regeneration in SOD-impacted redwood forests found no evidence that other broadleaf tree species are beginning to recruit. Instead, redwood was colonizing most mortality gaps. However, they also found inadequate regeneration in some areas and concluded that regeneration is continuing. Since this study only considered one site in Marin County, California, these results may not apply to other forests. Other impacts to the local ecology include, among others, the residual effects of spraying heavy pesticides (Agrifos) to treat SOD symptoms, and the heavy mortality of the native pollinator community that occurs as a result. Bee hives situated in areas of heavy Agrifos spraying have incurred significant losses of population in direct correlation to the application of these chemicals. Counties such as Napa and Sonoma may be doing significant damage to their native pollinator populations by virtue of adopting broad-based prophylactic pesticide policies. Such damage to the pollinator populations may have tertiary negative effects on the entire local plant community, compounding the loss of biodiversity, and thus environmental value, attributable to SOD.

Control

Early Detection

Early detection of *P. ramorum* is essential for its control. On an individual-tree basis, preventive treatments, which are more effective than therapeutic treatments, depend on knowledge of the pathogen's movement through the landscape to know when it is nearing prized trees. On the landscape level, *P. ramorum's* fast and often undetectable movement means that any treatment hoping to slow its spread must happen very early in the development of an infestation. Since *P. ramorum's* discovery, researchers have been working on the development of early detection methods on scales ranging from diagnosis in individual infected plants to landscape-level detection efforts involving large numbers of people.

Detecting the presence of *Phytophthora* species requires laboratory confirmation. The traditional method of culturing is on a growth medium selective against fungi (and, in some cases, against other oomycetes such as *Pythium* species). Host material is removed from the leading edge of a plant tissue canker caused by the pathogen; resulting growth is examined under a microscope to confirm the unique morphology of *P. ramorum*. Successful isolation of the pathogen often depends on the type of host tissue and the time of year that detection is attempted.

Because of these difficulties, researchers have developed some other approaches for identifying *P. ramorum*. The enzyme-linked immunosorbent assay test can be the first step in nonculture methods of identifying *P. ramorum*, but it can only be a first step, because it detects the presence of proteins that are produced by all *Phytophthora* species. In other words, it can identify to the genus level, but not to the species level. ELISA tests can process large numbers of samples at once, so researchers often use it to screen out likely positive samples from those that are not when the total number of samples is very large. Some manufacturers produce small-scale ELISA "field kits" that the homeowner can use to determine if plant tissue is infected by *Phytophthora*.

Researchers have also developed numerous molecular techniques for *P. ramorum* identification. These include amplifying DNA sequences in the internal transcribed spacer region of the *P. ramorum* genome (ITS polymerase chain reaction, or ITS PCR); real-time PCR, in which DNA abundance is measured in real time during the PCR reaction, using dyes or probes such as SBYR-Green or TaqMan; multiplex PCR, which amplifies more than one region of DNA at the same time; and single strand conformation polymorphism (SSCP), which uses the ITS DNA sequence amplified by the PCR reaction to differentiate *Phytophthora* species according to their differential movement through a gel.

Additionally, researchers have begun using features of the DNA sequence of *P. ramorum* to pinpoint the minuscule differences of separate *P. ramorum* isolates from each other. Two techniques for doing this are amplified fragment length polymorphism, which through comparing differences between various fragments in the sequence has enabled researchers to differentiate correctly between EU and U.S. isolates, and the examination of microsatellites, which are areas on the sequence featuring repeating base pairs. When *P. ramorum* propagules arrive in a new geographic location and establish colonies, these microsatellites begin to display mutation in a relatively short time, and they mutate in a stepwise fashion. Based on this, researchers in California have been able to construct trees, based on microsatellite analyses of isolates collected from around the state, that trace the movement of *P. ramorum* from two likely initial points of establishment in Marin and Santa Cruz Counties and out to subsequent points.

Early detection of *P. ramorum* on a landscape scale begins with the observation of symptoms on individual plants. Systematic ground-based monitoring has been difficult within the range of *P. ramorum* because most infected trees stand on a complex mosaic of lands with various ownerships. In some areas, targeted ground-based surveys have been conducted in areas of heavy recreation or visitor use such as parks, trailheads, and boat ramps. In California, when conducting ground-based detection, looking for symptoms on bay laurel is the most effective strategy, since *P. ramorum* infection of true oaks and tanoaks is almost always highly associated with bay laurel, the main epidemiological springboard for the pathogen. Moreover, on many sites in California (though not all), *P. ramorum* can typically be detected from infected bay laurel tissues via culturing techniques year-round; this is not the case for most other hosts, nor is it the case in Oregon, where tanoak is the most reliable host.

As part of a nationwide USDA program, a ground-based detection survey was implemented from 2003 to 2006 in 39 U.S. states to determine whether the pathogen was established outside the West Coast areas already known to be infested. Sampling areas were stratified by environmental variables likely to be conducive to pathogen growth and by proximity to possible points of inoculum introduction such as nurseries. Samples were collected along transects established in potentially susceptible forests or outside the perimeters of nurseries. The only positive samples were collected in California, confirming that *P. ramorum* was not yet established in the environment outside the West Coast.

Aerial surveying has proven useful for detection of *P. ramorum* infestations across large landscapes, although it is not as "early" a technique as some others because it depends on spotting dead tanoak crowns from fixed-wing aircraft. Sophisticated GPS and sketch-mapping technology enable spotters to mark the locations of dead trees so that ground crews can return to the area to sample from nearby vegetation.

Detection of *P. ramorum* in watercourses has emerged as the earliest of early detection methods. This technique employs pear or rhododendron baits suspended in the watercourse using ropes, buckets, mesh bags, or other similar devices. If plants in the watershed are infected with *P. ramorum*, zoospores of the pathogen (as well as other *Phytophthora* spp.) are likely present in adjacent waterways. Under conducive weather conditions, the zoospores are attracted to the baits and infect them, causing lesions that can be isolated to culture the pathogen or analyzed via PCR assay. This method has detected *P. ramorum* at scales ranging from small, intermittent seasonal drainages to the Garcia, Chetco, and South Fork Eel Rivers in California and Oregon (144, 352, and 689 mi2 drainage areas, respectively). It can detect the existence of infected plants in watersheds before any mortality from the infections becomes evident. Of course, it cannot detect the exact locations of those infected plants: at the first sign of *P. ramorum* propagules in the stream, crews must scour the watershed using all available means to find symptomatic vegetation.

A less technical means of detecting *P. ramorum* at the landscape level involves engaging local landowners across the landscape in the search. Many local county agriculture departments and University of California Cooperative Extension offices in California have been able to keep track of the distribution of the pathogen in their regions through reports and samples brought to them by the public. In 2008, the Garbelotto Laboratory at University of California, Berkeley, along with local collaborators, hosted a series of educational events, called "SOD Blitzes", designed to give local landowners basic information about *P. ramorum* and how to identify its symptoms; each participant was provided with a sampling kit, sampled a certain number of trees on his or her property, and returned the samples to the lab for analysis. This kind of citizen science hopefully can help generate an improved map of *P. ramorum* distribution in the areas where the workshops are held.

Wildland Management

The course that *P. ramorum* management should take depends on a number of factors, including the scale of the landscape upon which one hopes to manage it. Management of *P. ramorum* has been undertaken at the landscape/ regional level in Oregon in the form of a campaign to completely eradicate the pathogen from the forests in which it has been found (mostly private, but also USDA Forest Service and USDI Bureau of Land Management ownership). The eradication campaign involves vigorous early detection by airplane and watercourse monitoring, a U.S. Department of Agriculture Animal and Plant Health Inspection Service (USDA APHIS) and Oregon Department of Agriculture-led quarantine to prevent movement of host materials out of the area where infected trees are found, and immediate removal of *P. ramorum* host vegetation, symptomatic or not, within a 300-foot (91 m) buffer around each infected tree.

The Oregon eradication effort, which began near the town of Brookings in southwest Oregon in 2001, has adapted its management efforts over the years in response to new information about *P. ramorum*. For example, after inoculation trials of various tree species more clearly delineated which hosts are susceptible, the Oregon cooperators began leaving nonhost species such as Douglas fir and red alder on site. In another example, after finding that a small percentage of tanoak stumps that were resprouting on the host removal sites were infected with the pathogen—whether these infections were systemic or reached the sprouts from the surrounding environment is unknown—the cooperators began pretreating trees with very small, targeted amounts of herbicide to kill the root systems of infected tanoaks before cutting them down. The effort has been successful in that while it has not yet completely eradicated the pathogen from Oregon forests, the epidemic in Oregon has not taken the explosive course that it has in California forests.

California, though, faces significant obstacles that preclude it from mounting the same kind of eradication effort. For one thing, the organism was too well established in forests in the Santa Cruz and San Francisco Bay areas by the time the cause of sudden oak death was discovered to enable any eradication effort to succeed. Even in still relatively uninfested areas of the north coast and southern Big Sur, regionally coordinated efforts to manage the pathogen face huge challenges of leadership, coordination, and funding. Nevertheless, land managers are still working to coordinate efforts between states, counties, and agencies to provide *P. ramorum* management in a more comprehensive manner.

Several options exist for landowners who want to limit the impacts of SOD death on their properties. None of these options is foolproof, guaranteed to eradicate *P. ramorum*, or guaranteed to prevent a tree from becoming infected. Some are still in the initial stage of testing. Nevertheless, when used thoughtfully and thoroughly, some of the treatments do improve the likelihood of either slowing the spread of the pathogen or of limiting its impacts on trees or stands of trees. Assuming that the landowner has correctly identified the host tree(s) and symptom(s), has submitted a sample to a local authority to send to an approved laboratory for testing, and has received confirmation that the tree(s) are indeed infected with *P. ramorum*—or, alternatively, assuming that the landowner knows that *P. ramorum*-infected trees are nearby and wants to protect the resources on his or her property—he or she can attempt control by cultural (individual-tree), chemical, or silvicultural (stand-level) means.

The best evidence that cultural techniques might help protect trees against *P. ramorum* comes from research that has established a correlation between disease risk in coast live oak trees and the trees' proximity to bay laurel. In particular, this research found that bay laurel trees growing within 5 m of the trunk of an oak tree were the best predictors of disease risk. This implies that strategic removal of bay laurel trees near coast live oaks might decrease the risk of oak infection. Wholesale removal of bay laurel trees would not be warranted, since the bay laurels close to the oak trees appear to provide the greatest risk factor. Whether the same pattern is true for other oaks or tanoaks has yet to be established. Research on this subject has been started for tanoak, but any eventual cultural recommendations will be more complicated, because tanoak twigs also serve as sources of *P. ramorum* inoculum.

A promising treatment for preventing infection of individual oak and tanoak trees—not for curing an already established infection—is a phosphonate fungicide marketed under the trade name Agri-fos. Phosphonate is a neutralized form of phosphorous acid that works not by direct antagonism of *Phytophthora*, but by stimulating various kinds of immune responses on the part of the tree. It is mostly environmentally benign if not applied to nontarget plants and can be applied either as an injection into the tree stem or as a spray to the bole. When applying Agri-fos as a spray, it must be combined with an organosilicate surfactant, Pentra-bark, to enable the product to adhere to the bole long enough to be absorbed by the tree. Agri-fos has been very effective in preventing tree infections, but it must be applied when visible symptoms of *P. ramorum* on other trees in the immediate neighborhood are still relatively distant; otherwise, the tree to be treated likely is already infected, but visible symptoms have not yet developed (especially true for tanoak).

Trials of silvicultural methods for treating *P. ramorum* began in Humboldt County in northwest coastal California in 2006. The trials have taken place on a variety of infested properties both private and public and have generally focused on varying levels and kinds of host removal. The largest (50 acres (200,000 m²)) and most replicated trials have involved removal of tanoak and bay laurel by chainsaw throughout the infested stand, both with and without subsequent underburning designed to eliminate small seedlings and infested leaf litter. Other treatments included host removal in a modified "shaded fuelbreak" design in which all bay laurel is removed, but not all tanoaks; bay and tanoak removal using herbicides; and removal of bay laurel alone. The results of these treatments are still being monitored, but repeated sampling has so far detected only very small amounts of *P. ramorum* in the soil or on vegetation in the treated sites.

Nursery Management

Research and development in managing *P. ramorum* in nursery settings extends from *P. ramorum* in the individual plant, to *P. ramorum* in the nursery environment, to the pathogen's movement across state and national borders in infected plants.

An array of studies have tested the curative and protective effects of various chemical compounds against *P. ramorum* in plants valued as ornamentals or Christmas trees. Many studies have focused on the four main ornamental hosts of *P. ramorum* (*Rhododendron, Camellia, Viburnum,* and *Pieris*). Several effective compounds have been found; some of the most effective include mefenoxam, metalaxyl, dimethomorph, and fenamidone. Many of these studies have converged upon the following conclusions: chemical compounds are, in general, more effective as preventives than in curatives; when used preventively, chemical compounds must be reapplied at various intervals; and chemical compounds can mask the symptoms of *P. ramorum* infection in the host plant, potentially interfering with inspections for quarantine efforts. In general, these compounds suppress but do not eradicate the pathogen, and some researchers are concerned that with repeated use the pathogen may become resistant to them. These studies and conclusions are summarized by Kliejunas.

Another area of research and evolving practice deals with eliminating *P. ramorum* from nursery environments in which it is established to prevent human-mediated pathogen movement within the ornamental plant trade. One way of approaching this is through a robust quarantine and inspection program, which the various federal and state regulatory agencies have implemented. Under the federal *P. ramorum* quarantine program implemented by USDA APHIS, nurseries in California, Oregon, and Washington are regulated and must participate in an annual inspection regimen; nurseries in the 14 infested counties in coastal California, plus the limited infested area in Curry County, Oregon, must participate in a more stringent inspection schedule when shipping out of this area.

Much of the research into disinfesting nurseries has focused on the voluntary best management practices (BMPs) that nurseries can implement to prevent *P. ramorum's* introduction into the nursery and movement from plant to plant. In 2008, a group of nursery industry organizations issued a list of BMPs that includes subsections on pest prevention/management, training, internal/external monitoring/audits, records/traceability, and documentation. The document includes such specific recommendations as "Avoid overhead irrigation of high-risk plants"; "After every crop rotation, disinfect propagation mist beds, sorting area, cutting benches, machines and tools to minimize the spread or introduction of pathogens"; and "Nursery personnel should attend one or more *P. ramorum* trainings conducted by qualified personnel or document self-training".

Research on control of *P. ramorum* in nurseries has also focused on disinfesting irrigation water containing *P. ramorum* inoculum. Irrigation water can become infested from bay trees in the forest (if the irrigation source is a stream), from bay trees overhanging irrigation ponds, from runoff from infested forests, or from recirculated irrigation water. Experiments in Germany with three types of filters—slow sand filters, lava filters, and constructed wetlands—showed that the first two removed *P. ramorum* from the irrigation water completely, while 37% of the post-treatment water samples from the constructed wetland still contained *P. ramorum*.

Since *P. ramorum* can persist for an undetermined period of time within the soil profile, management programs in nurseries should also deal with delineating the pathogen's distribution in nursery soil and eliminating it from infested areas. A variety of chemical options has been tested for soil disinfestation, including such chemicals as chloropicrin, metham sodium, iodomethane, and dazomet. Lab tests indicated that all of these chemicals were effective when applied to infested soil in glass jars. Additionally, tests on volunteer nurseries with infested soil demonstrated that dazomet (trade name Basamid) fumigation followed by a 14-day tarping period successfully removed *P. ramorum* from the soil profile. Other soil disinfestation practices under investigation, or in which interest has been expressed, include steam sterilization, solarization, and paving of infested areas.

General Sanitation in Infested Areas

One of the most important aspects of *P. ramorum* control involves interrupting the human-mediated movement of the pathogen by ensuring that infested materials do not move from location to location. While enforceable quarantines perform part of this function, basic cleanliness when working or recreating in infested areas is also important. In most cases, cleanliness practices involve ridding potentially infested surfaces—such as shoes, vehicles, and pets—of foliage and mud before leaving the infested area. The demands of implementing these practices become more complex when large numbers of people are working in infested areas, as in construction, timber harvesting, or wildfire suppression. The California Department of Forestry and Fire Protection and USDA Forest Service have implemented guidelines and mitigation requirements for the latter two situations; basic information about cleanliness in *P. ramorum*-infested areas can be found at the California Oak Mortality Task Force web site (www.suddenoakdeath.org) under the "Treatment and Management" section (subsection "Sanitation and Reducing Spread").

Government Agency Involvement

In England in 2009, the Forestry Commission, DEFRA, the Food and Environment Research Agency, Cornwall County Council, and Natural England are working together to record the locations and deal with this disease. Natural England is offering grant funding through its Environmental Stewardship, Countryside Stewardship, and Environmentally Sensitive Area schemes to clear rhododendron. In 2011, the Forestry Commission started felling 10,000 acres (40 km^2) of larch forest in the south-west of England, as an attempt to halt the spread of the disease. In Northern Ireland at the end of 2011, the Department of Agriculture and Rural Development's Forest Service began felling 14 hectares of affected Larch woodland at Moneyscalp, on the edge of Tollymore Forest Park in County Down.

Phytophthora Cinnamomi

Phytophthora cinnamomi is a soil-borne water mould that produces an infection which causes a condition in plants called "root rot" or "dieback". The plant pathogen is one of the world's most invasive species and is present in over 70 countries around the world.

Life Cycle and Effects on Plants

Phytophthora cinnamomi lives in the soil and in plant tissues, can take different shapes and can move in water. During periods of harsh environmental conditions, the organisms become dormant chlamydospores. When environmental conditions are suitable, the chlamydospores germinate, producing mycelia (or hyphae) and sporangia. The sporangia ripen and release zoospores, which infect plant roots by entering the root behind the root tip. Zoospores need water to swim through the soil, therefore infection is most likely in moist soils. Mycelia grow throughout the root absorbing carbohydrates and nutrients, destroying the structure of the root tissues, "rotting" the root, and preventing the plant from absorbing water and nutrients. Sporangia and chlamydospores form on the mycelia of the infected root, and the cycle of infection continues to the next plant.

Early symptoms of infection include wilting, yellowing and retention of dried foliage and darkening of root color. Infection often leads to death of the plant, especially in dry summer conditions when plants may be water stressed.

Sexual Reproduction

Phytophthora cinnamomi is a diploid and heterothallic species with two mating types, A1 and A2. Sexual reproduction in heterothallic *Phytophthora* species ordinarily occurs when gametangia of opposite mating type interact in host tissue. This interaction leads to the formation of oospores that can survive for long periods in or outside the host. *Phytophthora cinnamomi* is also capable of self-fertilization (i.e. it can be homothallic). *Phytophthora cinnamomi* mating type A2 cultures can be induced to undergo sexual reproduction by exposure to damaging conditions, that is by exposure to hydrogen peroxide or mechanical damage.

In the Wild

When phytophthora dieback spreads to native plant communities, it kills many susceptible plants, resulting in a permanent decline in the biodiversity and a disruption of ecosystem processes. It can also change the composition of the forest or native plant community by increasing the number of resistant plants and reducing the number of susceptible plant species. Native animals that rely on susceptible plants for survival are reduced in numbers or are eliminated from sites infested by Phytophthora dieback.

Damage to forests suspected to be caused by *Phytophthora cinnamomi* was first recorded in the United States about 200 years ago. Infection is the cause of sudden death of a number of native tree species, including American chestnut, littleleaf disease of shortleaf pine (*Pinus echinata*), Christmas tree disease in nursery grown Fraser fir (*Abies fraseri*), while oaks are affected from South Carolina to Texas.

A heath landscape in the Stirling Range, Western Australia, with a dieback-infested valley in the mid ground

In Australia, where it is known as phytophthora dieback, dieback, jarrah dieback or cinnamon fungus, Phytophthora cinnamomi infects a number of native plants, causing damage to forests and removing habitats for small mammals.

Of particular concern is the infection and dieback of large areas of forest and heathland which support threatened species in the south-west corner of Western Australia. Many plants from the genera *Banksia*, *Darwinia*, *Grevillea*, *Leucopogon*, *Verticordia* and *Xanthorrhoea* are susceptible. This in turn will impact on animals reliant on these plants for food and shelter, such as the south-western pygmy possum (*Cercartetus concinnus*) and the honey possum (*Tarsipes rostratus*).

Littleleaf disease in *Pinus* spp., the tree on the left shows no symptoms of infection while the tree on the right shows stunted leaf growth characteristic of *Phytophthora cinnamomi* infection

Phytophthora cinnamomi is also a problem in the Mexican state of Colima, killing several native oak species and other susceptible vegetation in the surrounding woodlands. It is implicated in the die-off of the rare endemic shrub Ione manzanita (*Arctostaphylos myrtifolia*) in California, as well.

In addition to damage to native woodlands, *Phytophthora cinnamomi* can also infect fruit trees, nut trees and other ornamental plants. Research has shown that *Phytophthora cinnamomi* can infect club mosses, ferns, cycads, conifers, cord rushes, grasses, lilies and a large number of species from many dicotyledonous families. This is a remarkable range for a plant pathogen and highlights the effectiveness of *Phytophthora cinnamomi* as an aggressive primary pathogen. The Invasive Species Specialist Group includes this species in its list of "100 of the World's Worst Invasive Alien Species".

In Gardens and Crops

Phytophthora dieback affects a large number of common garden species, natives and horticultural crops. This list of susceptible plants includes roses, azaleas and fruit trees. Since there is no known cure, once the disease has been introduced into a garden it cannot be easily eradicated, and can become a major problem.

Protocols to prevent the disease from entering gardens include sourcing plants from non-dieback infested areas (not local bushland), using sterilised potting mixes, and using only mulch that has been properly composted. Transplanting established plants from one garden to another can also spread the disease. Propagating from seed and cuttings is less likely to transmit the disease because there is no soil transported with stock.

Many nurseries are accredited under the Nursery Industry Accreditation Scheme Australia (NIASA) and use hygienic practices to prevent Phytophthora dieback from infecting their stock. Hygienic practices prevent the spread of the disease in contaminated potting mix, plant material and water sources. Other preventative measures include raised benches, regular testing for phytophthora dieback, and the placing of new stock in quarantine.

Plants typically die from phytophthora dieback at the end of summer when the plants are under the most stress. For this reason phytophthora dieback can often be confused with symptoms of drought. Phytophthora dieback will affect a range of different susceptible plants, but will not impact on resistant plant species. If the disease is suspected, a likely mode of disease transmission should be identified. The best method to confirm the presence of the disease is testing of soil and/or plant samples by a diagnostic laboratory.

Control of existing *Phytophthora* infestations includes injecting or spraying plants with phosphite (a fungicide), using well-composted mulch, and using pre-planting techniques such as solarisation or biofumigation. Composted mulch is highly suppressive to phytophthora dieback and can prevent healthy plants getting infected. It is most important to prevent the spread of infected soil, plants or water. Infested areas can be revegetated or landscaped with resistant plant species which are not affected by the disease.

Impact on Avocado Farming

Phytophthora cinnamomi is the leading cause of damage to avocado trees, and is commonly known as "root rot" amongst avocado farmers. Since the 1940s various breeds of root rot-resistant avocados have been developed to minimize tree damage. Damaged trees generally die or become unproductive within three to five years. A 1960 study of the Fallbrook, California area correlated higher levels of avocado root rot to soils with poorer drainage and greater clay content.

Phosphite Fungicide Treatment

Phosphite (phosphonate) salts can be used as a biodegradable fungicide to protect plants against phytophthora dieback. It is usually applied as potassium phosphite, derived from phosphorous acid neutralized with potassium hydroxide. Calcium and magnesium phosphite may also be used. Phosphite works by boosting the plant's own natural defences and thereby allowing susceptible plants to survive within phytophthora dieback infested sites. It is important to note that there is no treatment that will eradicate phytophthora dieback, including phosphite. However, an integrated approach can successfully control the spread and impact of the disease. An integrated approach may combine strategic phosphite treatment, fumigants, controlling access, correcting drainage problems, removal of host plants and implementing excellent hygiene protocols.

Phosphite is not toxic to people or animals and its toxicity has been compared to that of table salt. There is a very low pollution risk associated with phosphite. When phosphite is sprayed on to the foliage of plants, it is applied at a very low rate, so any phosphite that reaches the soil is bound to the soil and does not reach the water table.

Phosphite needs to enter a plant's water transport system in order for it to be effective. This can be done by injecting phosphite into trees, or spraying the leaves of understorey plants. Phosphite not only protects a plant from phytophthora dieback infection, it can also help a plant to recover if it is already infected.

Phytophthora Capsici

Phytophthora capsici is an oomycete plant pathogen that causes blight and fruit rot of peppers and other important commercial crops. It was first described by L. Leonian at a New Mexico Agricultural Research Station in Las Cruces in 1922 on a crop of chili peppers. In 1967, a study by Satour and Butler found 45 species of cultivated plants and weeds susceptible to *P. capsici* In Greek, *Phytophthora capsici* means "plant destroyer of capsicums". *P. capsici* has a wide range of hosts including members of the Solanaceae and Cucurbitaceae family as well as Fabaceae.

Hosts

Under field conditions, *P. capsici* has been found to affect a wide range of hosts in the Cucurbitaceae, Fabaceae, and Solanaceae families, including: cantaloupe, cucumber, watermelon, bell pepper, tomato, snap beans, and lima beans.

Although beans, lima beans, and soybeans were previously thought to be immune to *P. capsici*, in 2000 and 2001, "Phytophthora capsici was isolated from five commercial cultivars of lima bean in Delaware, Maryland, and New Jersey. It was also recently isolated from commercial snap beans in northern Michigan"

Symptoms

General Symptoms: General symptoms on the solanaceous crops and cucurbits include seed rot and seedling blight which discolors the roots and causes seedlings to topple over. Preemergence and postemergence damping-off are also possible symptoms that may occur.

Bean: Include water-soaked foliage, stem and pod necrosis.

Pepper: Infection of the pepper commonly starts at the soil line leading to symptoms of dark, water soaked areas on the stem. Dark lesions of the stem may girdle the plant resulting in death. Roots of the pepper plant appear brown and mushy. Leaf spots start out small and become water soaked, and as time progresses may enlarge turn tan and crack. Blighting of new leaves may also take place. The fruit of the pepper is infected through the stem giving way to water soaked areas on the fruit that are overgrown by signs of the pathogen which appear as, "white-gray, cottony, fungal-like growth"(hyphae). The fruit mummifies and stays attached to the stem.

Eggplant (*Solanum melongena*): Fruit rot is the primary symptom of the eggplant. A dark brown area of the fruit expands into a light tan region. Signs of fungal-like growth may be seen on the lesions.

Tomato (Solanum lycopersicum): P. capsici can cause crown infections, leaf spot, and foliar blight in tomato. The plant may eventually topple over from the crown rot. Fruit rot with patterns of concentric rings is another possible symptom.

Squash: Foliar blight with rapidly expanding water soaked regions and fruit rot are common symptoms on susceptible species of summer and winter squash varieties. These lead to dieback of shoot tips, wilting, shoot rot, and plant death. White fungal growth is also a sign of the pathogen in squash.

Watermelon: Foliar symptoms are less common in watermelon than squash, but the leaves are still susceptible. Fruit rot is more common eventually leading to a total decay of the fruit.

Pumpkin: *P. capsici* causes pre- and post-emergence damping off of seedlings. It also causes vine blight contributing to developing water soaked lesions on the vine which start off as dark olive-colored and soon turn dark brown. This leads to rapid collapse and death of foliage above the lesions. Similar lesions may appear on the leaves and petioles of the pumpkin. Fruit rot is also a very common symptom.

Cantelope (C. cantelopensis): Similar symptoms to that of the watermelon.

Cucumber: Symptoms of the cucumber are similar to that of other cucurbits, but do not include crown gall as a symptom.

P. capsici blight on lower stem of a bell pepper plant.

Disease Cycle

P. capsici is a heterothallic oomycete. The sexual types are designated as A1 and A2. Phytophthora capsici produces both a male and a female type gametangia called an antheridium (male) and an oogonium (female). The antheridium is amphigynous in the species, meaning that the antheridium may remain in this male form of the gametangia or develop into the female gametangia which is an oogonium. Karyogamy between these two types of gametangia one being from the A1 sexualtype and the other of the A2 sexual type results in the formation of an overwintering oospore. The oospores may directly germinate into a germ tube or indirectly germinate and give rise to sporangia which then indirectly germinates and gives rise to zoospores. Zoospores are biflagellate motile spores responsible for the polycyclic qualities of this disease.

Chlamydospores, found in other *Phytophthera* species, have not been documented on *P. capsici* in nature or formed on isolates that were collected from a range of hosts and locations.

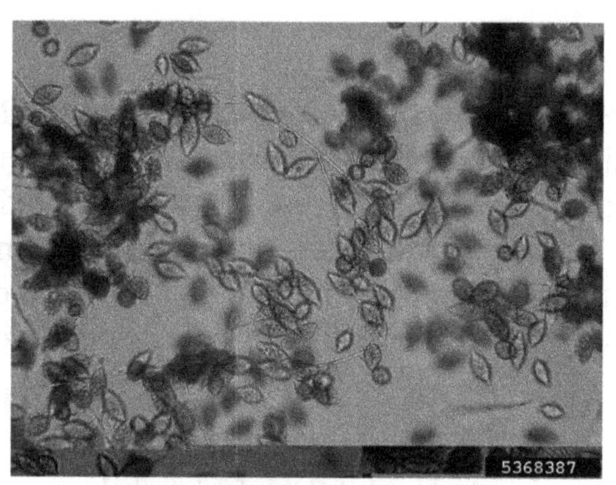

Detached sporangia of P. capsici

Environment

Disease initially occurs in low areas of fields where water accumulates, often leading growers to believe that stunting and death of the cultivar is due to water logging. *P. capsici* grows best at 80 degrees Fahrenheit. It rapidly spreads in warm wet conditions. The asexual spore bearing structures called sporangia are spread by irrigation water, drainage water, and rain. Theses indirectly germinate and release zoospores.

Management

Crop rotation may reduce the number of pathogens in the soil and, "a minimum of a 3 years crop rotation which alternates with non-host species is recommended to avoid build-up of *P. capsici* spores." Crops should also avoid conditions that would be conducive to the pathogen by using well drained soils and raised beds. As stated above, "Excess moisture is the single most important component to the initial infection and subsequent spreading of Phytophthora capsici." Overall, a study by K.H. Lamour and M.K. Hausbeck found that "crop rotation and mefenoxam are not likely to provide economic control". Mefenoxem is the active enantiomer contained in the racemic fungicide metalaxyl used to defend against *Phytophthora capsici*. Sexual recombination provides the genetic diversity to promote resistance towards fungicides in *P. capsici*. The failure of crop rotation as a means to control *P. capsici* may also be due to weeds playing the role of an alternative host in the absence of common hosts. According to a study done by the University of Florida, "In Florida, and perhaps elsewhere, weeds may contribute to pathogen survival in the absence of a host crop or when propagules may not readily survive in soil or plant debris." To avoid fruit rot of vegetable crops in the Cucurbitaceae family, trellising cucurbit fruits and other ways to keep the fruit off the ground is a way to control secondary inoculents (zoospores) from physically being splashed from the soil onto the fruit. Control of *Phytophthora capsici* is easier in drier climates with less rainfall such as California. In these areas, it is important to practice placing drip irrigation emitters away from the stems of pepper plants in order to reduce the incidence of crown rot in peppers. Although resistance has been developed in the cultivars Adra (Abbott and Cobb Seed Co.) and Emerald Isle (Harris Moran Seed Co.), they do not possess sufficient horticultural characteristics accepted by bell pepper growers in the U.S. Paladin (Novartis Seed Co.) has excellent resistance to the crown rot phase of Phytophthora rot and is acceptable to most growers. Paladin does not possess resistance to the foliar phase of this disease and one must use copper fungicides along with the resistant strain for control.

Importance

Phytophthora capsici was first described by Leon H. Leonian at the New Mexico Agricultural Research Station in Las Crucesin in 1922. After this,issues were documented in the Arkansas River Valley of Colorado in the 1930s and 40's. Major research was initiated by M.K. Hausbeck and K.H. Lamour when crop losses due to *P. capsici* threatened to bankrupt numerous vegetable producers in Michigan (which could potentially threaten 134 million dollars worth of vegetable crops). *P. capsici* is also important on a global scale. It is potentially the most destructive disease of peppers in Spain.

References

- Managing Phytophthora Dieback in Bushland: A Guide for Landholders and Community Conservation Groups (PDF) (5th ed.). Australia: Dieback Working Group. 2009. ISBN 9780646493046.
- Ruggiero; et al. (2015), "Higher Level Classification of All Living Organisms", PLoS ONE, 10 (4): e0119248, doi:10.1371/journal.pone.0119248, PMC 4418965, PMID 25923521
- Nowakowska, Marzena; et al. (3 Oct 2014), Appraisal of artificial screening techniques of tomato to accurately reflect field performance of the late blight resistance, PLOS ONE, doi:10.1371/journal.pone.0109328
- "Disease found in Japanese Larch Trees in Ireland". Department of Agriculture, Food & the Marine. 17 August 2010. Retrieved 17 February 2014.
- Grünwald, N. J.; Garbelotto, M.; Goss, E. M.; Heungens, K.; Prospero, S. (2012). "Emergence of the sudden oak death pathogen Phytophthora ramorum". Trends in Microbiology. 20 (3): 131–138. doi:10.1016/j.tim.2011.12.006. PMID 22326131.

4

Fungus: An Overview

Fungus is among the most widely distributed organisms on Earth. They are in large numbers and can be found in every part of our world. The types of fungus discussed in this section are ascomycota, Sclerotinia sclerotiorum, Magnaporthe grisea, sclerotium and armillaria. This text is an overview of the subject matter incorporating all the major aspects of fungus.

Fungus

A fungus is any member of the group of eukaryotic organisms that includes unicellular microorganisms such as yeasts and molds, as well as multicellular fungi that produce familiar fruiting forms known as mushrooms. These organisms are classified as a kingdom, Fungi, which is separate from the other eukaryotic life kingdoms of plants and animals.

A characteristic that places fungi in a different kingdom from plants, bacteria and some protists, is chitin in their cell walls. Similar to animals, fungi are heterotrophs; they acquire their food by absorbing dissolved molecules, typically by secreting digestive enzymes into their environment. Fungi do not photosynthesise. Growth is their means of mobility, except for spores (a few of which are flagellated), which may travel through the air or water. Fungi are the principal decomposers in ecological systems. These and other differences place fungi in a single group of related organisms, named the *Eumycota* (*true fungi* or *Eumycetes*), which share a common ancestor (is a *monophy-letic group*), an interpretation that is also strongly supported by molecular phylogenetics. This fungal group is distinct from the structurally similar myxomycetes (slime molds) and oomycetes (water molds). The discipline of biology devoted to the study of fungi is known as mycology (from the Greek, mukēs, meaning «fungus»). In the past, mycology was regarded as a branch of botany, although it is now known fungi are genetically more closely related to animals than to plants.

Abundant worldwide, most fungi are inconspicuous because of the small size of their structures, and their cryptic lifestyles in soil or on dead matter. Fungi include symbionts of plants, animals, or other fungi and also parasites. They may become noticeable when fruiting, either as mushrooms or as molds. Fungi perform an essential role in the decomposition of organic matter and have fundamental roles in nutrient cycling and exchange in the environment. They have long been used as a direct source of food, in the form of mushrooms and truffles; as a leavening agent for bread; and in the fermentation of various food products, such as wine, beer, and soy sauce. Since the 1940s, fungi have been used for the production of antibiotics, and, more recently, various enzymes produced by fungi are used industrially and in detergents. Fungi are also used as biological pesticides to control weeds, plant diseases and insect pests. Many species produce bioactive compounds called mycotoxins, such as alkaloids and polyketides, that are toxic to animals including humans.

The fruiting structures of a few species contain psychotropic compounds and are consumed recreationally or in traditional spiritual ceremonies. Fungi can break down manufactured materials and buildings, and become significant pathogens of humans and other animals. Losses of crops due to fungal diseases (e.g., rice blast disease) or food spoilage can have a large impact on human food supplies and local economies.

The fungus kingdom encompasses an enormous diversity of taxa with varied ecologies, life cycle strategies, and morphologies ranging from unicellular aquatic chytrids to large mushrooms. However, little is known of the true biodiversity of Kingdom Fungi, which has been estimated at 1.5 million to 5 million species, with about 5% of these having been formally classified. Ever since the pioneering 18th and 19th century taxonomical works of Carl Linnaeus, Christian Hendrik Persoon, and Elias Magnus Fries, fungi have been classified according to their morphology (e.g., characteristics such as spore color or microscopic features) or physiology. Advances in molecular genetics have opened the way for DNA analysis to be incorporated into taxonomy, which has sometimes challenged the historical groupings based on morphology and other traits. Phylogenetic studies published in the last decade have helped reshape the classification within Kingdom Fungi, which is divided into one subkingdom, seven phyla, and ten subphyla.

Characteristics

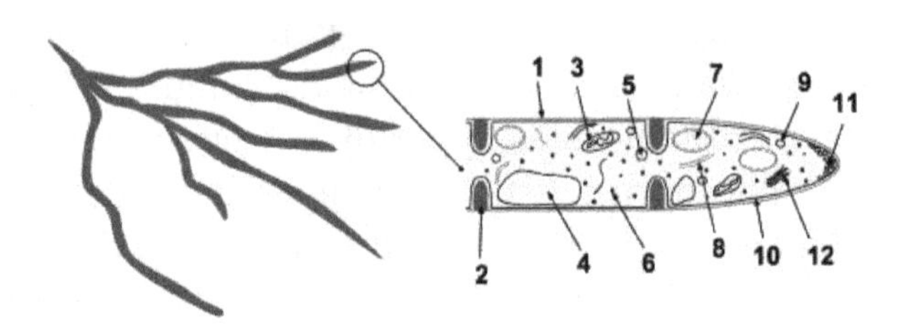

Fungal hyphae cells

1. Hyphal wall
2. Septum
3. Mitochondrion
4. Vacuole
5. Ergosterol crystal
6. Ribosome
7. Nucleus
8. Endoplasmic reticulum
9. Lipid body
10. Plasma membrane

11. Spitzenkörper

12. Golgi apparatus

Before the introduction of molecular methods for phylogenetic analysis, taxonomists considered fungi to be members of the plant kingdom because of similarities in lifestyle: both fungi and plants are mainly immobile, and have similarities in general morphology and growth habitat. Like plants, fungi often grow in soil and, in the case of mushrooms, form conspicuous fruit bodies, which sometimes resemble plants, such as mosses. The fungi are now considered a separate kingdom, distinct from both plants and animals, from which they appear to have diverged around one billion years ago. Some morphological, biochemical, and genetic features are shared with other organisms, while others are unique to the fungi, clearly separating them from the other kingdoms:

Shared features:

- With other eukaryotes: Fungal cells contain membrane-bound nuclei with chromosomes that contain DNA with noncoding regions called introns and coding regions called exons. Fungi have membrane-bound cytoplasmic organelles such as mitochondria, sterol-containing membranes, and ribosomes of the 80S type. They have a characteristic range of soluble carbohydrates and storage compounds, including sugar alcohols (e.g., mannitol), disaccharides, (e.g., trehalose), and polysaccharides (e.g., glycogen, which is also found in animals).

- With animals: Fungi lack chloroplasts and are heterotrophic organisms and so require preformed organic compounds as energy sources.

- With plants: Fungi have a cell wall and vacuoles. They reproduce by both sexual and asexual means, and like basal plant groups (such as ferns and mosses) produce spores. Similar to mosses and algae, fungi typically have haploid nuclei.

- With euglenoids and bacteria: Higher fungi, euglenoids, and some bacteria produce the amino acid L-lysine in specific biosynthesis steps, called the α-aminoadipate pathway.

- The cells of most fungi grow as tubular, elongated, and thread-like (filamentous) structures called hyphae, which may contain multiple nuclei and extend by growing at their tips. Each tip contains a set of aggregated vesicles—cellular structures consisting of proteins, lipids, and other organic molecules—called the Spitzenkörper. Both fungi and oomycetes grow as filamentous hyphal cells. In contrast, similar-looking organisms, such as filamentous green algae, grow by repeated cell division within a chain of cells. There are also single-celled fungi (yeasts) that do not form hyphae, and fungi with both hyphal and yeast forms.

- In common with some plant and animal species, more than 70 fungal species display bioluminescence.

Unique features:

- Some species grow as unicellular yeasts that reproduce by budding or binary fission. Dimorphic fungi can switch between a yeast phase and a hyphal phase in response to environmental conditions.

- The fungal cell wall is composed of glucans and chitin; while glucans are also found in plants and chitin in the exoskeleton of arthropods, fungi are the only organisms that combine these two structural molecules in their cell wall. Unlike those of plants and oomycetes, fungal cell walls do not contain cellulose.

Omphalotus nidiformis, a bioluminescent mushroom

Most fungi lack an efficient system for the long-distance transport of water and nutrients, such as the xylem and phloem in many plants. To overcome this limitation, some fungi, such as *Armillaria*, form rhizomorphs, which resemble and perform functions similar to the roots of plants. As eukaryotes, fungi possess a biosynthetic pathway for producing terpenes that uses mevalonic acid and pyrophosphate as chemical building blocks. Plants and some other organisms have an additional terpene biosynthesis pathway in their chloroplasts, a structure fungi and animals do not have. Fungi produce several secondary metabolites that are similar or identical in structure to those made by plants. Many of the plant and fungal enzymes that make these compounds differ from each other in sequence and other characteristics, which indicates separate origins and evolution of these enzymes in the fungi and plants.

Diversity

Fungi have a worldwide distribution, and grow in a wide range of habitats, including extreme environments such as deserts or areas with high salt concentrations or ionizing radiation, as well as in deep sea sediments. Some can survive the intense UV and cosmic radiation encountered during space travel. Most grow in terrestrial environments, though several species live partly or solely in aquatic habitats, such as the chytrid fungus *Batrachochytrium dendrobatidis*, a parasite that has been responsible for a worldwide decline in amphibian populations. This organism spends part of its life cycle as a motile zoospore, enabling it to propel itself through water and enter its amphibian host. Other examples of aquatic fungi include those living in hydrothermal areas of the ocean.

Bracket fungi on a tree stump

Around 100,000 species of fungi have been formally described by taxonomists, but the global biodiversity of the fungus kingdom is not fully understood. On the basis of observations of the ratio of the number of fungal species to the number of plant species in selected environments, the fungal kingdom has been estimated to contain about 1.5 million species. A recent (2011) estimate suggests there may be over 5 million species. In mycology, species have historically been distinguished by a variety of methods and concepts. Classification based on morphological characteristics, such as the size and shape of spores or fruiting structures, has traditionally dominated fungal taxonomy. Species may also be distinguished by their biochemical and physiological characteristics, such as their ability to metabolize certain biochemicals, or their reaction to chemical tests. The biological species concept discriminates species based on their ability to mate. The application of molecular tools, such as DNA sequencing and phylogenetic analysis, to study diversity has greatly enhanced the resolution and added robustness to estimates of genetic diversity within various taxonomic groups.

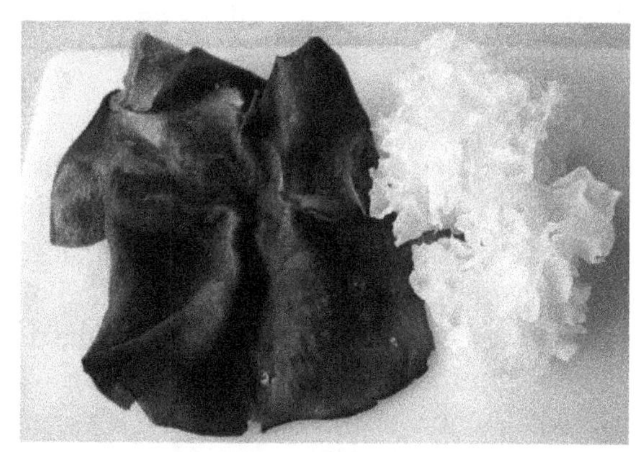

Two types of edible fungi

Mycology

Mycology is the branch of biology concerned with the systematic study of fungi, including their genetic and biochemical properties, their taxonomy, and their use to humans as a source of medicine, food, and psychotropic substances consumed for religious purposes, as well as their dangers, such as poisoning or infection. The field of phytopathology, the study of plant diseases, is closely related because many plant pathogens are fungi.

The use of fungi by humans dates back to prehistory; Ötzi the Iceman, a well-preserved mummy of a 5,300-year-old Neolithic man found frozen in the Austrian Alps, carried two species of polypore mushrooms that may have been used as tinder (*Fomes fomentarius*), or for medicinal purposes (*Piptoporus betulinus*). Ancient peoples have used fungi as food sources–often unknowingly–for millennia, in the preparation of leavened bread and fermented juices. Some of the oldest written records contain references to the destruction of crops that were probably caused by pathogenic fungi.

In 1729, Pier A. Micheli first published descriptions of fungi.

History

Mycology is a relatively new science that became systematic after the development of the microscope in the 16th century. Although fungal spores were first observed by Giambattista della Porta in 1588, the seminal work in the development of mycology is considered to be the publication of Pier Antonio Micheli's 1729 work *Nova plantarum genera*. Micheli not only observed spores but also showed that, under the proper conditions, they could be induced into growing into the same species of fungi from which they originated. Extending the use of the binomial system of nomenclature introduced by Carl Linnaeus in his *Species plantarum* (1753), the Dutch Christian Hendrik Persoon (1761–1836) established the first classification of mushrooms with such skill so as to be considered a founder of modern mycology. Later, Elias Magnus Fries (1794–1878) further elaborated the classification of fungi, using spore color and various microscopic characteristics, methods still used by taxonomists today. Other notable early contributors to mycology in the 17th–19th and early 20th centuries include Miles Joseph Berkeley, August Carl Joseph Corda, Anton de Bary, the brothers Louis René and Charles Tulasne, Arthur H. R. Buller, Curtis G. Lloyd, and Pier Andrea Saccardo. The 20th century has seen a modernization of mycology that has come from advances in biochemistry, genetics, molecular biology, and biotechnology. The use of DNA sequencing technologies and phylogenetic analysis has provided new insights into fungal relationships and biodiversity, and has challenged traditional morphology-based groupings in fungal taxonomy.

Morphology

Microscopic Structures

Most fungi grow as hyphae, which are cylindrical, thread-like structures 2–10 μm in diameter and up to several centimeters in length. Hyphae grow at their tips (apices); new hyphae are typically formed by emergence of new tips along existing hyphae by a process called *branching*, or occasionally growing hyphal tips fork, giving rise to two parallel-growing hyphae. The combination of apical growth and branching/forking leads to the development of a mycelium, an interconnected network of hyphae. Hyphae can be either septate or coenocytic. Septate hyphae are divided into compartments separated by cross walls (internal cell walls, called septa, that are formed at right angles to the cell wall giving the hypha its shape), with each compartment containing one or more nuclei; coenocytic hyphae are not compartmentalized. Septa have pores that allow cytoplasm, organelles, and sometimes nuclei to pass through; an example is the dolipore septum in fungi of the phylum Basidiomycota. Coenocytic hyphae are in essence multinucleate supercells.

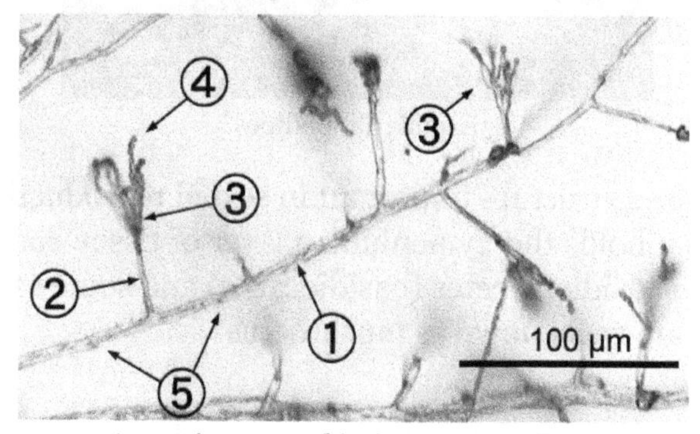

An environmental isolate of Penicillium

1. hypha
2. conidiophore
3. phialide
4. conidia
5. septa

Many species have developed specialized hyphal structures for nutrient uptake from living hosts; examples include haustoria in plant-parasitic species of most fungal phyla, and arbuscules of several mycorrhizal fungi, which penetrate into the host cells to consume nutrients.

Although fungi are opisthokonts—a grouping of evolutionarily related organisms broadly characterized by a single posterior flagellum—all phyla except for the chytrids have lost their posterior flagella. Fungi are unusual among the eukaryotes in having a cell wall that, in addition to glucans (e.g., β-1,3-glucan) and other typical components, also contains the biopolymer chitin.

Macroscopic Structures

Fungal mycelia can become visible to the naked eye, for example, on various surfaces and substrates, such as damp walls and spoiled food, where they are commonly called molds. Mycelia grown on solid agar media in laboratory petri dishes are usually referred to as colonies. These colonies can exhibit growth shapes and colors (due to spores or pigmentation) that can be used as diagnostic features in the identification of species or groups. Some individual fungal colonies can reach extraordinary dimensions and ages as in the case of a clonal colony of *Armillaria solidipes*, which extends over an area of more than 900 ha (3.5 square miles), with an estimated age of nearly 9,000 years.

Armillaria solidipes

The apothecium—a specialized structure important in sexual reproduction in the ascomycetes—is a cup-shaped fruit body that holds the hymenium, a layer of tissue containing the spore-bearing cells. The fruit bodies of the basidiomycetes (basidiocarps) and some ascomycetes can sometimes grow very large, and many are well known as mushrooms.

Growth and Physiology

Mold growth covering a decaying peach. The frames were taken approximately 12 hours apart over a period of six days.

The growth of fungi as hyphae on or in solid substrates or as single cells in aquatic environments is adapted for the efficient extraction of nutrients, because these growth forms have high surface area to volume ratios. Hyphae are specifically adapted for growth on solid surfaces, and to invade substrates and tissues. They can exert large penetrative mechanical forces; for example, the plant pathogen *Magnaporthe grisea* forms a structure called an appressorium that evolved to puncture plant tissues. The pressure generated by the appressorium, directed against the plant epidermis, can exceed 8 megapascals (1,200 psi). The filamentous fungus *Paecilomyces lilacinus* uses a similar structure to penetrate the eggs of nematodes.

The mechanical pressure exerted by the appressorium is generated from physiological processes that increase intracellular turgor by producing osmolytes such as glycerol. Adaptations such as these are complemented by hydrolytic enzymes secreted into the environment to digest large organic molecules—such as polysaccharides, proteins, and lipids—into smaller molecules that may then be absorbed as nutrients. The vast majority of filamentous fungi grow in a polar fashion—i.e., by extension into one direction—by elongation at the tip (apex) of the hypha. Other forms of fungal growth include intercalary extension (longitudinal expansion of hyphal compartments that are below the apex) as in the case of some endophytic fungi, or growth by volume expansion during the development of mushroom stipes and other large organs. Growth of fungi as multicellular structures consisting of somatic and reproductive cells—a feature independently evolved in animals and plants—has several functions, including the development of fruit bodies for dissemination of sexual spores and biofilms for substrate colonization and intercellular communication.

The fungi are traditionally considered heterotrophs, organisms that rely solely on carbon fixed by other organisms for metabolism. Fungi have evolved a high degree of metabolic versatility that allows them to use a diverse range of organic substrates for growth, including simple compounds such as nitrate, ammonia, acetate, or ethanol. In some species the pigment melanin may play a role in extracting energy from ionizing radiation, such as gamma radiation. This form of "radiotrophic" growth has been described for only a few species, the effects on growth rates are small, and the underlying biophysical and biochemical processes are not well known. This process might bear similarity to CO_2 fixation via visible light, but instead uses ionizing radiation as a source of energy.

Reproduction

Polyporus squamosus

Fungal reproduction is complex, reflecting the differences in lifestyles and genetic makeup within this diverse kingdom of organisms. It is estimated that a third of all fungi reproduce using more than one method of propagation; for example, reproduction may occur in two well-differentiated stages within the life cycle of a species, the teleomorph and the anamorph. Environmental conditions trigger genetically determined developmental states that lead to the creation of specialized structures for sexual or asexual reproduction. These structures aid reproduction by efficiently dispersing spores or spore-containing propagules.

Asexual Reproduction

Asexual reproduction occurs via vegetative spores (conidia) or through mycelial fragmentation. Mycelial fragmentation occurs when a fungal mycelium separates into pieces, and each component grows into a separate mycelium. Mycelial fragmentation and vegetative spores maintain clonal populations adapted to a specific niche, and allow more rapid dispersal than sexual reproduction. The "Fungi imperfecti" (fungi lacking the perfect or sexual stage) or Deuteromycota comprise all the species that lack an observable sexual cycle.

Sexual Reproduction

Sexual reproduction with meiosis exists in all fungal phyla except Glomeromycota. It differs in many aspects from sexual reproduction in animals or plants. Differences also exist between fungal groups and can be used to discriminate species by morphological differences in sexual structures and reproductive strategies. Mating experiments between fungal isolates may identify species on the basis of biological species concepts. The major fungal groupings have initially been delineated based on the morphology of their sexual structures and spores; for example, the spore-containing structures, asci and basidia, can be used in the identification of ascomycetes and basidiomycetes, respectively. Some species may allow mating only between individuals of opposite mating type, whereas others can mate and sexually reproduce with any other individual or itself. Species of the former mating system are called heterothallic, and of the latter homothallic.

Most fungi have both a haploid and a diploid stage in their life cycles. In sexually reproducing fungi, compatible individuals may combine by fusing their hyphae together into an interconnected network; this process, anastomosis, is required for the initiation of the sexual cycle. Ascomycetes and basidiomycetes go through a dikaryotic stage, in which the nuclei inherited from the two parents do not combine immediately after cell fusion, but remain separate in the hyphal cells.

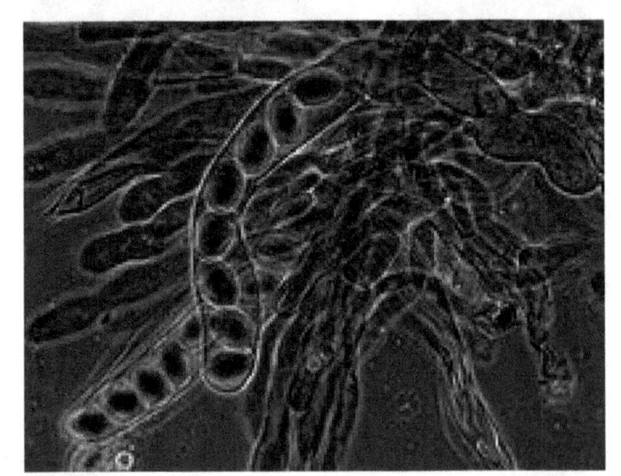

The 8-spore asci of *Morchella elata*, viewed with phase contrast microscopy

In ascomycetes, dikaryotic hyphae of the hymenium (the spore-bearing tissue layer) form a characteristic *hook* at the hyphal septum. During cell division, formation of the hook ensures proper distribution of the newly divided nuclei into the apical and basal hyphal compartments. An ascus (plural *asci*) is then formed, in which karyogamy (nuclear fusion) occurs. Asci are embedded in an ascocarp, or fruiting body. Karyogamy in the asci is followed immediately by meiosis and the production of ascospores. After dispersal, the ascospores may germinate and form a new haploid mycelium.

Sexual reproduction in basidiomycetes is similar to that of the ascomycetes. Compatible haploid hyphae fuse to produce a dikaryotic mycelium. However, the dikaryotic phase is more extensive in the basidiomycetes, often also present in the vegetatively growing mycelium. A specialized anatomical structure, called a clamp connection, is formed at each hyphal septum. As with the structurally similar hook in the ascomycetes, the clamp connection in the basidiomycetes is required for controlled transfer of nuclei during cell division, to maintain the dikaryotic stage with two genetically different nuclei in each hyphal compartment. A basidiocarp is formed in which club-like structures known as basidia generate haploid basidiospores after karyogamy and meiosis. The most commonly known basidiocarps are mushrooms, but they may also take other forms.

In glomeromycetes (formerly zygomycetes), haploid hyphae of two individuals fuse, forming a gametangium, a specialized cell structure that becomes a fertile gamete-producing cell. The gametangium develops into a zygospore, a thick-walled spore formed by the union of gametes. When the zygospore germinates, it undergoes meiosis, generating new haploid hyphae, which may then form asexual sporangiospores. These sporangiospores allow the fungus to rapidly disperse and germinate into new genetically identical haploid fungal mycelia.

Spore Dispersal

Both asexual and sexual spores or sporangiospores are often actively dispersed by forcible ejection from their reproductive structures. This ejection ensures exit of the spores from the reproductive structures as well as traveling through the air over long distances.

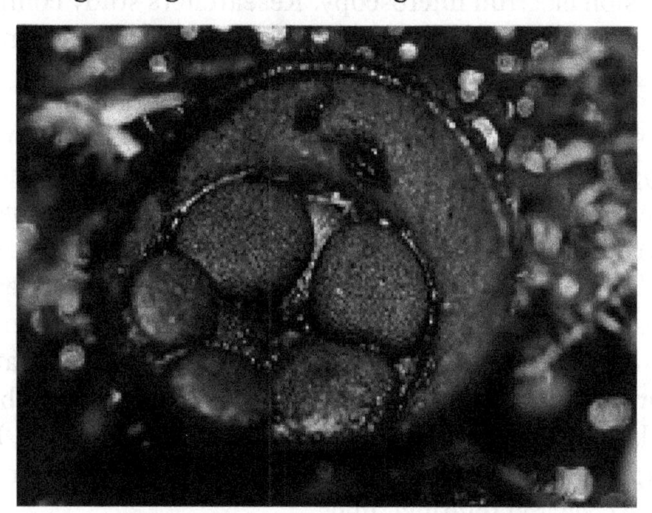

The bird's nest fungus Cyathus stercoreus

Specialized mechanical and physiological mechanisms, as well as spore surface structures (such as hydrophobins), enable efficient spore ejection. For example, the structure of the spore-bearing cells in some ascomycete species is such that the buildup of substances affecting cell volume and fluid balance enables the explosive discharge of spores into the air. The forcible discharge of single spores termed *ballistospores* involves formation of a small drop of water (Buller's drop), which upon contact with the spore leads to its projectile release with an initial acceleration of more than 10,000 g; the net result is that the spore is ejected 0.01–0.02 cm, sufficient distance for it to fall through the gills or pores into the air below. Other fungi, like the puffballs, rely on alternative mechanisms for spore release, such as external mechanical forces. The bird's nest fungi use the force of falling water drops to liberate the spores from cup-shaped fruiting bodies. Another strategy is seen in the stinkhorns, a group of fungi with lively colors and putrid odor that attract insects to disperse their spores.

Other Sexual Processes

Besides regular sexual reproduction with meiosis, certain fungi, such as those in the genera *Penicillium* and *Aspergillus*, may exchange genetic material via parasexual processes, initiated by anastomosis between hyphae and plasmogamy of fungal cells. The frequency and relative importance of parasexual events is unclear and may be lower than other sexual processes. It is known to play a role in intraspecific hybridization and is likely required for hybridization between species, which has been associated with major events in fungal evolution.

Evolution

In contrast to plants and animals, the early fossil record of the fungi is meager. Factors that likely contribute to the under-representation of fungal species among fossils include the nature of fungal fruiting bodies, which are soft, fleshy, and easily degradable tissues and the microscopic dimensions of most fungal structures, which therefore are not readily evident. Fungal fossils are difficult to distinguish from those of other microbes, and are most easily identified when they resemble extant fungi. Often recovered from a permineralized plant or animal host, these samples are typically studied by making thin-section preparations that can be examined with light microscopy or transmission electron microscopy. Researchers study compression fossils by dissolving the surrounding matrix with acid and then using light or scanning electron microscopy to examine surface details.

The earliest fossils possessing features typical of fungi date to the Proterozoic eon, some 1,430 million years ago (Ma); these multicellular benthic organisms had filamentous structures with septa, and were capable of anastomosis. More recent studies (2009) estimate the arrival of fungal organisms at about 760–1060 Ma on the basis of comparisons of the rate of evolution in closely related groups.For much of the Paleozoic Era (542–251 Ma), the fungi appear to have been aquatic and consisted of organisms similar to the extant chytrids in having flagellum-bearing spores. The evolutionary adaptation from an aquatic to a terrestrial lifestyle necessitated a diversification of ecological strategies for obtaining nutrients, including parasitism, saprobism, and the development of mutualistic relationships such as mycorrhiza and lichenization. Recent (2009) studies suggest that the ancestral ecological state of the Ascomycota was saprobism, and that independent lichenization events have occurred multiple times.

It is presumed that the fungi colonized the land during the Cambrian (542–488.3 Ma), long before land plants. Fossilized hyphae and spores recovered from the Ordovician of Wisconsin (460 Ma) resemble modern-day Glomerales, and existed at a time when the land flora likely consisted of only non-vascular bryophyte-like plants. Prototaxites, which was probably a fungus or lichen, would have been the tallest organism of the late Silurian. Fungal fossils do not become common and uncontroversial until the early Devonian (416–359.2 Ma), when they occur abundantly in the

Rhynie chert, mostly as Zygomycota and Chytridiomycota. At about this same time, approximately 400 Ma, the Ascomycota and Basidiomycota diverged, and all modern classes of fungi were present by the Late Carboniferous (Pennsylvanian, 318.1–299 Ma).

Lichen-like fossils have been found in the Doushantuo Formation in southern China dating back to 635–551 Ma. Lichens formed a component of the early terrestrial ecosystems, and the estimated age of the oldest terrestrial lichen fossil is 400 Ma; this date corresponds to the age of the oldest known sporocarp fossil, a *Paleopyrenomycites* species found in the Rhynie Chert. The oldest fossil with microscopic features resembling modern-day basidiomycetes is *Palaeoancistrus*, found permineralized with a fern from the Pennsylvanian. Rare in the fossil record are the Homobasidiomycetes (a taxon roughly equivalent to the mushroom-producing species of the Agaricomycetes). Two amber-preserved specimens provide evidence that the earliest known mushroom-forming fungi (the extinct species *Archaeomarasmius leggetti*) appeared during the late Cretaceous, 90 Ma.

Some time after the Permian–Triassic extinction event (251.4 Ma), a fungal spike (originally thought to be an extraordinary abundance of fungal spores in sediments) formed, suggesting that fungi were the dominant life form at this time, representing nearly 100% of the available fossil record for this period. However, the relative proportion of fungal spores relative to spores formed by algal species is difficult to assess, the spike did not appear worldwide, and in many places it did not fall on the Permian–Triassic boundary.

Taxonomy

Although commonly included in botany curricula and textbooks, fungi are more closely related to animals than to plants and are placed with the animals in the monophyletic group of opisthokonts. Analyses using molecular phylogenetics support a monophyletic origin of the Fungi. The taxonomy of the Fungi is in a state of constant flux, especially due to recent research based on DNA comparisons. These current phylogenetic analyses often overturn classifications based on older and sometimes less discriminative methods based on morphological features and biological species concepts obtained from experimental matings.

There is no unique generally accepted system at the higher taxonomic levels and there are frequent name changes at every level, from species upwards. Efforts among researchers are now underway to establish and encourage usage of a unified and more consistent nomenclature. Fungal species can also have multiple scientific names depending on their life cycle and mode (sexual or asexual) of reproduction. Web sites such as Index Fungorum and ITIS list current names of fungal species (with cross-references to older synonyms).

The 2007 classification of Kingdom Fungi is the result of a large-scale collaborative research effort involving dozens of mycologists and other scientists working on fungal taxonomy. It recognizes seven phyla, two of which—the Ascomycota and the Basidiomycota—are contained within a branch representing subkingdom Dikarya. The accompanying cladogram depicts the major fungal taxa and their relationship to opisthokont and unikont organisms, based on the work of Philippe Silar and "The Mycota: A Comprehensive Treatise on Fungi as Experimental Systems for Basic and Applied Research". The lengths of the branches are not proportional to evolutionary distances.

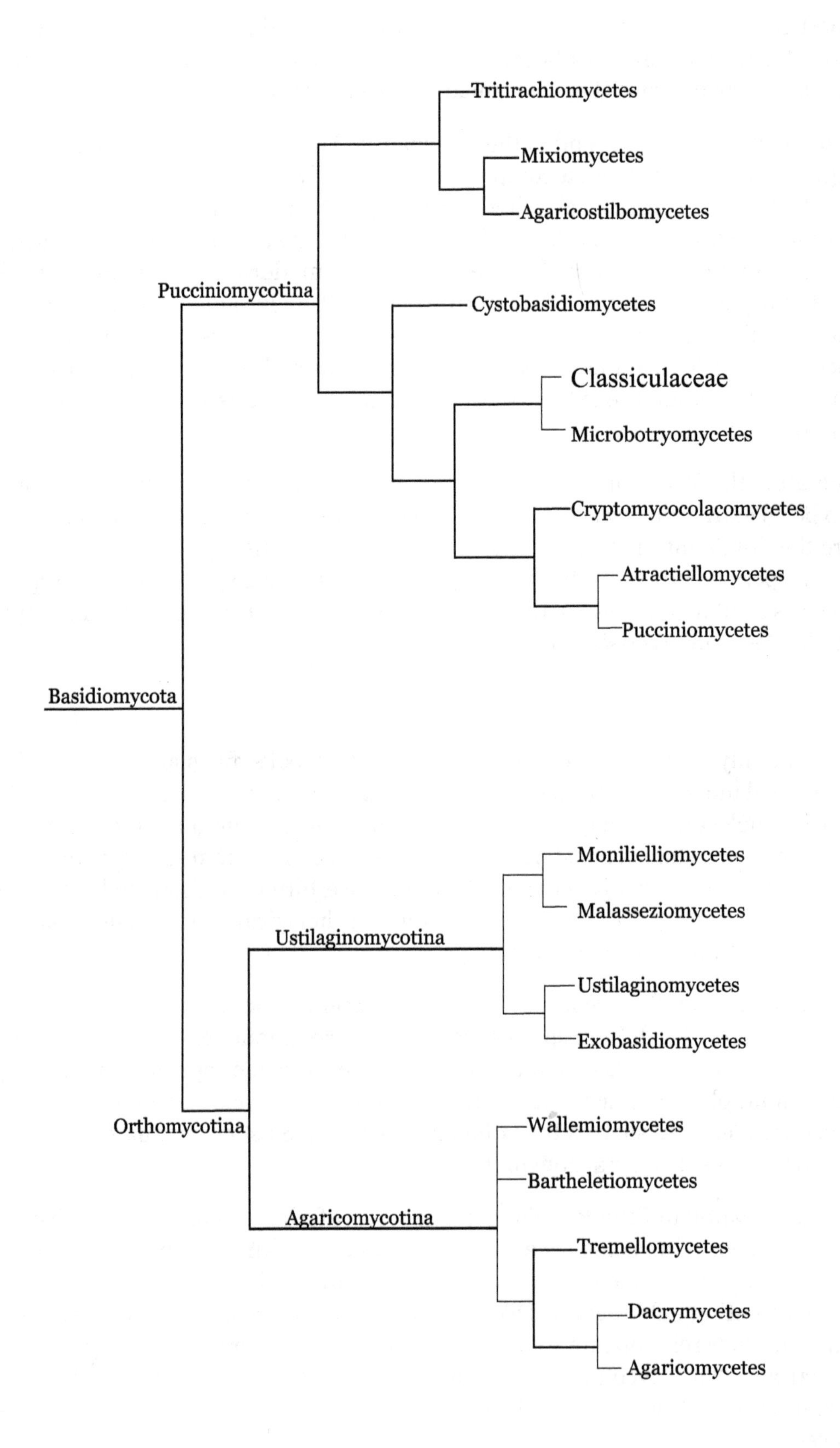
Tritirachiomycetes
Mixiomycetes
Agaricostilbomycetes
Pucciniomycotina
Cystobasidiomycetes
Classiculaceae
Microbotryomycetes
Cryptomycocolacomycetes
Atractiellomycetes
Pucciniomycetes
Basidiomycota
Moniliellomycetes
Malasseziomycetes
Ustilaginomycotina
Ustilaginomycetes
Exobasidiomycetes
Orthomycotina
Wallemiomycetes
Bartheletiomycetes
Agaricomycotina
Tremellomycetes
Dacrymycetes
Agaricomycetes

Taxonomic Groups

The major phyla (sometimes called divisions) of fungi have been classified mainly on the basis of characteristics of their sexual reproductive structures. Currently, seven phyla are proposed: Microsporidia, Chytridiomycota, Blastocladiomycota, Neocallimastigomycota, Glomeromycota, Ascomycota, and Basidiomycota.

Arbuscular mycorrhiza seen under microscope. Flax root cortical cells containing paired arbuscules.

Phylogenetic analysis has demonstrated that the Microsporidia, unicellular parasites of animals and protists, are fairly recent and highly derived endobiotic fungi (living within the tissue of another species). One 2006 study concludes that the Microsporidia are a sister group to the true fungi; that is, they are each other's closest evolutionary relative. Hibbett and colleagues suggest that this analysis does not clash with their classification of the Fungi, and although the Microsporidia are elevated to phylum status, it is acknowledged that further analysis is required to clarify evolutionary relationships within this group.

The Chytridiomycota are commonly known as chytrids. These fungi are distributed worldwide. Chytrids produce zoospores that are capable of active movement through aqueous phases with a single flagellum, leading early taxonomists to classify them as protists. Molecular phylogenies, inferred from rRNA sequences in ribosomes, suggest that the Chytrids are a basal group divergent from the other fungal phyla, consisting of four major clades with suggestive evidence for paraphyly or possibly polyphyly.

The Blastocladiomycota were previously considered a taxonomic clade within the Chytridiomycota. Recent molecular data and ultrastructural characteristics, however, place the Blastocladiomycota as a sister clade to the Zygomycota, Glomeromycota, and Dikarya (Ascomycota and Basidiomycota). The blastocladiomycetes are saprotrophs, feeding on decomposing organic matter, and they are parasites of all eukaryotic groups. Unlike their close relatives, the chytrids, most of which exhibit zygotic meiosis, the blastocladiomycetes undergo sporic meiosis.

The Neocallimastigomycota were earlier placed in the phylum Chytridomycota. Members of this small phylum are anaerobic organisms, living in the digestive system of larger herbivorous mammals and in other terrestrial and aquatic environments enriched in cellulose (e.g., domestic waste landfill sites). They lack mitochondria but contain hydrogenosomes of mitochondrial origin. As the related chrytrids, neocallimastigomycetes form zoospores that are posteriorly uniflagellate or polyflagellate.

Members of the Glomeromycota form arbuscular mycorrhizae, a form of symbiosis wherein fungal hyphae invade plant root cells and both species benefit from the resulting increased supply of nutrients. All known Glomeromycota species reproduce asexually. The symbiotic association between the Glomeromycota and plants is ancient, with evidence dating to 400 million years ago. Formerly part of the Zygomycota (commonly known as 'sugar' and 'pin' molds), the Glomeromycota were elevated to phylum status in 2001 and now replace the older phylum Zygomycota. Fungi that were placed in the Zygomycota are now being reassigned to the Glomeromycota, or the subphyla incertae sedis Mucoromycotina, Kickxellomycotina, the Zoopagomycotina and the Entomophthoromycotina. Some well-known examples of fungi formerly in the Zygomycota include black bread mold (*Rhizopus stolonifer*), and *Pilobolus* species, capable of ejecting spores several meters through the air. Medically relevant genera include *Mucor*, *Rhizomucor*, and *Rhizopus*.

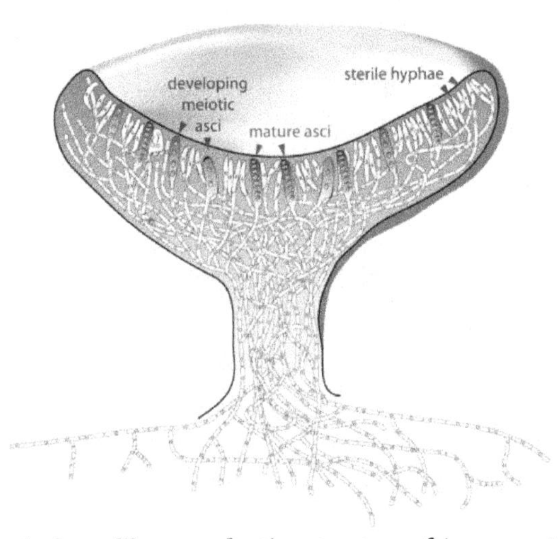

Diagram of an apothecium (the typical cup-like reproductive structure of Ascomycetes) showing sterile tissues as well as developing and mature asci.

The Ascomycota, commonly known as sac fungi or ascomycetes, constitute the largest taxonomic group within the Eumycota. These fungi form meiotic spores called ascospores, which are enclosed in a special sac-like structure called an ascus. This phylum includes morels, a few mushrooms and truffles, unicellular yeasts (e.g., of the genera *Saccharomyces*, *Kluyveromyces*, *Pichia*, and *Candida*), and many filamentous fungi living as saprotrophs, parasites, and mutualistic symbionts. Prominent and important genera of filamentous ascomycetes include *Aspergillus*, *Penicillium*, *Fusarium*, and *Claviceps*. Many ascomycete species have only been observed undergoing asexual reproduction (called anamorphic species), but analysis of molecular data has often been able to identify their closest teleomorphs in the Ascomycota. Because the products of meiosis are retained within the sac-like ascus, ascomycetes have been used for elucidating principles of genetics and heredity (e.g., *Neurospora crassa*).

Members of the Basidiomycota, commonly known as the club fungi or basidiomycetes, produce meiospores called basidiospores on club-like stalks called basidia. Most common mushrooms belong to this group, as well as rust and smut fungi, which are major pathogens of grains. Other important basidiomycetes include the maize pathogen *Ustilago maydis*, human commensal species of the genus *Malassezia*, and the opportunistic human pathogen, *Cryptococcus neoformans*.

Fungus-like Organisms

Because of similarities in morphology and lifestyle, the slime molds (mycetozoans, plasmodiophorids, acrasids, *Fonticula* and labyrinthulids, now in Amoebozoa, Rhizaria, Excavata, Opisthokonta and Stramenopiles, respectively), water molds (oomycetes) and hyphochytrids (both Stramenopiles) were formerly classified in the kingdom Fungi, in groups like Mastigomycotina, Gymnomycota and Phycomycetes. The slime molds were studied also as protozoans, leading to a ambiregnal, duplicated taxonomy.

Unlike true fungi, the cell walls of oomycetes contain cellulose and lack chitin. Hyphochytrids have both chitin and cellulose. Slime molds lack a cell wall during the assimilative phase (except labyrinthulids, which have a wall of scales), and ingest nutrients by ingestion (phagocytosis, except labyrinthulids) rather than absorption (osmotrophy, as fungi, labyrinthulids, oomycetes and hyphochytrids). Neither water molds nor slime molds are closely related to the true fungi, and, therefore, taxonomists no longer group them in the kingdom Fungi. Nonetheless, studies of the oomycetes and myxomycetes are still often included in mycology textbooks and primary research literature.

The Eccrinales and Amoebidiales are opisthokont protists, previously thought to be zygomycete fungi. Other groups now in Opisthokonta (e.g., *Corallochytrium*, Ichthyosporea) were also at given time classified as fungi. The genus *Blastocystis*, now in Stramenopiles, was originally classified as a yeast. *Ellobiopsis*, now in Alveolata, was considered a chytrid. The bacteria were also included in fungi in some classifications, as the group Schizomycetes.

The Rozellida clade, including the "ex-chytrid" *Rozella*, is a genetically disparate group known mostly from environmental DNA sequences that is a sister group to fungi. Members of the group that have been isolated lack the chitinous cell wall that is characteristic of fungi.

The nucleariids, protists currently grouped in the Choanozoa (Opisthokonta), may be the next sister group to the eumycete clade, and as such could be included in an expanded fungal kingdom.

Many Actinomycetales (Actinobacteria), a group with many filamentous bacteria, were also long believed to be fungi.

Ecology

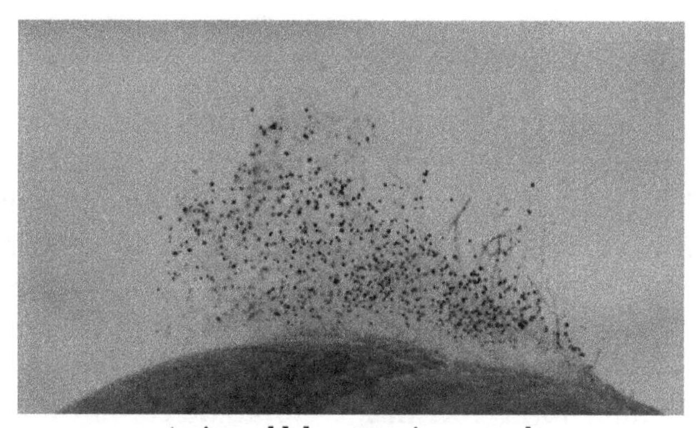

A pin mold decomposing a peach

Although often inconspicuous, fungi occur in every environment on Earth and play very important roles in most ecosystems. Along with bacteria, fungi are the major decomposers in most terrestrial (and some aquatic) ecosystems, and therefore play a critical role in biogeochemical cycles and in many food webs. As decomposers, they play an essential role in nutrient cycling, especially as saprotrophs and symbionts, degrading organic matter to inorganic molecules, which can then re-enter anabolic metabolic pathways in plants or other organisms.

Symbiosis

Many fungi have important symbiotic relationships with organisms from most if not all Kingdoms. These interactions can be mutualistic or antagonistic in nature, or in the case of commensal fungi are of no apparent benefit or detriment to the host.

With Plants

Mycorrhizal symbiosis between plants and fungi is one of the most well-known plant–fungus associations and is of significant importance for plant growth and persistence in many ecosystems; over 90% of all plant species engage in mycorrhizal relationships with fungi and are dependent upon this relationship for survival.

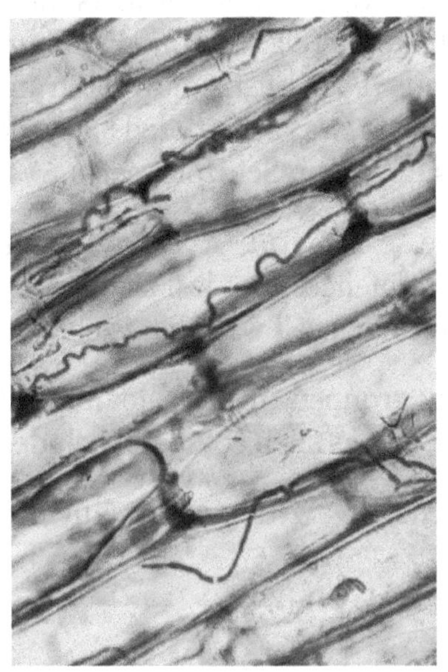

The dark filaments are hyphae of the endophytic fungus *Neotyphodium coenophialum* in the intercellular spaces of tall fescue leaf sheath tissue

The mycorrhizal symbiosis is ancient, dating to at least 400 million years ago. It often increases the plant's uptake of inorganic compounds, such as nitrate and phosphate from soils having low concentrations of these key plant nutrients. The fungal partners may also mediate plant-to-plant transfer of carbohydrates and other nutrients. Such mycorrhizal communities are called "common mycorrhizal networks". A special case of mycorrhiza is myco-heterotrophy, whereby the plant parasitizes the fungus, obtaining all of its nutrients from its fungal symbiont. Some fungal species inhabit the tissues inside roots, stems, and leaves, in which case they are called endophytes. Similar to mycorrhiza, endophytic colonization by fungi may benefit both symbionts; for example, endophytes of grasses impart to their host increased resistance to herbivores and other environmental stresses and receive food and shelter from the plant in return.

With Algae and Cyanobacteria

The lichen Lobaria pulmonaria, a symbiosis of fungal, algal, and cyanobacterial species

Lichens are a symbiotic relationship between fungi and algae or cyanobacteria. The algae partner in the relationship is referred to in lichen terminology as a "photobiont". The fungi part of the relationship are composed mostly of various species of ascomycetes and a few basidiomycetes. Lichens occur in every ecosystem on all continents, play a key role in soil formation and the initiation of biological succession, and are the dominating life forms in extreme environments, including polar, alpine, and semiarid desert regions. They are able to grow on inhospitable surfaces, including bare soil, rocks, tree bark, wood, shells, barnacles and leaves. As in mycorrhizas, the photobiont provides sugars and other carbohydrates via photosynthesis to the fungus, while the fungus provides minerals and water to the photobiont. The functions of both symbiotic organisms are so closely intertwined that they function almost as a single organism; in most cases the resulting organism differs greatly from the individual components. Lichenization is a common mode of nutrition for fungi; around 20% of fungi—between 17,500 and 20,000 described species—are lichenized. Characteristics common to most lichens include obtaining organic carbon by photosynthesis, slow growth, small size, long life, long-lasting (seasonal) vegetative reproductive structures, mineral nutrition obtained largely from airborne sources, and greater tolerance of desiccation than most other photosynthetic organisms in the same habitat.

With Insects

Many insects also engage in mutualistic relationships with fungi. Several groups of ants cultivate fungi in the order Agaricales as their primary food source, while ambrosia beetles cultivate various species of fungi in the bark of trees that they infest. Likewise, females of several wood wasp species (genus *Sirex*) inject their eggs together with spores of the wood-rotting fungus *Amylostereum areolatum* into the sapwood of pine trees; the growth of the fungus provides ideal nutritional conditions for the development of the wasp larvae. At least one species of stingless bee has a relationship with a fungus in the genus *Monascus*, where the larvae consume and depend on fungus transferred from old to new nests. Termites on the African savannah are also known to cultivate fungi, and yeasts of the genera *Candida* and *Lachancea* inhabit the gut of a wide range of insects, including neuropterans, beetles, and cockroaches; it is not known whether these fungi benefit their hosts. Fungi ingrowing dead wood are essential for xylophagous insects (e.g. woodboring beetles). They deliver nutrients needed by xylophages to nutritionally scarce dead wood. Thanks to this nutritional enrichment the larvae of woodboring insect is able to grow and develop to adulthood. The larvae of many families of fungicolous flies, particularly those within the superfamily Sciaroidea such as the Mycetophilidae and some Keroplatidae feed on fungal fruiting bodies and sterile mycorrhizae.

As Pathogens and Parasites

The plant pathogen Aecidium magellanicum causes calafate rust, seen here on a Berberis shrub in Chile.

Many fungi are parasites on plants, animals (including humans), and other fungi. Serious pathogens of many cultivated plants causing extensive damage and losses to agriculture and forestry include the rice blast fungus *Magnaporthe oryzae*, tree pathogens such as *Ophiostoma ulmi* and *Ophiostoma novo-ulmi* causing Dutch elm disease, and *Cryphonectria parasitica* responsible for chestnut blight, and plant pathogens in the genera *Fusarium*, *Ustilago*, *Alternaria*, and *Cochliobolus*. Some carnivorous fungi, like *Paecilomyces lilacinus*, are predators of nematodes, which they capture using an array of specialized structures such as constricting rings or adhesive nets.

Some fungi can cause serious diseases in humans, several of which may be fatal if untreated. These include aspergillosis, candidiasis, coccidioidomycosis, cryptococcosis, histoplasmosis, mycetomas, and paracoccidioidomycosis. Furthermore, persons with immuno-deficiencies are particularly susceptible to disease by genera such as *Aspergillus*, *Candida*, *Cryptoccocus*, *Histoplasma*, and *Pneumocystis*. Other fungi can attack eyes, nails, hair, and especially skin, the so-called dermatophytic and keratinophilic fungi, and cause local infections such as ringworm and athlete's foot. Fungal spores are also a cause of allergies, and fungi from different taxonomic groups can evoke allergic reactions.

Mycotoxins

Many fungi produce biologically active compounds, several of which are toxic to animals or plants and are therefore called mycotoxins. Of particular relevance to humans are mycotoxins produced by molds causing food spoilage, and poisonous mushrooms. Particularly infamous are the lethal amatoxins in some *Amanita* mushrooms, and ergot alkaloids, which have a long history of causing serious epidemics of ergotism (St Anthony's Fire) in people consuming rye or related cereals contaminated with sclerotia of the ergot fungus, *Claviceps purpurea*. Other notable mycotoxins include the aflatoxins, which are insidious liver toxins and highly carcinogenic metabolites produced by certain *Aspergillus* species often growing in or on grains and nuts consumed by humans, ochratoxins, patulin, and trichothecenes (e.g., T-2 mycotoxin) and fumonisins, which have significant impact on human food supplies or animal livestock.

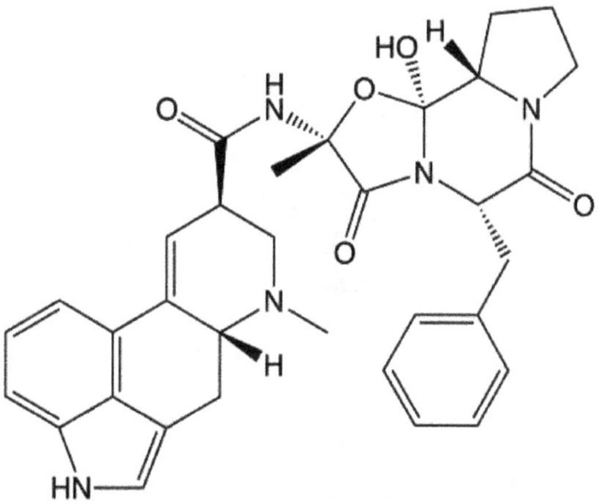

Ergotamine, a major mycotoxin produced by *Claviceps* species, which if ingested can cause gangrene, convulsions, and hallucinations

Mycotoxins are secondary metabolites (or natural products), and research has established the existence of biochemical pathways solely for the purpose of producing mycotoxins and other natural products in fungi. Mycotoxins may provide fitness benefits in terms of physiological adaptation, competition with other microbes and fungi, and protection from consumption (fungivory).

Pathogenic Mechanisms

Ustilago maydis is a pathogenic plant fungus that causes smut disease in maize and teosinte. Plants have evolved efficient defense systems against pathogenic microbes such as *U. maydis*. A rapid defense reaction after pathogen attack is the oxidative burst where the plant produces reactive oxygen species at the site of the attempted invasion. *U. maydis* can respond to the oxidative burst with an oxidative stress response, regulated by the gene *YAP1*. The response protects *U. maydis* from the host defense, and is necessary for the pathogen's virulence. Furthermore, *U. maydis* has a well-established recombinational DNA repair system which acts during mitosis and meiosis. The system may assist the pathogen in surviving DNA damage arising from the host plant's oxidative defensive response to infection.

Cryptococcus neoformans is an encapsulated yeast that can live in both plants and animals. *C. neoformans* usually infects the lungs, where it is phagocytosed by alveolar macrophages. Some *C. neoformans* can survive inside macrophages, which appears to be the basis for latency, disseminated disease, and resistance to antifungal agents. One mechanism by which *C. neoformans* survives the hostile macrophage environment is by up-regulating the expression of genes involved in the oxidative stress response. Another mechanism involves meiosis. The majority of *C. neoformans* are mating "type a". Filaments of mating "type a" ordinarily have haploid nuclei, but they can become diploid (perhaps by endoduplication or by stimulated nuclear fusion) to form blastospores. The diploid nuclei of blastospores can undergo meiosis, including recombination, to form haploid basidiospores that can be dispersed. This process is referred to as monokaryotic fruiting. this process requires a gene called *DMC1*, which is a conserved homologue of genes *recA* in bacteria and *RAD51* in eukaryotes, that mediates homologous chromosome pairing during meiosis and repair of DNA double-strand breaks. Thus, *C. neoformans* can undergo a meiosis, monokaryotic fruiting, that promotes recombinational repair in the oxidative, DNA damaging environment of the host macrophage, and the repair capability may contribute to its virulence.

Human Use

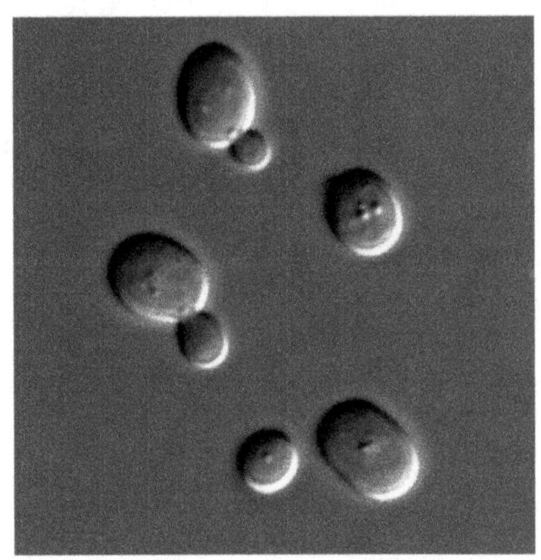

Saccharomyces cerevisiae cells shown with DIC microscopy

The human use of fungi for food preparation or preservation and other purposes is extensive and has a long history. Mushroom farming and mushroom gathering are large industries in many countries. The study of the historical uses and sociological impact of fungi is known as ethnomycology. Because of the capacity of this group to produce an enormous range of natural products with antimicrobial or other biological activities, many species have long been used or are being developed for industrial production of antibiotics, vitamins, and anti-cancer and cholesterol-lowering drugs. More recently, methods have been developed for genetic engineering of fungi, enabling metabolic engineering of fungal species. For example, genetic modification of yeast species—which are easy to grow at fast rates in large fermentation vessels—has opened up ways of pharmaceutical production that are potentially more efficient than production by the original source organisms.

Therapeutic Uses

Modern Chemotherapeutics

Many species produce metabolites that are major sources of pharmacologically active drugs. Particularly important are the antibiotics, including the penicillins, a structurally related group of β-lactam antibiotics that are synthesized from small peptides. Although naturally occurring penicillins such as penicillin G (produced by *Penicillium chrysogenum*) have a relatively narrow spectrum of biological activity, a wide range of other penicillins can be produced by chemical modification of the natural penicillins. Modern penicillins are semisynthetic compounds, obtained initially from fermentation cultures, but then structurally altered for specific desirable properties. Other antibiotics produced by fungi include: ciclosporin, commonly used as an immunosuppressant during transplant surgery; and fusidic acid, used to help control infection from methicillin-resistant *Staphylococcus aureus* bacteria. Widespread use of antibiotics for the treatment of bacterial diseases, such as tuberculosis, syphilis, leprosy, and others began in the early 20th century and continues to date. In nature, antibiotics of fungal or bacterial origin appear to play a dual role: at high concentrations they act as chemical defense against competition with other microorganisms in species-rich environments, such as the rhizosphere, and at low concentrations as quorum-sensing molecules for intra- or interspecies signaling. Other drugs produced by fungi include griseofulvin isolated from *Penicillium griseofulvum*, used to treat fungal infections, and statins (HMG-CoA reductase inhibitors), used to inhibit cholesterol synthesis. Examples of statins found in fungi include mevastatin from *Penicillium citrinum* and lovastatin from *Aspergillus terreus* and the oyster mushroom.

Traditional and Folk Medicine

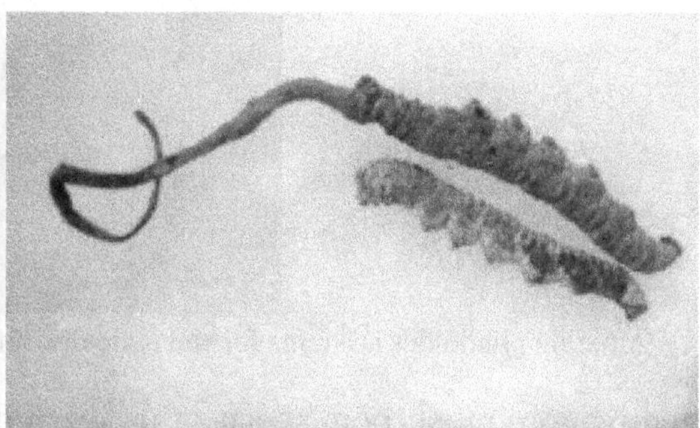

The medicinal fungi Ganoderma lucidum (left) and Ophiocordyceps sinensis (right)

Certain mushrooms enjoy usage as therapeutics in folk medicines, such as Traditional Chinese medicine. Notable medicinal mushrooms with a well-documented history of use include *Agaricus subrufescens*, *Ganoderma lucidum*, and *Ophiocordyceps sinensis*. Research has identified compounds produced by these and other fungi that have inhibitory biological effects against viruses and cancer cells. Specific metabolites, such as polysaccharide-K, ergotamine, and β-lactam antibiotics, are routinely used in clinical medicine. The shiitake mushroom is a source of lentinan, a clinical drug approved for use in cancer treatments in several countries, including Japan. In Europe and Japan, polysaccharide-K (brand name Krestin), a chemical derived from *Trametes versicolor*, is an approved adjuvant for cancer therapy.

Cultured Foods

Baker's yeast or *Saccharomyces cerevisiae*, a unicellular fungus, is used to make bread and other wheat-based products, such as pizza dough and dumplings. Yeast species of the genus *Saccharomyces* are also used to produce alcoholic beverages through fermentation. Shoyu koji mold (*Aspergillus oryzae*) is an essential ingredient in brewing Shoyu (soy sauce) and sake, and the preparation of miso, while *Rhizopus* species are used for making tempeh. Several of these fungi are domesticated species that were bred or selected according to their capacity to ferment food without producing harmful mycotoxins, which are produced by very closely related *Aspergilli*. Quorn, a meat substitute, is made from *Fusarium venenatum*.

Edible and Poisonous Species

Edible mushrooms are well-known examples of fungi. Many are commercially raised, but others must be harvested from the wild. *Agaricus bisporus*, sold as button mushrooms when small or Portobello mushrooms when larger, is a commonly eaten species, used in salads, soups, and many other dishes. Many Asian fungi are commercially grown and have increased in popularity in the West. They are often available fresh in grocery stores and markets, including straw mushrooms (*Volvariella volvacea*), oyster mushrooms (*Pleurotus ostreatus*), shiitakes (*Lentinula edodes*), and enokitake (*Flammulina* spp.).

Amanita phalloides accounts for the majority of fatal mushroom poisonings worldwide.

There are many more mushroom species that are harvested from the wild for personal consumption or commercial sale. Milk mushrooms, morels, chanterelles, truffles, black trumpets, and *porcini* mushrooms (*Boletus edulis*) (also known as king boletes) demand a high price on the market. They are often used in gourmet dishes.

Certain types of cheeses require inoculation of milk curds with fungal species that impart a unique flavor and texture to the cheese. Examples include the blue color in cheeses such as Stilton or Roquefort, which are made by inoculation with *Penicillium roqueforti*. Molds used in cheese production are non-toxic and are thus safe for human consumption; however, mycotoxins (e.g., aflatoxins, roquefortine C, patulin, or others) may accumulate because of growth of other fungi during cheese ripening or storage.

Stilton cheese veined with Penicillium roqueforti

Many mushroom species are poisonous to humans, with toxicities ranging from slight digestive problems or allergic reactions as well as hallucinations to severe organ failures and death. Genera with mushrooms containing deadly toxins include *Conocybe*, *Galerina*, *Lepiota*, and, the most infamous, *Amanita*. The latter genus includes the destroying angel *(A. virosa)* and the death cap *(A. phalloides)*, the most common cause of deadly mushroom poisoning. The false morel (*Gyromitra esculenta*) is occasionally considered a delicacy when cooked, yet can be highly toxic when eaten raw. *Tricholoma equestre* was considered edible until it was implicated in serious poisonings causing rhabdomyolysis. Fly agaric mushrooms (*Amanita muscaria*) also cause occasional non-fatal poisonings, mostly as a result of ingestion for its hallucinogenic properties. Historically, fly agaric was used by different peoples in Europe and Asia and its present usage for religious or shamanic purposes is reported from some ethnic groups such as the Koryak people of north-eastern Siberia.

As it is difficult to accurately identify a safe mushroom without proper training and knowledge, it is often advised to assume that a wild mushroom is poisonous and not to consume it.

Pest Control

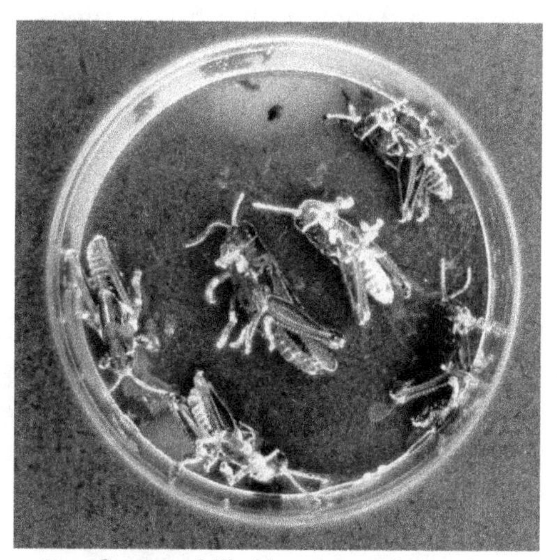

Grasshoppers killed by Beauveria bassiana

In agriculture, fungi may be useful if they actively compete for nutrients and space with pathogenic microorganisms such as bacteria or other fungi via the competitive exclusion principle, or if they are parasites of these pathogens. For example, certain species may be used to eliminate or suppress the growth of harmful plant pathogens, such as insects, mites, weeds, nematodes, and other fungi that cause diseases of important crop plants. This has generated strong interest in practical applications that use these fungi in the biological control of these agricultural pests. Entomopathogenic fungi can be used as biopesticides, as they actively kill insects. Examples that have been used as biological insecticides are *Beauveria bassiana*, *Metarhizium* spp, *Hirsutella* spp, *Paecilomyces* (*Isaria*) spp, and *Lecanicillium lecanii*. Endophytic fungi of grasses of the genus *Neotyphodium*, such as *N. coenophialum*, produce alkaloids that are toxic to a range of invertebrate and vertebrate herbivores. These alkaloids protect grass plants from herbivory, but several endophyte alkaloids can poison grazing animals, such as cattle and sheep. Infecting cultivars of pasture or forage grasses with *Neotyphodium* endophytes is one approach being used in grass breeding programs; the fungal strains are selected for producing only alkaloids that increase resistance to herbivores such as insects, while being non-toxic to livestock.

Bioremediation

Certain fungi, in particular "white rot" fungi, can degrade insecticides, herbicides, pentachlorophenol, creosote, coal tars, and heavy fuels and turn them into carbon dioxide, water, and basic elements. Fungi have been shown to biomineralize uranium oxides, suggesting they may have application in the bioremediation of radioactively polluted sites.

Model Organisms

Several pivotal discoveries in biology were made by researchers using fungi as model organisms, that is, fungi that grow and sexually reproduce rapidly in the laboratory. For example, the one gene-one enzyme hypothesis was formulated by scientists using the bread mold *Neurospora crassa* to test their biochemical theories. Other important model fungi are *Aspergillus nidulans* and the yeasts *Saccaromyces cerevisiae* and *Schizosaccharomyces pombe*, each of which with a long history of use to investigate issues in eukaryotic cell biology and genetics, such as cell cycle regulation, chromatin structure, and gene regulation. Other fungal models have more recently emerged that address specific biological questions relevant to medicine, plant pathology, and industrial uses; examples include *Candida albicans*, a dimorphic, opportunistic human pathogen, *Magnaporthe grisea*, a plant pathogen, and *Pichia pastoris*, a yeast widely used for eukaryotic protein production.

Others

Fungi are used extensively to produce industrial chemicals like citric, gluconic, lactic, and malic acids, and industrial enzymes, such as lipases used in biological detergents, cellulases used in making cellulosic ethanol and stonewashed jeans, and amylases, invertases, proteases and xylanases. Several species, most notably *Psilocybin mushrooms* (colloquially known as *magic mushrooms*), are ingested for their psychedelic properties, both recreationally and religiously.

Ascomycota

Ascomycota is a division or phylum of the kingdom Fungi that, together with the Basidiomycota, form the subkingdom Dikarya. Its members are commonly known as the sac fungi or ascomycetes. They are the largest phylum of Fungi, with over 64,000 species. The defining feature of this fungal group is the "ascus" , a microscopic sexual structure in which nonmotile spores, called ascospores, are formed. However, some species of the Ascomycota are asexual, meaning that they do not have a sexual cycle and thus do not form asci or ascospores. Previously placed in the Deuteromycota along with asexual species from other fungal taxa, asexual (or anamorphic) ascomycetes are now identified and classified based on morphological or physiological similarities to ascus-bearing taxa, and by phylogenetic analyses of DNA sequences.

The ascomycetes are a monophyletic group, i.e. it contains all descendants of one common ancestor. This group is of particular relevance to humans as sources for medicinally important compounds, such as antibiotics and for making bread, alcoholic beverages, and cheese, but also as pathogens of humans and plants. Familiar examples of sac fungi include morels, truffles, brewer's yeast and baker's yeast, dead man's fingers, and cup fungi. The fungal symbionts in the majority of lichens (loosely termed "ascolichens") such as *Cladonia* belong to the Ascomycota. There are many plant-pathogenic ascomycetes, including apple scab, rice blast, the ergot fungi, black knot, and the powdery mildews. Several species of ascomycetes are biological model organisms in laboratory research. Most famously, *Neurospora crassa*, several species of yeasts, and *Aspergillus* species are used in many genetics and cell biology studies. *Penicillium* species on cheeses and those producing antibiotics for treating bacterial infectious diseases are examples of taxa that belong to the Ascomycota.

Asexual Reproduction in Ascomycetes and their Characteristics

Ascomycetes:

Ascomycetes are 'spore shooters'. They are fungi which produce microscopic spores inside special, elongated cells or sacs, known as 'asci', which give the group its name.

Asexual Reproduction:

Asexual reproduction is the dominant form of propagation in the Ascomycota, and is responsible for the rapid spread of these fungi into new areas. Asexual reproduction of ascomycetes is very diverse from both structural and functional points of view. The most important and general is production of conidia, but chlamydospores are also frequently produced. Furthermore, Ascomycota also reproduce asexually through budding.

1) Conidia Formation:

Asexual reproduction may occur through vegetative reproductive spores, the conidia. Asexual, non-motile haploid spore of a fungus, which is named after the Greek word for dust; conia and hence also known as conidiospores and mitospores. The conidiospores commonly contain one nucleus and are products of mitotic cell divisions and thus are sometimes call mitospores, which are genetically identical to the mycelium from which they originate. They are typically formed at the ends of specialized hyphae, the conidiophores. Depending on the species they may be dispersed by wind or water, or by animals. Conidiophores may simply branch off from the mycelia or they may be formed in fruiting bodies.

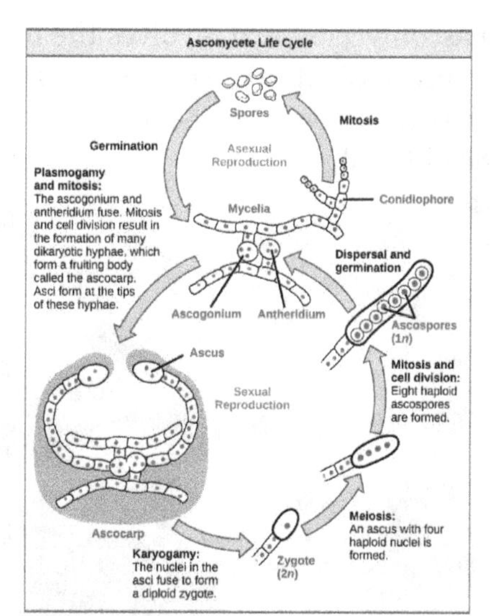

The hypha that creates the sporing (conidiating) tip can be very similar to the normal hyphal tip, or it can be differentiated. The most common differentiation is the formation of a bottle shaped cell called a phialide, from which the spores are produced. As all of these asexual structures are not single hyphae. In some groups, the conidiophores (the structures that bear the conidia) are aggregated to form a thick structure.

E.g. In the order *Moniliales,* all of them are single hyphae with the exception of the aggregations, termed as coremia or synnema. These produce structures rather like corn-stokes, with many conidia being produced in a mass from the aggregated conidiophores.

The diverse conidia and conidiophores sometimes develop in asexual sporocarps with different characteristics (e.g. aecervulus, pycnidium, sporodochium). Some species of *Ascomycetes* form their structures within plant tissue, either as parasite or saprophytes. These fungi have evolved more complex asexual sporing structures, probably influenced by the cultural conditions of plant tissue as a substrate. These structures are called the sporodochium. This is a cushion of conidiophores created from a pseudoparenchymatous stoma in plant tissue. The pycnidium is a globose to flask-shaped parenchymatous structure, lined on its inner wall with conidiophores. The acervulus is a flat saucer shaped bed of conidiophores produced under a plant cuticle, which eventually erupt through the cuticle for dispersal.

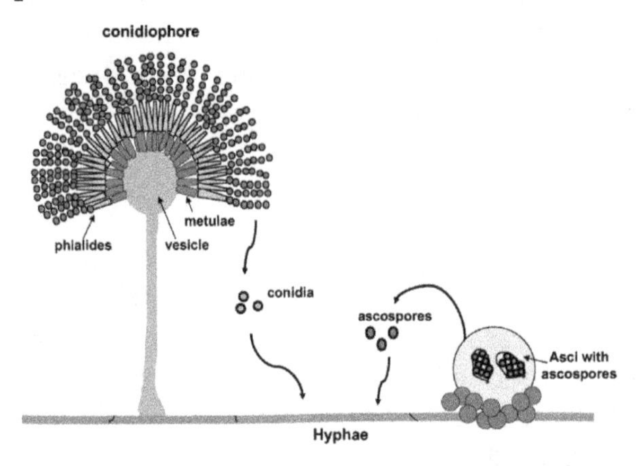

2) Budding:

Asexual reproduction process in ascomycetes also involves the budding which we clearly observe in yeast. This is termed a "blastic process". It involves the blowing out or blebbing of the hyphal tip wall. The blastic process can involve all wall layers, or there can be a new cell wall synthesized which is extruded from within the old wall.

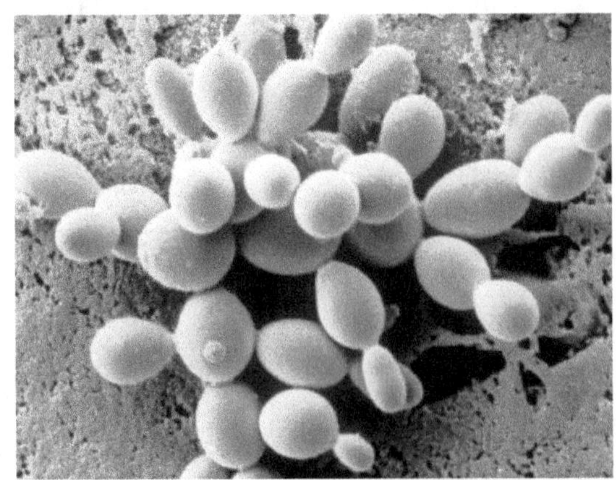

The initial events of budding can be seen as the development of a ring of chitin around the point where the bud is about to appear. This reinforces and stabilizes the cell wall. Enzymatic activity and turgor pressure act to weaken and extrude the cell wall. New cell wall material is incorporated during this phase. Cell contents are forced into the progeny cell, and as the final phase of mitosis ends a cell plate, the point at which a new cell wall will grow inwards from, forms.

Characteristics of Ascomycetes:

· Ascomycota are morphologically diverse. The group includes organisms from unicellular yeasts to complex cup fungi.

· There are 2000 identified genera and 30,000 species of Ascomycota.

· The unifying characteristic among these diverse groups is the presence of a reproductive structure known as the ascus, though in some cases it has a reduced role in the life cycle.

· Many ascomycetes are of commercial importance. Some play a beneficial role, such as the yeasts used in baking, brewing, and wine fermentation, plus truffles and morels, which are held as gourmet delicacies.

· Many of them cause tree diseases, such a

s Dutch elm disease and apple blights.

· Some of the plant pathogenic ascomycetes are apple scab, rice blast, the ergot fungi, black knot, and the powdery mildews.

· The yeasts are used to produce alcoholic beverages and breads. The mold Penicillium is used to produce the anti-biotic penicillin.

· Almost half of all members of the phylum Ascomycota form symbiotic associations with algae to form lichens.

· Others, such as morels (a highly prized edible fungi), form important mychorrhizal relationships with plants, thereby providing enhanced water and nutrient uptake and, in some cases, protection from insects.

· Almost all ascomycetes are terrestrial or parasitic. However, a few have adapted to marine or freshwater environments.

· The cell walls of the hyphae are variably composed of chitin and β-glucans, just as in Basidiomycota. However, these fibers are set in a matrix of glycoprotein containing the sugars galactose and mannose.

· The mycelium of ascomycetes is usually made up of septate hyphae. However, there is not necessarily any fixed number of nuclei in each of the divisions.

· The septal walls have septal pores which provide cytoplasmic continuity throughout the individual hyphae. Under appropriate conditions, nuclei may also migrate between septal compartments through the septal pores.

· A unique character of the Ascomycota (but not present in all ascomycetes) is the presence of Woronin bodies on each side of the septa separating the hyphal segments which control the septal pores. If an adjoining hypha is ruptured, the Woronin bodies block the pores to prevent loss of cytoplasm into the ruptured compartment. The Woronin bodies are spherical, hexagonal, or rectangular membrane bound structures with a crystalline protein matrix.

Modern Classification of Ascomycota

There are three subphyla that are described and accepted:

- The *Pezizomycotina* are the largest subphylum and contains all ascomycetes that produce ascocarps (fruiting bodies), except for one genus, *Neolecta*, in the Taphrinomycotina. It is roughly equivalent to the previous taxon, *Euascomycetes*. The Pezizomycotina includes most macroscopic "ascos" such as truffles, ergot, ascolichens, cup fungi (discomycetes), pyrenomycetes, lorchels, and caterpillar fungus. It also contains microscopic fungi such as powdery mildews, dermatophytic fungi, and Laboulbeniales.
- The *Saccharomycotina* comprise most of the "true" yeasts, such as baker's yeast and *Candida*, which are single-celled (unicellular) fungi, which reproduce vegetatively by budding. Most of these species were previously classified in a taxon called *Hemiascomycetes*.
- The *Taphrinomycotina* include a disparate and basal group within the Ascomycota that was recognized following molecular (DNA) analyses. The taxon was originally named *Archiascomycetes* (or *Archaeascomycetes*). It includes both hyphal fungi (*Neolecta, Taphrina, Archaeorhizomyces*), fission yeasts (*Schizosaccharomyces*), and the mammalian lung parasite, *Pneumocystis*.

Outdated Taxon Names

Several outdated taxon names—based on morphological features—are still occasionally used for species of the Ascomycota. These include the following sexual (teleomorphic) groups, defined by the structures of their sexual fruiting bodies: the Discomycetes, which included all species forming apothecia; the Pyrenomycetes, which included all sac fungi that formed perithecia or pseudothecia, or any structure resembling these morphological structures; and the Plectomycetes, which included those species that form cleistothecia. Hemiascomycetes included the yeasts and yeast-like fungi that have now been placed into the Saccharomycotina or Taphrinomycotina, while the Euascomycetes included the remaining species of the Ascomycota, which are now in the Pezizomycotina, and the Neolecta, which are in the Taphrinomycotina.

Some ascomycetes do not reproduce sexually or are not known to produce asci and are therefore anamorphic species. Those anamorphs that produce conidia (mitospores) were previously described as Mitosporic Ascomycota. Some taxonomists placed this group into a separate artificial phylum, the Deuteromycota (or "Fungi Imperfecti"). Where recent molecular analyses have identified close relationships with ascus-bearing taxa, anamorphic species have been grouped into the Ascomycota, despite the absence of the defining ascus. Sexual and asexual isolates of the same species commonly carry different binomial species names, as, for example, *Aspergillus nidulans* and *Emericella nidulans*, for asexual and sexual isolates, respectively, of the same species.

Species of the Deuteromycota were classified as Coelomycetes if they produced their conidia in minute flask- or saucer-shaped conidiomata, known technically as *pycnidia* and *acervuli*. The Hyphomycetes were those species where the conidiophores (*i.e.*, the hyphal structures that carry conidia-forming cells at the end) are free or loosely organized. They are mostly isolated but sometimes also appear as bundles of cells aligned in parallel (described as *synnematal*) or as cushion-shaped masses (described as *sporodochial*).

Morphology

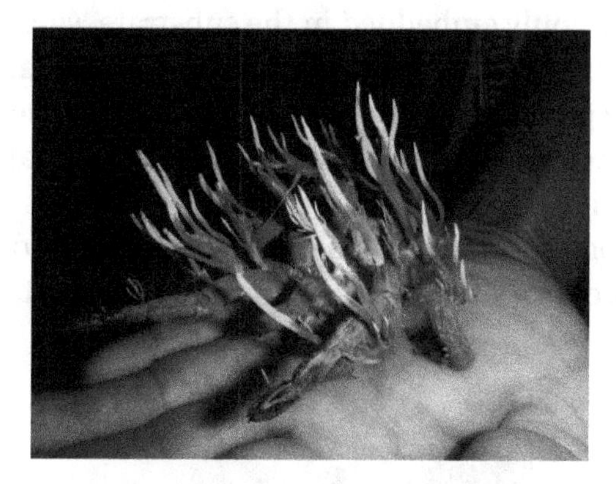

A member of the Cordyceps genus which is parasitic on arthropods. Note the elongated ascocarps. Species unknown, perhaps *Cordyceps ignota*.

Most species grow as filamentous, microscopic structures called hyphae. Many interconnected hyphae form a mycelium, which—when visible to the naked eye (macroscopic)—is commonly called mold (or, in botanical terminology, thallus). During sexual reproduction, many Ascomycota typically produce large numbers of asci. The asci is often contained in a multicellular, occasionally readily visible fruiting structure, the ascocarp (also called an *ascoma*). Ascocarps come in a very large variety of shapes: cup-shaped, club-shaped, potato-like, spongy, seed-like, oozing and pimple-like, coral-like, nit-like, golf-ball-shaped, perforated tennis ball-like, cushion-shaped, plated and feathered in miniature (Laboulbeniales), microscopic classic Greek shield-shaped, stalked or sessile. They can appear solitary or clustered. Their texture can likewise be very variable, including fleshy, like charcoal (carbonaceous), leathery, rubbery, gelatinous, slimy, powdery, or cob-web-like. Ascocarps come in multiple colors such as red, orange, yellow, brown, black, or, more rarely, green or blue. Some ascomyceous fungi, such as *Saccharomyces cerevisiae*, grow as single-celled yeasts, which—during sexual reproduction—develop into an ascus, and do not form fruiting bodies.

The "candlesnuff fungus", Xylaria hypoxylon

In lichenized species, the thallus of the fungus defines the shape of the symbiotic colony. Some dimorphic species, such as *Candida albicans*, can switch between growth as single cells and as filamentous, multicellular hyphae. Other species are pleomorphic, exhibiting asexual (anamorphic) as well as a sexual (teleomorphic) growth forms.

Except for lichens, the non-reproductive (vegetative) mycelium of most ascomycetes is usually inconspicuous because it is commonly embedded in the substrate, such as soil, or grows on or inside a living host, and only the ascoma may be seen when fruiting. Pigmentation, such as melanin in hyphal walls, along with prolific growth on surfaces can result in visible mold colonies; examples include *Cladosporium* species, which form black spots on bathroom caulking and other moist areas. Many ascomycetes cause food spoilage, and, therefore, the pellicles or moldy layers that develop on jams, juices, and other foods are the mycelia of these species or occasionally Mucoromycotina and almost never Basidiomycota. Sooty molds that develop on plants, especially in the tropics are the thalli of many species.

The ascocarp of a morel contains numerous apothecia.

Large masses of yeast cells, asci or ascus-like cells, or conidia can also form macroscopic structures. For example. *Pneumocystis* species can colonize lung cavities (visible in x-rays), causing a form of pneumonia. Asci of *Ascosphaera* fill honey bee larvae and pupae causing mummification with a chalk-like appearance, hence the name “chalkbrood”. Yeasts for small colonies in vitro and in vivo, and excessive growth of *Candida* species in the mouth or vagina causes “thrush”, a form of candidiasis.

The cell walls of the ascomycetes almost always contain chitin and β-glucans, and divisions within the hyphae, called “septa”, are the internal boundaries of individual cells (or compartments). The cell wall and septa give stability and rigidity to the hyphae and may prevent loss of cytoplasm in case of local damage to cell wall and cell membrane. The septa commonly have a small opening in the center, which functions as a cytoplasmic connection between adjacent cells, also sometimes allowing cell-to-cell movement of nuclei within a hypha. Vegetative hyphae of most ascomycetes contain only one nucleus per cell (*uninucleate* hyphae), but multinucleate cells—especially in the apical regions of growing hyphae—can also be present.

Metabolism

In common with other fungal phyla, the Ascomycota are heterotrophic organisms that require organic compounds as energy sources. These are obtained by feeding on a variety of organic substrates including dead matter, foodstuffs, or as symbionts in or on other living organisms. To obtain these nutrients from their surroundings, ascomycetous fungi secrete powerful digestive enzymes that break down organic substances into smaller molecules, which are then taken up into the cell. Many species live on dead plant material such as leaves, twigs, or logs. Several species colonize plants, animals, or other fungi as parasites or mutualistic symbionts and derive all their metabolic energy in form of nutrients from the tissues of their hosts.

Owing to their long evolutionary history, the Ascomycota have evolved the capacity to break down almost every organic substance. Unlike most organisms, they are able to use their own enzymes to digest plant biopolymers such as cellulose or lignin. Collagen, an abundant structural protein in animals, and keratin—a protein that forms hair and nails—, can also serve as food sources. Unusual examples include *Aureobasidium pullulans*, which feeds on wall paint, and the kerosene fungus *Amorphotheca resinae*, which feeds on aircraft fuel (causing occasional problems for the airline industry), and may sometimes block fuel pipes. Other species can resist high osmotic stress and grow, for example, on salted fish, and a few ascomycetes are aquatic.

The Ascomycota is characterized by a high degree of specialization; for instance, certain species of Laboulbeniales attack only one particular leg of one particular insect species. Many Ascomycota engage in symbiotic relationships such as in lichens—symbiotic associations with green algae or cyanobacteria—in which the fungal symbiont directly obtains products of photosynthesis. In common with many basidiomycetes and Glomeromycota, some ascomycetes form symbioses with plants by colonizing the roots to form mycorrhizal associations. The Ascomycota also represents several carnivorous fungi, which have developed hyphal traps to capture small protists such as amoebae, as well as roundworms (*Nematoda*), rotifers, tardigrades, and small arthropods such as springtails (*Collembola*).

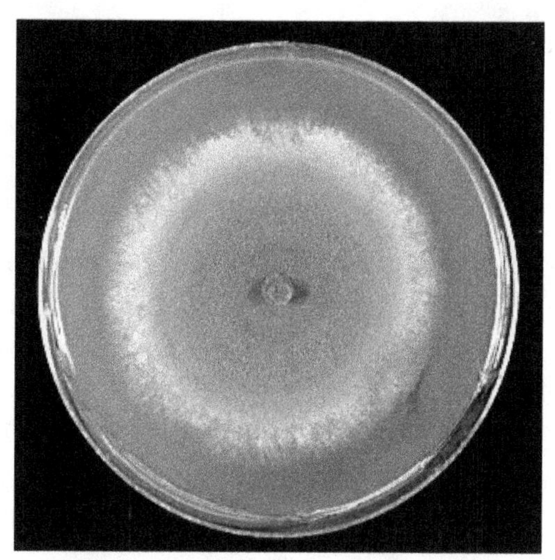

Hypomyces completus on culture medium

Distribution and Living Environment

The Ascomycota are represented in all land ecosystems worldwide, occurring on all continents including Antarctica. Spores and hyphal fragments are dispersed through the atmosphere and freshwater environments, as well as ocean beaches and tidal zones. The distribution of species is variable; while some are found on all continents, others, as for example the white truffle *Tuber magnatum*, only occur in isolated locations in Italy and Eastern Europe. The distribution of plant-parasitic species is often restricted by host distributions; for example, *Cyttaria* is only found on *Nothofagus* (Southern Beech) in the Southern Hemisphere.

Reproduction

Asexual Reproduction

Asexual reproduction is the dominant form of propagation in the Ascomycota, and is responsible for the rapid spread of these fungi into new areas. It occurs through vegetative reproductive spores, the conidia. The conidiospores commonly contain one nucleus and are products of mitotic cell divisions and thus are sometimes called mitospores, which are genetically identical to the mycelium from which they originate. They are typically formed at the ends of specialized hyphae, the *conidiophores*. Depending on the species they may be dispersed by wind or water, or by animals.

Asexual Spores

Different types of asexual spores can be identified by colour, shape, and how they are released as individual spores. Spore types can be used as taxonomic characters in the classification within the Ascomycota. The most frequent types are the single-celled spores, which are designated *amerospores*. If the spore is divided into two by a cross-wall (septum), it is called a *didymospore*.

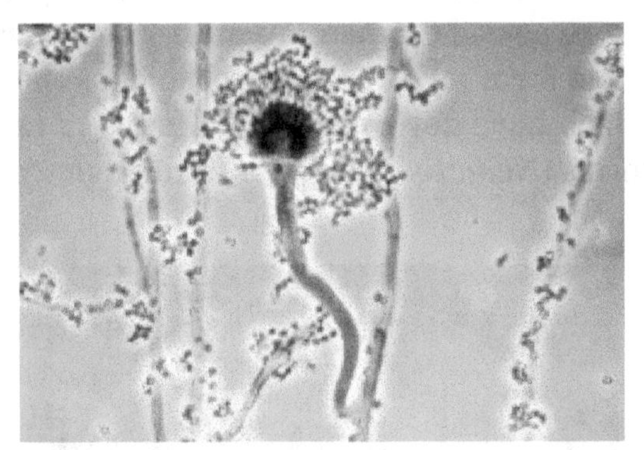

Conidiophores of molds of the genus Aspergillus, conidiogenesis is blastic-phialidic

When there are two or more cross-walls, the classification depends on spore shape. If the septae are *transversal*, like the rungs of a ladder, it is a *phragmospore*, and if they possess a net-like structure it is a *dictyospore*. In *staurospores* ray-like arms radiate from a central body; in others (*helicospores*) the entire spore is wound up in a spiral like a spring. Very long worm-like spores with a length-to-diameter ratio of more than 15:1, are called *scolecospores*.

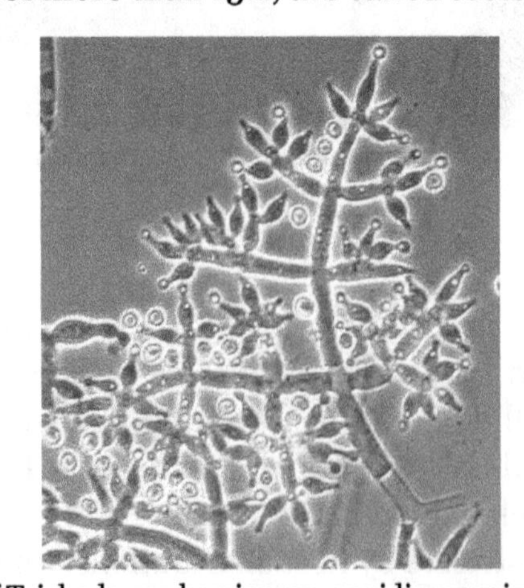

Conidiophores of Trichoderma harzianum, conidiogenesis is blastic-phialidic

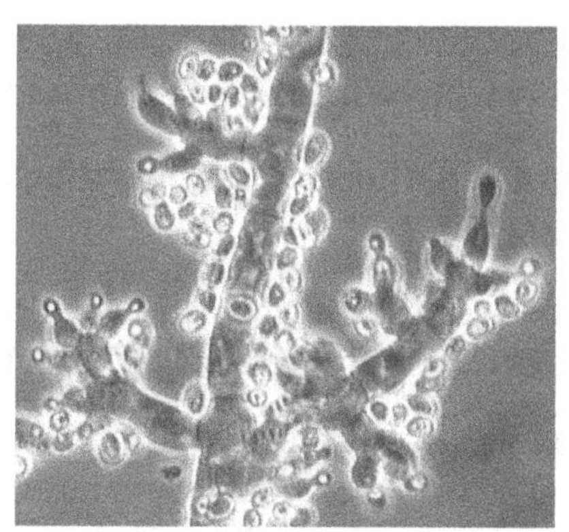

Conidiophores of Trichoderma fertile with vase-shaped phialides and newly formed conidia on their ends (bright points)

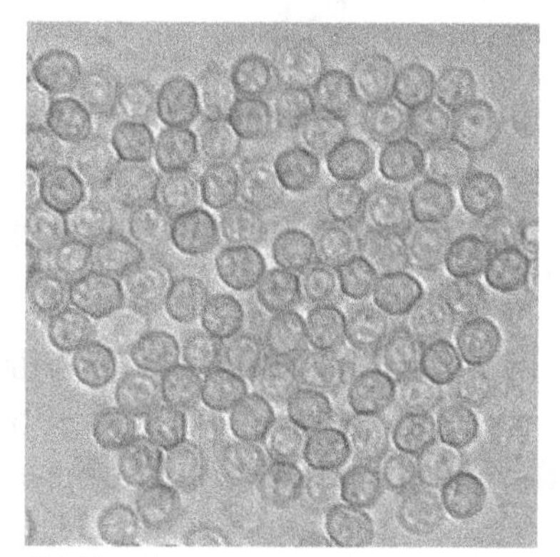

Conidiospores of Trichoderma aggressivum, Diameter approx. 3μm

Conidiogenesis and Dehiscence

Important characteristics of the anamorphs of the Ascomycota are *conidiogenesis*, which includes spore formation and dehiscence (separation from the parent structure). Conidiogenesis corresponds to Embryology in animals and plants and can be divided into two fundamental forms of development: *blastic* conidiogenesis, where the spore is already evident before it separates from the conidiogenic hypha, and *thallic* conidiogenesis, during which a cross-wall forms and the newly created cell develops into a spore. The spores may or may not be generated in a large-scale specialized structure that helps to spread them.

These two basic types can be further classified as follows:

- *blastic-acropetal* (repeated budding at the tip of the conidiogenic hypha, so that a chain of spores is formed with the youngest spores at the tip),
- *blastic-synchronous* (simultaneous spore formation from a central cell, sometimes with secondary acropetal chains forming from the initial spores),
- *blastic-sympodial* (repeated sideways spore formation from behind the leading spore, so that the oldest spore is at the main tip),
- *blastic-annellidic* (each spore separates and leaves a ring-shaped scar inside the scar left by the previous spore),

- *blastic-phialidic* (the spores arise and are ejected from the open ends of special conidiogenic cells called phialides, which remain constant in length),
- *basauxic* (where a chain of conidia, in successively younger stages of development, is emitted from the mother cell),
- *blastic-retrogressive* (spores separate by formation of crosswalls near the tip of the conidiogenic hypha, which thus becomes progressively shorter),
- *thallic-arthric* (double cell walls split the conidiogenic hypha into cells that develop into short, cylindrical spores called *arthroconidia*; sometimes every second cell dies off, leaving the arthroconidia free),
- *thallic-solitary* (a large bulging cell separates from the conidiogenic hypha, forms internal walls, and develops to a *phragmospore*).

Sometimes the conidia are produced in structures visible to the naked eye, which help to distribute the spores. These structures are called "conidiomata" (singular: conidioma), and may take the form of *pycnidia* (which are flask-shaped and arise in the fungal tissue) or *acervuli* (which are cushion-shaped and arise in host tissue).

Dehiscence happens in two ways. In *schizolytic* dehiscence, a double-dividing wall with a central lamella (layer) forms between the cells; the central layer then breaks down thereby releasing the spores. In *rhexolytic* dehiscence, the cell wall that joins the spores on the outside degenerates and releases the conidia.

Heterokaryosis and Parasexuality

Several Ascomycota species are not known to have a sexual cycle. Such asexual species may be able to undergo genetic recombination between individuals by processes involving *heterokaryosis* and *parasexual* events.

Parasexuality refers to the process of heterokaryosis, caused by merging of two hyphae belonging to different individuals, by a process called *anastomosis*, followed by a series of events resulting in genetically different cell nuclei in the mycelium. The merging of nuclei is not followed by meiotic events, such as gamete formation and results in an increased number of chromosomes per nuclei. *Mitotic crossover* may enable recombination, i.e., an exchange of genetic material between homologous chromosomes. The chromosome number may then be restored to its haploid state by nuclear division, with each daughter nuclei being genetically different from the original parent nuclei. Alternatively, nuclei may lose some chromosomes, resulting in aneuploid cells. *Candida albicans* (class Saccharomycetes) is an example of a fungus that has a parasexual cycle.

Sexual Reproduction

Sexual reproduction in the Ascomycota leads to the formation of the *ascus*, the structure that defines this fungal group and distinguishes it from other fungal phyla. The ascus is a tube-shaped vessel, a *meiosporangium*, which contains the sexual spores produced by meiosis and which are called *ascospores*.

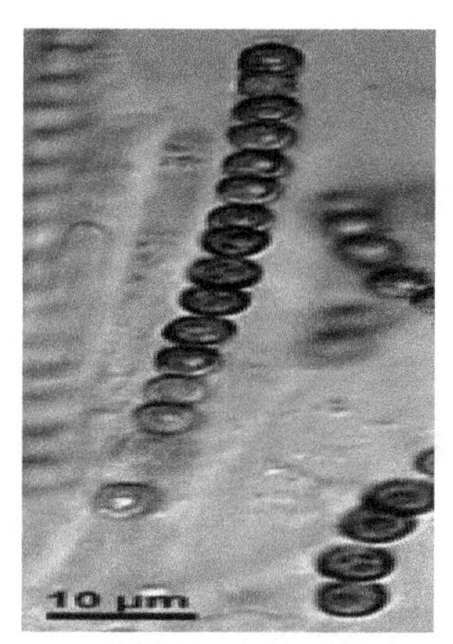

Ascus of Hypocrea virens with eight two-celled Ascospores

Apart from a few exceptions, such as *Candida albicans*, most ascomycetes are haploid, i.e., they contain one set of chromosomes per nucleus. During sexual reproduction there is a diploid phase, which commonly is very short, and meiosis restores the haploid state. The sexual cycle of one well-studied representative species of Ascomycota is described in greater detail in Neurospora crassa.

Formation of Sexual Spores

The sexual part of the life cycle commences when two hyphal structures mate. In the case of *homothallic* species, mating is enabled between hyphae of the same fungal clone, whereas in *heterothallic* species, the two hyphae must originate from fungal clones that differ genetically, i.e., those that are of a different mating type. Mating types are typical of the fungi and correspond roughly to the sexes in plants and animals; however one species may have more than two mating types, resulting in sometimes complex vegetative incompatibility systems. The adaptive function of mating type is discussed in Neurospora crassa.

Gametangia are sexual structures formed from hyphae, and are the generative cells. A very fine hypha, called trichogyne emerges from one gametangium, the *ascogonium*, and merges with a gametangium (the *antheridium*) of the other fungal isolate. The nuclei in the antheridium then migrate into the ascogonium, and plasmogamy—the mixing of the cytoplasm—occurs. Unlike in animals and plants, plasmogamy is not immediately followed by the merging of the nuclei (called *karyogamy*). Instead, the nuclei from the two hyphae form pairs, initiating the *dikaryophase* of the sexual cycle, during which time the pairs of nuclei synchronously divide. Fusion of the paired nuclei leads to mixing of the genetic material and recombination and is followed by meiosis. A similar sexual cycle is present in the red algae (Rhodophyta). A discarded hypothesis held that a second karyogamy event occurred in the ascogonium prior to ascogeny, resulting in a tetraploid nucleus which divided into four diploid nuclei by meiosis and then into eight haploid nuclei by a supposed process called brachymeiosis, but this hypothesis was disproven in the 1950s.

From the fertilized ascogonium, *dinucleate* hyphae emerge in which each cell contains two nuclei. These hyphae are called *ascogenous* or fertile hyphae. They are supported by the vegetative mycelium containing uni– (or mono–) nucleate hyphae, which are sterile. The mycelium containing both sterile and fertile hyphae may grow into fruiting body, the *ascocarp*, which may contain millions of fertile hyphae.

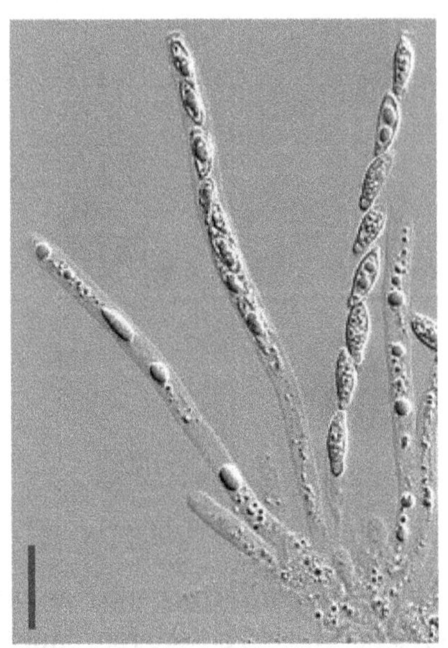

Unitunicate-inoperculate Asci of Hypomyces chrysospermus

The sexual structures are formed in the fruiting layer of the ascocarp, the hymenium. At one end of ascogenous hyphae, characteristic U-shaped hooks develop, which curve back opposite to the growth direction of the hyphae. The two nuclei contained in the apical part of each hypha divide in such a way that the threads of their mitotic spindles run parallel, creating two pairs of genetically different nuclei. One daughter nucleus migrates close to the hook, while the other daughter nucleus locates to the basal part of the hypha. The formation of two parallel cross-walls then divides the hypha into three sections: one at the hook with one nucleus, one at the basal of the original hypha that contains one nucleus, and one that separates the U-shaped part, which contains the other two nuclei.

Fusion of the nuclei (karyogamy) takes place in the U-shaped cells in the hymenium, and results in the formation of a diploid zygote. The zygote grows into the ascus, an elongated tube-shaped or cylinder-shaped capsule. Meiosis then gives rise to four haploid nuclei, usually followed by a further mitotic division that results in eight nuclei in each ascus. The nuclei along with some cytoplasma become enclosed within membranes and a cell wall to give rise to ascospores that are aligned inside the ascus like peas in a pod.

Upon opening of the ascus, ascospores may be dispersed by the wind, while in some cases the spores are forcibly ejected form the ascus; certain species have evolved spore cannons, which can eject ascospores up to 30 cm. away. When the spores reach a suitable substrate, they germinate, form new hyphae, which restarts the fungal life cycle.

The form of the ascus is important for classification and is divided into four basic types: unitunicate-operculate, unitunicate-inoperculate, bitunicate, or prototunicate.

Ecology

The Ascomycota fulfil a central role in most land-based ecosystems. They are important decomposers, breaking down organic materials, such as dead leaves and animals, and helping the detritivores (animals that feed on decomposing material) to obtain their nutrients. Ascomycetes along with other fungi can break down large molecules such as cellulose or lignin, and thus have important roles in nutrient cycling such as the carbon cycle.

The fruiting bodies of the Ascomycota provide food for many animals ranging from insects and slugs and snails (*Gastropoda*) to rodents and larger mammals such as deer and wild boars.

Many ascomycetes also form symbiotic relationships with other organisms, including plants and animals.

Lichens

Probably since early in their evolutionary history, the Ascomycota have formed symbiotic associations with green algae (*Chlorophyta*), and other types of algae and cyanobacteria. These mutualistic associations are commonly known as lichens, and can grow and persist in terrestrial regions of the earth that are inhospitable to other organisms and characterized by extremes in temperature and humidity, including the Arctic, the Antarctic, deserts, and mountaintops. While the photoautotrophic algal partner generates metabolic energy through photosynthesis, the fungus offers a stable, supportive matrix and protects cells from radiation and dehydration. Around 42% of the Ascomycota (about 18,000 species) form lichens, and almost all the fungal partners of lichens belong to the Ascomycota.

Mycorrhizal Fungi and Endophytes

Members of the Ascomycota form two important types of relationship with plants: as mycorrhizal fungi and as endophytes. Mycorrhiza are symbiotic associations of fungi with the root systems of the plants, which can be of vital importance for growth and persistence for the plant. The fine mycelial network of the fungus enables the increased uptake of mineral salts that occur at low levels in the soil. In return, the plant provides the fungus with metabolic energy in the form of photosynthetic products.

Endophytic fungi live inside plants, and those that form mutualistic or commensal associations with their host, do not damage their hosts. The exact nature of the relationship between endophytic fungus and host depends on the species involved, and in some cases fungal colonization of plants can bestow a higher resistance against insects, roundworms (nematodes), and bacteria; in the case of grass endophytes the fungal symbiont produces poisonous alkaloids, which can affect the health of plant-eating (herbivorous) mammals and deter or kill insect herbivores.

Symbiotic Relationships with Animals

Several ascomycetes of the genus *Xylaria* colonize the nests of leafcutter ants and other fungus-growing ants of the tribe Attini, and the fungal gardens of termites (Isoptera). Since they do not generate fruiting bodies until the insects have left the nests, it is suspected that, as confirmed in several cases of Basidiomycota species, they may be cultivated.

Bark beetles (family Scolytidae) are important symbiotic partners of ascomycetes. The female beetles transport fungal spores to new hosts in characteristic tucks in their skin, the *mycetangia*. The beetle tunnels into the wood and into large chambers in which they lay their eggs. Spores released from the mycetangia germinate into hyphae, which can break down the wood.

The beetle larvae then feed on the fungal mycelium, and, on reaching maturity, carry new spores with them to renew the cycle of infection. A well-known example of this is Dutch elm disease, caused by *Ophiostoma ulmi*, which is carried by the European elm bark beetle, *Scolytus multistriatus*.

Importance for Humans

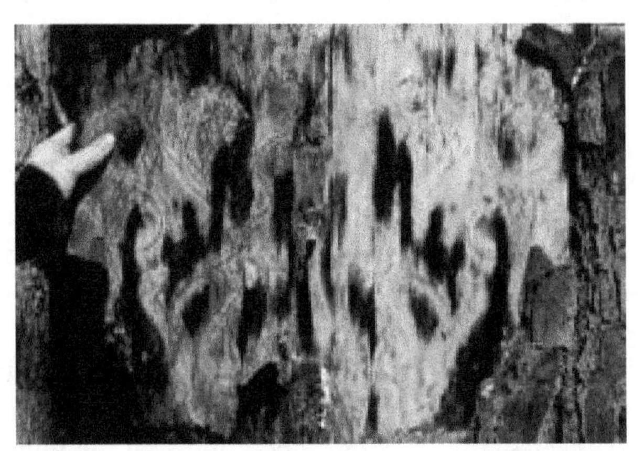

Tree attacked by the Bluestain fungus, Ophiostoma minus

Ascomycetes make many contributions to the good of humanity, and also have many ill effects.

Harmful Interactions

One of their most harmful roles is as the agent of many plant diseases. For instance:

- Dutch Elm Disease, caused by the closely related species *Ophiostoma ulmi* and *Ophiostoma novo-ulmi*, has led to the death of many elms in Europe and North America.

Claviceps purpurea on rye (Secale cereale)

- The originally Asian *Cryphonectria parasitica* is responsible for attacking Sweet Chestnuts (*Castanea sativa*), and virtually eliminated the once-widespread American Chestnut (*Castanea dentata*),

- A disease of maize (*Zea mays*), which is especially prevalent in North America, is brought about by *Cochliobolus heterostrophus*.
- *Taphrina deformans* causes leaf curl of peach.
- *Uncinula necator* is responsible for the disease powdery mildew, which attacks grapevines.
- Species of *Monilinia* cause brown rot of stone fruit such as peaches (*Prunus persica*) and sour cherries (*Prunus ceranus*).
- Members of the Ascomycota such as *Stachybotrys chartarum* are responsible for fading of woollen textiles, which is a common problem especially in the tropics.
- Blue-green, red and brown molds attack and spoil foodstuffs - for instance *Penicillium italicum* rots oranges.
- Cereals infected with *Fusarium graminearum* contain mycotoxins like deoxynivalenol (DON), which can lead to skin and mucous membrane lesions when eaten by pigs.
- Ergot (*Claviceps purpurea*) is a direct menace to humans when it attacks wheat or rye and produces highly poisonous and carcinogenic alkaloids, causing ergotism if consumed. Symptoms include hallucinations, stomach cramp, and a burning sensation in the limbs ("Saint Anthony's Fire").
- *Aspergillus flavus*, which grows on peanuts and other hosts, generates aflatoxin, which damages the liver and is highly carcinogenic.
- *Candida albicans*, a yeast that attacks the mucous membranes, can cause an infection of the mouth or vagina called thrush or candidiasis, and is also blamed for "yeast allergies".
- Fungi like *Epidermophyton* cause skin infections but are not very dangerous for people with healthy immune systems. However, if the immune system is damaged they can be life-threatening; for instance, *Pneumocystis jirovecii* is responsible for severe lung infections that occur in AIDS patients.

Positive Effects

On the other hand, ascus fungi have brought some important benefits to humanity.

- The most famous case may be that of the mould *Penicillium chrysogenum* (formerly *Penicillium notatum*), which, probably to attack competing bacteria, produces an antibiotic that, under the name of penicillin, triggered a revolution in the treatment of bacterial infectious diseases in the 20th century.
- The medical importance of *Tolypocladium niveum* as an immunosuppressor can hardly be exaggerated. It excretes Ciclosporin, which, as well as being given during Organ transplantation to prevent rejection, is also prescribed for auto-immune diseases such as multiple sclerosis, although there is some doubt over the long-term side-effects of the treatment.
- Some ascomycete fungi can be altered relatively easily through genetic engineering proce-

dures. They can then produce useful proteins such as insulin, human growth hormone, or TPa, which is employed to dissolve blood clots.

- Several species are common model organisms in biology, including *Saccharomyces cerevisiae*, *Schizosaccharomyces pombe*, and *Neurospora crassa*. The genomes of a number of ascomycete fungi have been fully sequenced.
- Baker's Yeast (*Saccharomyces cerevisiae*) is used to make bread, beer and wine, during which process sugars such as glucose or sucrose are fermented to make ethanol and carbon dioxide. Bakers use the yeast for carbon dioxide production, causing the bread to rise, with the ethanol boiling off during cooking. Most vintners use it for ethanol production, with the carbon dioxide being released into the atmosphere during fermentation. Brewers and traditional producers of sparkling wine use both, with a primary fermentation for the alcohol and a secondary one to produce the carbon dioxide bubbles that provide the drinks with "sparkling" texture in the case of wine and the desirable foam in the case of beer.
- Enzymes of *Penicillium camemberti* play a role in the manufacture of the cheeses Camembert and Brie, while those of *Penicillium roqueforti* do the same for Gorgonzola, Roquefort and Stilton.
- In Asia *Aspergillus oryzae* is added to a pulp of soaked soya beans to make soy sauce.
- Finally, some members of the Ascomycota are choice edibles; morels (*Morchella spp.*), truffles (*Tuber spp.*), and lobster mushroom (*Hypomyces lactifluorum*) are some of the most sought-after fungal delicacies.

Sclerotinia Sclerotiorum

Sclerotinia sclerotiorum is a plant pathogenic fungus and can cause a disease called white mold if conditions are correct. S. scler*otiorum* can also be known as cottony rot, watery soft rot, stem rot, drop, crown rot and blossom blight. A key characteristic of this pathogen is its ability to produce black resting structures known as sclerotia and white fuzzy growths of mycelium on the plant it infects. These sclerotia give rise to a fruiting body in the spring that produces spores in a sac which is why fungi in this class are called sac fungi (Ascomycetes). This pathogen can occur on many continents and has a wide host range of plants. When *S. sclerotiorum* is onset in the field by favorable environmental conditions, losses can be great and control measures should be considered.

Hosts and Symptoms

Common hosts of white mold are herbaceous, succulent plants, particularly flowers and vegetables. It can also affect woody ornamentals occasionally, usually on juvenile tissue. White mold can affect their hosts at any stage of growth, including seedlings, mature plants, and harvested products. It can usually be found on tissues with high water content and in close proximity to the soil. One of the first symptoms noticed is an obvious area of white, fluffy mycelial growth. Usually this is preceded by pale to dark brown lesions on the stem at the soil line. The mycelium then cover this necrotic area. Once the xylem is affected, other symptoms occur higher up in the plant. These can include chlorosis, wilting, leaf drop, and death quickly follows. On fruits, the initial dark lesions occur on the tissue that comes in contact with the soil. Next, white fungal mycelium covers the fruit and it decays. This can occur when the fruit is in the field or when in storage.

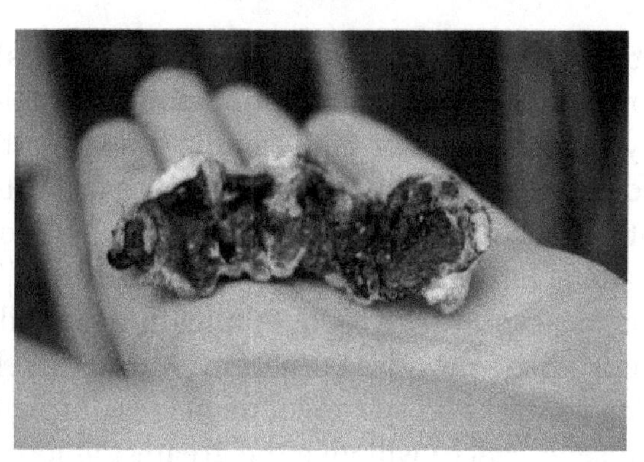

Importance

White mold affects a wide range of hosts. It is known to infect 408 plant species. Its diverse host range and ability to infect plants at any stage of growth makes white mold a very serious disease. The fungus can survive on infected tissues, in the soil, and on living plants. It affects young seedlings, mature plants, and fruit in the field or in storage. White mold can spread quickly in the field from plant to plant. It can also spread in a storage facility throughout the harvested crop. Some crops it affects commonly are soybeans, green beans, sunflowers, canola, and peanuts.

Environment

The pathogenic fungus *Sclerotinia sclerotioum* proliferates in moist environments. Under moist field conditions, *S. sclerotiorum* is capable of completely invading a plant host, colonizing nearly all of the plant's tissues with mycelium. Optimal temperatures for growth range from 15 to 21 degrees Celsius. Under wet conditions, *S. sclerotiorum* will produce an abundance of mycelium and sclerotia. The fungus can survive in the soil mainly on the previous year's plant debris. Like most fungi, *S. sclerotiorum* prefers darker, shadier conditions as opposed to direct exposure to sunlight.

Life Cycle

The lifecycle of *Sclerotinia sclerotiorum* can be described as monocyclic, as there are no secondary inoculums produced. During late summer/ early fall the fungus will produce a survival structure called a sclerotium either on or inside the tissues of a host plant. The following spring the dormant sclerotia will germinate to produce fruiting bodies called apothecia, which are small, thin stalks ending with a cup-like structure about 5-15mm in diameter (similar in shape to a small mushroom). The cup of the apothecium is lined with asci, in which the ascospores are contained. When the ascospores are released from the asci, they are carried by the wind until they land on a suitable host. The ascospores will then germinate on the host and begin to invade the host's tissues via mycelium, causing infection. *S. sclerotiorum* is capable of invading nearly all tissue types including stems, foliage, flowers, fruits, and roots. Eventually white, fluffy mycelium will begin to grow on the surface of the infected tissues. At the end of the growing season, *S. sclerotiorum* will once again produce sclerotia. The sclerotia will then remain on the surface of the ground or in the soil, on either living or dead plant parts until the next season. The lifecycle will then continue respectively.

Control

Control of white mold on crops can depend greatly on specific cultural practices and the application of chemical sprays. Crops that are highly susceptible to white mold should be planted in well drained soils. Also, properly spaced plants or maintaining growth in the fields will help to maintain good air circulation and create microclimates that are less favorable for disease development (Pohronezny 25). Fields with heavy disease pressure may also be flooded for a period of four to five weeks so as the sclerotia may lose their viability(Pohronezny 25). This may not be a practical measure and may be why chemical sprays are used. Fumigating the soils with broad spectrum fungicides can be effective in controlling sclerotia (Pernezyny, Momol and lopes 23). During the growing season, however, applying contact and systemic fungicides during times of the plants greatest susceptibility (flowering and senescence) will give the greatest amount of protection against disease. Crop rotation with cereals in fields with high levels of disease incidence can help in reduction of inoculums. Rotation should last for at least three years since sclerotia can be viable for at least this long.

Sclerotinia sclerotiorum on bushbean

Magnaporthe Grisea

Magnaporthe grisea, also known as rice blast fungus, rice rotten neck, rice seedling blight, blast of rice, oval leaf spot of graminea, pitting disease, ryegrass blast, and Johnson spot, is a plant-pathogenic fungus that causes a serious disease affecting rice. It is now known that M. grisea consists of a cryptic species complex containing at least two biological species that have clear genetic differences and do not interbreed. Complex members isolated from *Digitaria* have been more narrowly defined as *M. grisea*. The remaining members of the complex isolated from rice and a variety of other hosts have been rena*med Magnaporthe oryzae. Confu*sion on which of these two names to use for the rice blast pathogen remains, as both are now used by different authors.

Members of the *Magnaporthe grisea* complex can also infect other agriculturally important cere-als including wheat, rye, barley, and pearl millet causing diseases called blast disease or blight disease. Rice blast causes economically significant crop losses annually. Each year it is estimated to destroy enough rice to feed more than 60 million people. The fungus is known to occur in 85 countries worldwide.

Hosts and Symptoms

Lesions on rice leaves caused by infection with *M. grisea*

M. grisea is an ascomycete fungus. It is an extremely effective plant pathogen as it can reproduce both sexually and asexually to produce specialized infectious structures known as appressoria that infect aerial tissues and hyphae that can infect root tissues.

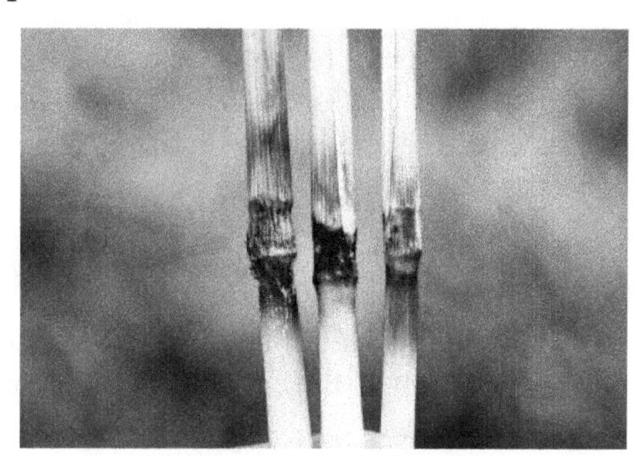

Rice blast lesions on plant nodes

Rice blast has been observed on rice strains M-201, M-202, M-204, M-205, M-103, M-104, S-102, L-204, Calmochi-101, with M-201 being the most vulnerable. Initial symptoms are white to gray-green lesions or spots with darker borders produced on all parts of the shoot, while older lesions are elliptical or spindle-shaped and whitish to gray with necrotic borders. Lesions may enlarge and coalesce to kill the entire leaf. Symptoms are observed on all above-ground parts of the plant. Lesions can be seen on the leaf collar, culm, culm nodes, and panicle neck node. Internodal infection of the culm occurs in a banded pattern. Nodal infection causes the culm to break at the infected node (rotten neck). It also affects reproduction by causing the host to produce fewer seeds. This is caused by the disease preventing maturation of the actual grain.

Disease Cycle

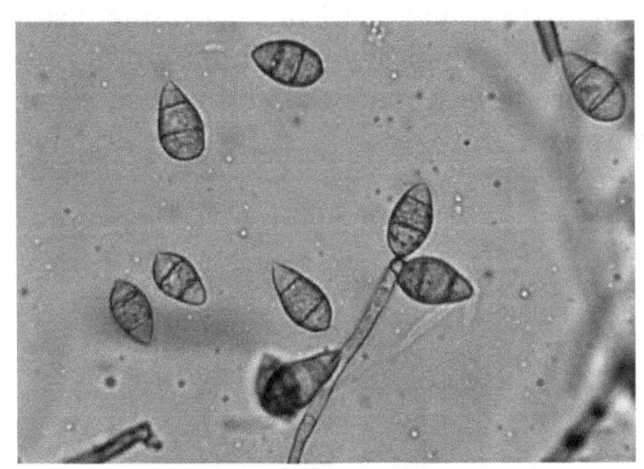

Spores of M. grisea

The pathogen infects as a spore that produces lesions or spots on parts of the rice plant such as the leaf, leaf collar, panicle, culm and culm nodes. Using a structure called an appressorium, the pathogen penetrates the plant. *M. grisea* then sporulates from the diseased rice tissue to be dispersed as conidiospores. After overwintering in sources such as rice straw and stubble, the cycle repeats.

A single cycle can be completed in about a week under favorable conditions where one lesion can generate up to thousands of spores in a single night. With the ability to continue to produce the spores for over 20 days, rice blast lesions can be devastating to susceptible rice crops.

Environment

Rice blast is a significant problem in temperate regions and can be found in areas such as irrigated lowland and upland. Conditions conducive for rice blast include long periods of free moisture where leaf wetness is required for infection and high humidity is common. Sporulation increases with high relative humidity and at 77-82 degrees F, spore germination, lesion formation, and sporulation are at optimum levels.

In terms of control, excessive use of nitrogen fertilization as well as drought stress increase rice susceptibility to the pathogen as the plant is placed in a weakened state and its defenses are low. Extended drain periods also favor infection as they aerate the soil, converting ammonium to nitrate and thus causing stress to rice crops, as well.

Management

The fungus has been able to establish resistance to both chemical treatments and genetic resistance in some types of rice developed by plant breeders. It is thought that the fungus can achieve this by genetic change through mutation. In order to most effectively control infection by *M. grisea*, an integrated management program should be implemented to avoid overuse of a single control method and fight against genetic resistance. For example, eliminating crop residue could reduce the occurrence of overwintering and discourage inoculation in subsequent seasons. Another strategy would be to plant resistant rice varieties that are not as susceptible to infection by *M. grisea*. Knowledge of the pathogenicity of *M. grisea* and its need for free moisture suggest other control strategies such as regulated irrigation and a combination of chemical treatments with different modes of action. Managing the amount of water supplied to the crops limits spore mobility thus dampening the opportunity for infection. Chemical controls such as Carpropamid have been shown to prevent penetration of the appressoria into rice epidermal cells, leaving the grain unaffected.

J. Sendra rice affected by Magnaporthe grisea.

Importance

Rice blast is the most important disease concerning the rice crop in the world. Since rice is an important food source for much of the world, its effects have a broad range. It has been found in over 85 countries across the world and reached the United States in 1996. Every year the amount of crops lost to rice blast could feed 60 million people. Although there are some resistant strains of rice, the disease persists wherever rice is grown. The disease has never been eradicated from a region.

Sirococcus Clavigignenti-juglandacearum

Sirococcus clavigignenti-juglandacearum is a mitosporic fungus that causes a lethal canker disease of Butternut trees (Juglans cinerea). Known in the vernacular as butternut canker, it is also known to parasitize other members of the Juglans genus on occasion, and very rarely other related trees including hickories. The fungus is found throughout North America, occurring on up to 91% of butternut trees, and may be threatening the viability of butternut as a species.

Distribution

Butternut, the primary host of *S. clavigignenti-juglandacearum*, is found in mixed hardwood forests throughout central North America, from New Brunswick to North Carolina.

Butternut canker was identified as an invasive species in 1967. It was first discovered in Wisconsin, but has since spread to other states and into Canada. Its native origin is unknown, but possibly in

Asia given the resistance of Asian walnuts to the disease. The United States Forest Service found that 84% of all butternuts in Michigan as well as 58% of all trees from Wisconsin have been affected; later surveys by the Wisconsin Department of Natural Resources revealed that 91% of all living trees in Wisconsin were diseased or cankered. In Virginia and North Carolina, the butternut population has been reduced from 7.5 million to 2.5 million.

Symptoms

Broad dead areas known as cankers form on the main stem, branches, young twigs, and exposed roots. Most cankers are covered with bark cracks. The fungus forms a dark mat of branching mycelium below the bark, from which arise peg-like hypha that lift and rupture the bark. In the later stages of infection, the bark above the canker is shredded.

Life Cycle

Butternut trees killed by Butternut canker

S. clavigignenti-juglandacearum produces its spores asexually; its sexual form of reproduction has never been observed.

Pycnidiospores are released during rainy periods. When the spores make contact with wounds or broken branches, they germinate and penetrate deep into the tree to produce cankers. Infection hyphae typically penetrates through the parenchyma phloem intracellularly but they can also penetrate intercellularly through uni and multiceliate xylem ray cells and paranchyma cells. Later, the fungus will produce mycelial mats of stroma and mycelial pegs.

Stroma mats will produce uni or multilocular pycnidia. Inside the pycnidia are branched and unbranched conidiophores with two-celled pycniospores, which later are ejected from the pycnidial ostiole.

Additionally, the stroma will produce a peg of interwoven mycelium. These pegs put pressure on the outer peridium of the host bark, which exposes the pycnidia below. These pegs also produce pycnidia that are smaller than the pycnidia in the stroma. While different in size, the spores produced are identical.

Resistance

Many species of tree show varying degrees of resistance, such as the heartnut, Butternut, and the

Japanese, Black, and Persian Walnuts. It is sometimes claimed that higher levels of resistance result from thicker bark; however, since the disease enters through breaks in the bark, it is unlikely that bark thickness influences resistance. Additionally, both trees produce phenolics immediately upon attack, later producing gums and tyloses to surround the pathogen.

Control

Breeding for resistance is important for fighting butternut canker. While standard practice has been that infected trees should be removed to prevent further spread, there is a growing opinion that the time for this is past. The disease has now been found in virtually all parts of the butternut range. Additionally, it is suggested that "removing diseased trees" is a guarantee that infected, but not dying trees, i.e. those that are specifically "partially resistant" to the fungus, will be killed; eliminating any chance of increased resistance in progeny. Instances are known of long-term survival of pure butternuts infected by the canker.

Recent reports have shown that the fungus can be internally seed-borne, so seeds should be subjected to intense quarantine protocols; most especially if destined for plantings where the disease is not already established.

Claviceps Purpurea

Claviceps purpurea is an ergot fungus that grows on the ears of rye and related cereal and forage plants. Consumption of grains or seeds contaminated with the survival structure of this fungus, the ergot sclerotium, can cause ergotism in humans and other mammals. *C. purpurea* most commonly affects outcrossing species such as rye (its most common host), as well as triticale, wheat and barley. It affects oats only rarely.

Life Cycle

fruiting bodies with head and stipe on Sclerotium

An *ergot kernel* called *Sclerotium clavus* develops when a floret of flowering grass or cereal is infected by an ascospore of *C. purpurea*. The infection process mimics a pollen grain growing into an ovary during fertilization. Because infection requires access of the fungal spore to the stigma, plants infected by *C. purpurea* are mainly outcrossing species with open flowers, such as rye (*Secale cereale*) and Alopecurus.

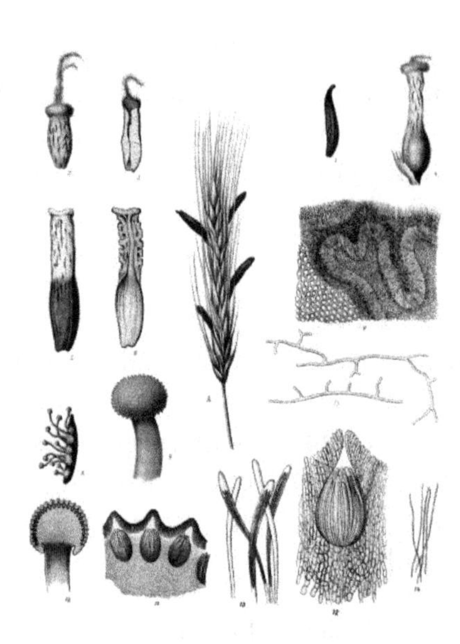

Various stages in the life cycle of *Claviceps purpurea*

The proliferating fungal mycelium then destroys the plant ovary and connects with the vascular bundle originally intended for feeding the developing seed. The first stage of ergot infection manifests itself as a white soft tissue (known as *Sphacelia segetum*) producing sugary honeydew, which often drops out of the infected grass florets. This honeydew contains millions of asexual spores (conidia) which are dispersed to other florets by insects or rain. Later, the *Sphacelia segetum* convert into a hard dry *Sclerotium clavus* inside the husk of the floret. At this stage, alkaloids and lipids (e.g. ricinoleic acid) accumulate in the Sclerotium.

When a mature *Sclerotium* drops to the ground, the fungus remains dormant until proper conditions trigger its fruiting phase (onset of spring, rain period, need of fresh temperatures during winter, etc.). It germinates, forming one or several fruiting bodies with head and stipe, variously colored (resembling a tiny mushroom). In the head, threadlike sexual spores (ascospores) are formed in perithecia, which are ejected simultaneously, when suitable grass hosts are flowering. Ergot infection causes a reduction in the yield and quality of grain and hay produced, and if infected grain or hay is fed to livestock it may cause a disease called ergotism.

Polistes dorsalis, a species of social wasps, have been recorded as a vector of the spread of this particular fungus. During their foraging behavior, particles of the fungal conidia get bound to parts of this wasp's body. As *P. dorsalis* travels from source to source, it leaves the fungal infection behind. Insects, including flies and moths, have also been shown to carry conidia of *Claviceps* species, but if insects play a role in spreading the fungus from infected to healthy plants is unknown.

Intraspecific Variations

Early scientists have observed *Claviceps purpurea* on other Poaceae as *Secale cereale*. 1855, Grandclement described ergot on *Triticum aestivum*. During more than a century scientists aimed to describe specialized species or specialized varieties inside the species *Claviceps purpurea*. That's how the species are created.

- *Claviceps microcephala* Tul. (1853)

- *Claviceps wilsonii* Cooke (1884)

Later scientists tried to determine host varieties as

- Claviceps purpurea var. agropyri
- Claviceps purpurea var. purpurea
- Claviceps purpurea var. spartinae
- Claviceps purpurea var.wilsonii.

But molecular biology hasn't confirmed this hypothesis but has distinguished three groups differing in their ecological specificity

- G1 — land grasses of open meadows and fields;
- G2 — grasses from moist, forest, and mountain habitats;
- G3 (C. purpurea var. spartinae) — salt marsh grasses (Spartina, Distichlis).

Sclerotium of Claviceps purpurea on Alopecurus myosuroides

Morphological criteria to distinguish different groups: The shape and the size of sclerotia are not good indicators because they strongly depend on the size and shape of the host floret. The size of conidia can be an indication but it is weak and it is necessary to pay attention to that, due to osmotic pressure, it varies significantly if the spores are observed in honeydew or in water. The sclerotial density can be used as the groups G2 and G3 float in water.

The compound of alkaloids is also used to differentiate the strains.

Host Range

Model of *Claviceps purpurea*, Botanical Museum Greifswald

Pooideae

Agrostis canina, Alopecurus myosuroides (G2), Alopecurus arundinaceus (G2), Alopecurus pratense, Bromus arvensis, Bromus commutatus, Bromus hordeaceus (G2), Bromus inermis, Bromus marginatus, Elymus tsukushiense, Festuca arundinacea, Elytrigia repens (G1), Nardus stricta, Poa annua (G2), Phleum pratense, Phalaris arundinacea (G2), Poa pratensis (G1), Stipa.

Arundinoideae

Danthonia, Molinia caerulea.

Chloridoideae

Spartina, Distichlis (G3)

Panicoideae

Setaria

Epidemiology

Claviceps purpurea has been known to mankind for a long time, and its appearance has been linked to extremely cold winters that were followed by rainy springs.

The sclerotial stage of *C. purpurea* conspicuous on the heads of ryes and other such grains is known as ergot. Sclerotia germinate in spring after a period of low temperature. A temperature of 0-5 °C for at least 25 days is required. Water before the cold period is also necessary. Favorable temperatures for stroma production are in the range of 10-25 °C. Favorable temperatures for mycelial growth are in the range of 20-30 °C with an optimum at 25 °C.

Sunlight has a chromogenic effect on the mycelium with intense coloration.

Effects

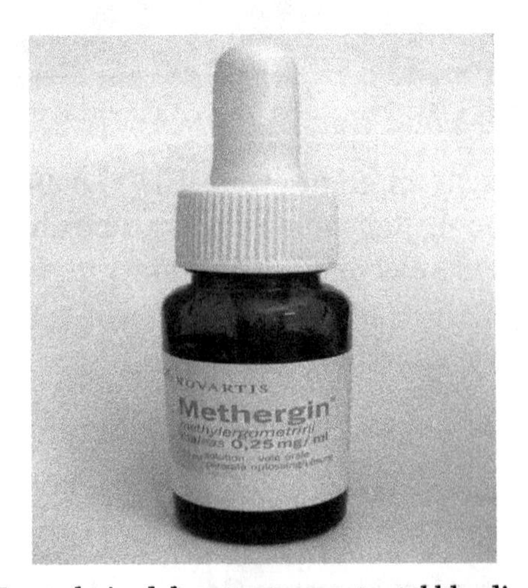

Ergot-derived drug to stop postnatal bleeding

The disease cycle of the ergot fungus was first described in 1853, but the connection with ergot and epidemics among people and animals was reported already in a scientific text in 1676. The ergot sclerotium contains high concentrations (up to 2% of dry mass) of the alkaloid ergotamine, a complex molecule consisting of a tripeptide-derived cyclol-lactam ring connected via amide linkage to a lysergic acid (ergoline) moiety, and other alkaloids of the ergoline group that are biosynthesized by the fungus. Ergot alkaloids have a wide range of biological activities including effects on circulation and neurotransmission.

Ergotism is the name for sometimes severe pathological syndromes affecting humans or animals that have ingested ergot alkaloid-containing plant material, such as ergot-contaminated grains. Monks of the order of St. Anthony the Great specialized in treating ergotism victims with balms containing tranquilizing and circulation-stimulating plant extracts; they were also skilled in amputations. The common name for ergotism is "St. Anthony's Fire", in reference to monks who cared for victims as well as symptoms, such as severe burning sensations in the limbs. These are caused by effects of ergot alkaloids on the vascular system due to vasoconstriction of blood vessels, sometimes leading to gangrene and loss of limbs due to severely restricted blood circulation.

The neurotropic activities of the ergot alkaloids may also cause hallucinations and attendant irrational behaviour, convulsions, and even death. Other symptoms include strong uterine contractions, nausea, seizures, and unconsciousness. Since the Middle Ages, controlled doses of ergot were used to induce abortions and to stop maternal bleeding after childbirth. Ergot alkaloids are also used in products such as Cafergot (containing caffeine and ergotamine or ergoline) to treat migraine headaches. Ergot extract is no longer used as a pharmaceutical preparation.

Ergot contains no lysergic acid diethylamide (LSD) but rather ergotamine, which is used to synthesize lysergic acid, an analog of and precursor for synthesis of LSD. Moreover, ergot sclerotia naturally contain some amounts of lysergic acid.

Culture

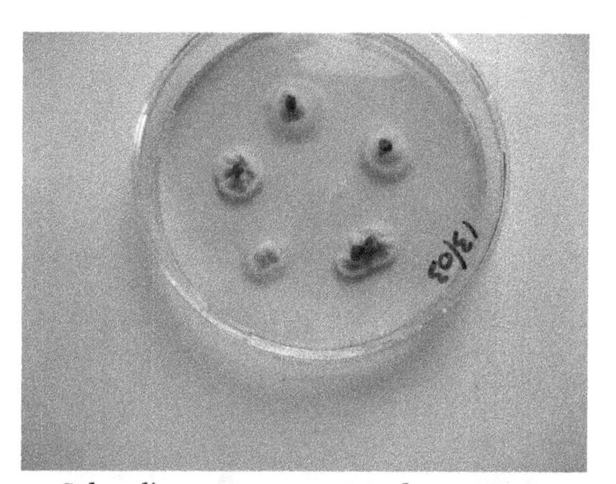

Sphacelia segetum on potato dextrose agar

Potato dextrose agar, wheat seeds or oat flour are suitable substrates for growth of the fungus in the laboratory. Agricultural production of Claviceps purpurea on rye is used to produce ergot alkaloids. Biological production of ergot alkaloids is also carried out by saprophytic cultivations.

Speculations

Human poisoning due to the consumption of rye bread made from ergot-infected grain was common in Europe in the Middle Ages. The epidemic was known as Saint Anthony's fire, or *ignis sacer*.

Gordon Wasson proposed that the psychedelic effects were the explanation behind the festival of Demeter at the Eleusinian Mysteries.

Linnda R. Caporael posited in 1976 that the hysterical symptoms of young women that had spurred the Salem witch trials had been the result of consuming ergot-tainted rye. However, her conclusions were later disputed by Nicholas P. Spanos and Jack Gottlieb, after a review of the historical and medical evidence. Other authors have likewise cast doubt on ergotism having been the cause of the Salem witch trials.

The Great Fear in France during the Revolution has also been linked by some historians to the influence of ergot.

British author John Grigsby claims that the presence of ergot in the stomachs of some of the so-called 'bog-bodies' (Iron Age human remains from peat bogs N E Europe such as Tollund Man), reveals that ergot was once a ritual drink in a prehistoric fertility cult akin to the Eleusinian Mysteries cult of ancient Greece. In his book Beowulf and Grendel he argues that the Anglo-Saxon poem Beowulf is based on a memory of the quelling of this fertility cult by followers of Odin. He states that Beowulf, which he translates as *barley-wolf*, suggests a connection to ergot which in German was known as the 'tooth of the wolf'.

In 1951 at Pont St. Esprit in the south of France there was an outbreak of violent hallucinations among hundreds of residents.

Sclerotium

Ergot sclerotia developing on wheat spikes

A sclerotium, plural sclerotia, is a compact mass of hardened fungal mycelium containing food reserves. One role of sclerotia is to survive environmental extremes. In some higher fungi such as ergot, sclerotia become detached and remain dormant until favorable growth conditions return. Sclerotia initially were mistaken for individual organisms and described as separate species until Louis René Tulasne proved in 1853 that sclerotia are only a stage in the life cycle of some fungi. Further investigation showed that this stage appears in many fungi belonging to many diverse groups. Sclerotia are important in the understanding of the life cycle and reproduction of fungi, as a food source, as medicine (for example, ergotamine), and in agricultural blight management.

Examples of fungi that form sclerotia are ergot (*Claviceps purpurea*), *Polyporus tuberaster*, *Psilocybe mexicana*, *Sclerotium delphinii* and many species in Sclerotiniaceae. The plasmodium of slime molds can form sclerotia in adverse environmental conditions.

Description

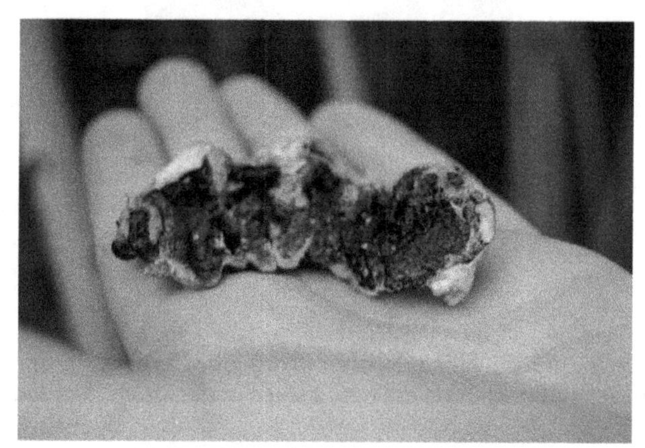

Sclerotinia sclerotiorum sclerotia

Sclerotia are often composed of a thick, dense shell with thick and dark cells and a core of thin colorless cells. Sclerotia are rich in hyphae emergency supplies, especially oil. They contain a very small amount of water (5-10%) and can survive in a dry environment for several years without losing the ability to grow. In most cases, the sclerotium consists exclusively of fungal hyphae, whereas some may consist partly of fungal hyphae plexus and partly in between tissues of the substrate (ergot, Sclerotinia). In favorable conditions, sclerotia germinate to form fruiting bodies (Basidiomycetes) or mycelium with conidia (in imperfect fungi). Sclerotia sizes can range from a fraction of a millimeter to a few tens of centimeters as, for example *Laccocephalum mylittae*, which has sclerotia with diameters up to 30 cm and weighing up to 20 kg.

Sclerotia resemble cleistothecia in both their morphology and the genetic control of their development. This suggests the two structures may be homologous, sclerotia being vestigial cleistothecia that lost the capacity to produce ascospores.

History

In the Middle Ages *Claviceps purpurea* sclerotia contaminated rye grain used in bread and led to ergot poisoning by way of which thousands of people were killed and mutilated. *Claviceps purpurea* sclerotia contain alkaloids that, when consumed, can cause ergotism which is a disease that causes paranoia and hallucinations, twitches, spasms, loss of peripheral sensation, edema and loss of affected tissues.

Louis Rene Tulasne discovered the relationship between infected rye plants and ergotism in the 19th century. With this discovery, more efforts were developed to reduce sclerotia from growing on rye and ergotism became rare. However, in 1879–1881 an outbreak developed in Germany, in 1926–1927 Russia was infected, and in 1977–1978 Ethiopia was infected.

Pleurotus tuber-regium, which forms edible sclerotia up to 30 cm wide, has a history of economic importance in Africa as food and as a medicinal mushroom.

Sclerotia initially were mistaken for individual organisms and described as separate species until Louis René Tulasne proved in 1853 that sclerotia are only a stage in the life cycle of some fungi.

Sclerotia as Part of Fungal Life Cycles

For example, *Claviceps purpurea* sclerotia form and begin regrowth in the spring, infecting grass and rye plants by way of releasing their ascospores from perithecia. *Claviceps purpurea* can infect a wide variety of plants by infecting the ovaries. The fungal spores germinate at the anthesis and grow down the pollen tube without branching any hyphae outward. When the fungus reaches the bottom of the ovary, it leaves the pollen tube path and enters the vascular tissues where it branches its hypha. Approximately seven days into the infection, the mycelium produces conidia. The conidia are then secreted out of the plant in a sugary liquid that insects, attracted by the sugars, transfer to other plants. After two weeks of being infected by the fungus, the plant no longer generates the sugary liquid, and the fungus produces sclerotia. The sclerotium is an overwinter structure, which contains ergot alkaloids. *Claviceps purpurea*'s life cycle is an interesting model for plant pathologists and cell biologists because:

- Strict organ specificity (ovaries)
- The plant lacks defense reactions
- Strict polar, oriented growth in the first infection stage
- Biotrophic life style

Formation

In fungi, there are three stages in the development of sclerotia:

1. Initial aggregation of hyphae;
2. Increase in size due to the growth and branching of hyphae;

3. Maturation with the formation of an outer coating that isolates the sclerotia from the surrounding environment, with the progressive dehydration of the hyphae and accumulation of reserve substances and pigments.

Sclerotia as Food

Wolfiporia Extensa

Edible Wolfiporia extensa sclerotium

This sclerotium of *Wolfiporia extensa* (called "Tuckahoe", or Indian bread) was used by Native Americans as a source of food in times of scarcity. It is a wood-decay fungus but has a terrestrial growth habit. It is notable in the development of a large, long-lasting underground sclerotium, which resembles a small coconut.

Pleurotus Tuber-regium

Pleurotus tuber-regium, which forms edible sclerotia up to 30 cm wide, has a history of economic importance as food in Africa.

Sclerotia as Medicine and Hallucinogenic Drug

Inonotus obliquus (chaga) sclerotium growing on a birch tree

Over billions of years of Earth's history, organisms have acquired the ability to produce secondary metabolites, that is chemical compounds that afford protection from pathogens and ultraviolet light damage from the sun. Fungi are no exception, and due to their exposure to a wide variety of environments, they have developed the ability to produce a large number of such chemical compounds that are very valuable in medicine.

Claviceps Purpurea

In early times, ergot alkaloids have been used for medicinal purposes. For example, ergot was used as a form of abortion in Europe, but it led to hyper-contraction. In the 19th century, it was used to aid in the prevention of bleeding in after childbirth and treatment for migraines and Parkinson's disease.

Acid hydrolysis is used to convert alkaloids, produced by the fungus *Claviceps purpurea,* into D-lysergic acid which is the starting material for many pharmaceutical and illegal drugs. In 1938 Albert Hofmann synthesized one of the strongest known hallucinogens, lysergic acid diethylamide (LSD), from ergot alkaloid. Despite side effects of the drug such as paranoia, loss of judgment and flashbacks, psychotherapists and psychiatrists used it to treat patients with neuroses, sexual dysfunctions and anxiety. The secret service may have also used it for interrogation purposes. In 1966 the United States government made LSD illegal. Recently, clinics have shown an interest in ergoline to treat patients with autism.

Ophiocordyceps Sinensis

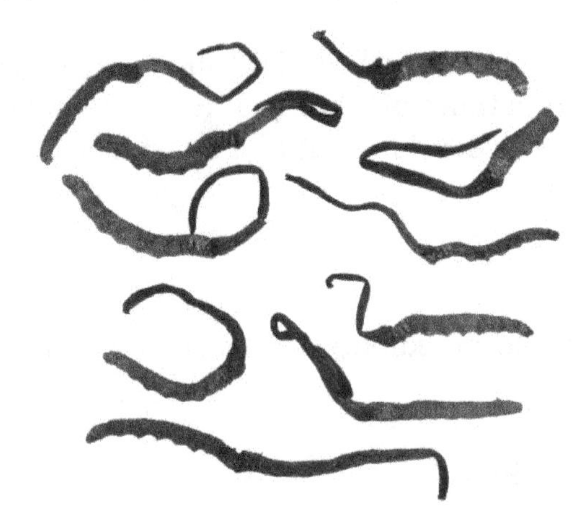

Caterpillars with emerging Ophiocordyceps sinensis

Ophiocordyceps sinensis (syn. *Cordyceps sinensis*) is a fungus which infects a caterpillar and uses nutrients out of it to create mycelia and replaces its body with a sclerotium. The fungus then sprouts out of the head of the caterpillar. In Chinese the fungus is known as *Dōng chóng xià cǎo.*

Inonotus Obliquus

Inonotus obliquus (chaga mushroom) is a sclerotium growing mostly on birch trees in northern climates. It has been used as a tonic and a remedy for thousands of years in Canada, Russia, Japan, etc. The tree sclerotium develops over the years as the mycelium sucks the energy of the living tree.

Psilocybe Galindoi

Certain grassland *Psilocybe* species have sclerotia to protect them from fire and from other disturbances. The sclerotia forming species contain, as many *Psilocybe* species do, the organic compounds psilocin and psilocybin, which are actively being researched to treat cluster headaches, depression, and to help the mental health of people with fatal cancer.

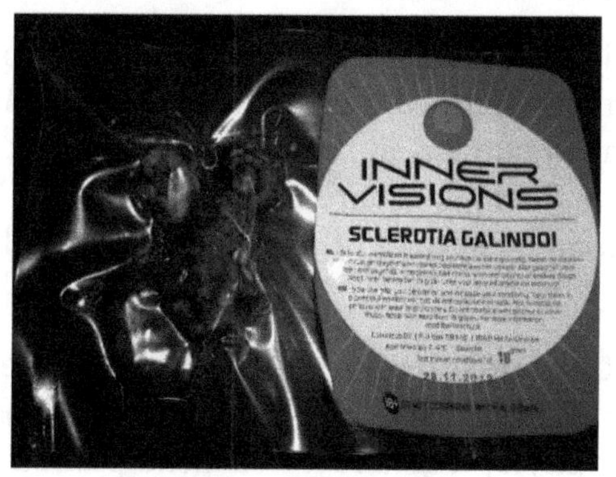

Psilocybe galindoi sclerotia

Psilocybe Mexicana and Psilocybe Tampanensis

Sclerotia from *Psilocybe mexicana* and *Psilocybe tampanensis* also contain the active metabolites psilocin and psilocybin. These sclerotia can be bought at smartshops under different trade names such as "Philosopher's Stone", "truffles" or "tripknollen" (Dutch for "hallucinogenic bulbs"), and have the same hallucinogenic affect as magic mushrooms.

Wolfiporia Extensa

Wolfiporia extensa is used as a medicinal mushroom in Chinese medicine.

Common names for it include *hoelen*, poria, *tuckahoe*, China root, *fu ling*, *fu shen* (or *fus-hen*) and *matsuhodo*.

Some Species with Sclerotia as Agricultural Pests

Many methods have been created to reduce the growth of agriculturally pathogenic sclerotia like changes in crop rotation, deeper ploughing and sifting out sclerotia. Fungicides, breeding disease resistance rye and cross breeding natural rye with hybrid rye have reduced *C. purpurea* infections. Other fungi that produce sclerotia are prominent pathogens for canola crops. These and related fungi are generally controlled through the use of fungicides and crop rotation.

Sclerotium delphinii sclerotia on infected host

Armillaria

Honey fungus, or Armillaria is a genus of parasitic fungi that live on trees and woody shrubs. It includes about 10 species formerly lumped together as *A. mellea*. *Armillarias* are long lived and form some of the largest living organisms in the world. The largest organism (of the species *Armillaria solidipes*) covers more than 3.4 square miles (8.8 km^2) in Malheur National Forest and is more than 2,400 years old. Some species of *Armillaria* display bioluminescence, resulting in foxfire.

Armillaria can be a destructive forest pathogen. It causes "white rot" root disease of forests, which is distinguishes it from *Tricholoma*, a mycorrhizal (non-parasitic) genus. Because *Armillaria* is a facultative saprophyte, it also feeds on dead plant material, allowing it to kill its host, unlike parasites that must moderate its growth to avoid host death.

In the Canadian Prairies (particularly Manitoba), *Armillaria* is not referred to as "honey fungus" but instead often as *pidpenky* (Ukrainian: підпеньки), meaning "beneath the stump" in Ukrainian.

Description

Armillaria mellea

The basidiocarp (reproductive structure) of the fungus is a mushroom that grows on wood, typically in small dense clumps or tufts. Their caps (mushroom tops) are typically yellow-brown, somewhat sticky to touch when moist, and, depending on age, may range in shape from conical to convex to depressed in the center. The stipe (stalk) may or may not have a ring. All *Armillaria* species have a white spore print and none have a volva (cup at base) (compare *Amanita*).

Similar species include *Pholiota* spp. which also grow in cespitose (mat-like) clusters on wood and fruit in the fall. *Pholiota* spp. are separated from Armillaria by its yellowish to greenish-yellow tone and a dark brown to grey-brown spore print. Mushroom hunters need to be wary of *Galerina* spp. which can grow side-by-side with *Armillaria* spp. on wood. *Galerina* have a dark brown spore print and are deadly poisonous (alpha-amanitin).

Armillaria hinnulea

Plant Pathology

Honey fungus is a "white rot" fungus, which is a pathogenic organism that affects trees, shrubs, woody climbers and, rarely, woody herbaceous perennial plants. Honey fungus can grow on living, decaying, and dead plant material.

Honey fungus spreads from living trees, dead and live roots and stumps by means of reddish-brown to black rhizomorphs (root-like structures) at the rate of approximately 1 m a year, but infection by root contact is possible. Infection by spores is rare. Rhizomorphs grow close to the soil surface (in the top 20 cm) and invade new roots, or the root collar (where the roots meet the stem) of plants. An infected tree will die once the fungus has girdled it, or when significant root damage has occurred. This can happen rapidly, or may take several years. Infected plants will deteriorate, although may exhibit prolific flower or fruit production shortly before death.

Initial symptoms of honey fungus infection include dieback or shortage of leaves in spring. Rhizomorphs appear under the bark and around the tree, and mushrooms grow in clusters from the infected plant in autumn and die back after the first frost. However these symptoms and signs do not necessarily mean that the pathogenic strains of honey fungus are the cause, so other identification methods are advised before diagnosis. Thin sheets of cream colored mycelium, beneath the bark at the base of the trunk or stem indicated that honey fungus is likely the pathogen. It will give off a strong mushroom scent and the mushrooms sometimes extend upward. On conifers honey fungus often exudes a gum or resin from cracks in the bark.

The linkage of morphological, genetic, and molecular characters of *Armillaria* over the past few decades has led to the recognition of intersterile groups designated as "biological species". Data from such studies, especially those using molecular diagnostic tools, have removed much uncertainty for mycologists and forest pathologists. New questions remain unanswered regarding the phylogeny of North American *Armillaria* species and their relationships to their European counterparts, particularly within the "*Armillaria mellea* complex". Some data suggest that North American and European *A. gallica* isolates are not monophyletic. Although North American and European isolates of *A. gallica* may be interfertile, some North American isolates of *A. gallica* are more closely related to the North American taxon *A. calvescens* than to European isolates of

A. gallica. The increase in genetic divergence has not necessarily barred inter-sterility between isolated populations of *A. gallica*. Although the relationships among some groups in the genus seem clearer, the investigation of geographically diverse isolates has revealed that the relationship between some North American species is still unclear (Hughes et al. 2003).

Intersterile species of *Armillaria* occurring in North America (North American Biological Species = NABS) were listed by Mallett (1992):

- I *Armillaria ostoyae* (Romagn.) Herink
- II *Armillaria gemina* Bérubé & Dessureault
- III *Armillaria calvescens* Bérubé & Dessureault
- V *Armillaria sinapina* Bérubé & Dessureault
- VI *Armillaria mellea* (Vahl.:Fries) Kummer
- VII *Armillaria gallica* (Marxmüller & Romagn.)
- IX, X, and XI taxonomically undescribed

NABS I, V, VII, IX, X, and XI have been found in British Columbia; I, III, V have been found in the Prairie Provinces, with I and V occurring in both the boreal and subalpine regions; I, III, V, and VII have been found in Ontario; and I, II, III, V, and VI have been found in Quebec. *Armillaria ostoyae* is the species most commonly found in all Canadian provinces surveyed (Mallett 1990). *Armillaria* root rot occurs in the Northwest Territories, and was identified on white spruce at Pine Point on Great Slave Lake prior to NABS findings.

Edibility

Honey Fungus are regarded in Ukraine, Russia, Poland, Germany and other European countries as one of the best wild mushrooms. They are commonly ranked above morels and chanterelles and only the cep / porcini is more highly prized. However, honey fungus must be thoroughly cooked as they are mildly poisonous raw. One of the four UK species can cause sickness when ingested with alcohol. For those unfamiliar with the species, it is advisable not to drink alcohol for 12 hours before and 24 after eating this mushroom to avoid any possible nausea and vomiting. However, if these rules are followed this variety of mushroom is a delicacy with a distinctive mushroomy and nutty flavor. Reference texts for identification are *Collins Complete British Mushrooms and Toadstools* for the variety of field pictures in it, and Roger Philips' *Mushrooms* for the quality of his out of field pictures and descriptions.

Norway used to consider Honey Fungus edible, but because the health department is moving away from parboiling, they are now considered poisonous.

Hosts

Potential hosts include conifers and various monocotyledonous and dicotyledonous trees, shrubs, and herbaceous species, ranging from asparagus and strawberry to large forest trees (Patton and

Vasquez Bravo 1967). *Armillaria* root rot enters hosts through the roots. In Alberta, 75% of trap logs (Mallett and Hiratsuka 1985) inserted into the soil between planted spruce became infected with the distinctive white mycelium of *Armillaria* within one year. Of the infestations, 12% were *A. ostoyae*, and 88% were *A. sinapina* (Blenis et al. 1995). Reviews of the biology, diversity, pathology, and control of *Armillaria* in Fox (2000) are useful.

References

- According to one 2001 estimate, some 10,000 fungal diseases are known. Struck C (2006). "Infection strategies of plant parasitic fungi". In Cooke BM, Jones DG, Kaye B. The Epidemiology of Plant Diseases. Berlin, Germany: Springer. p. 117. ISBN 1-4020-4580-8.
- Chang S-T, Miles PG (2004). Mushrooms: Cultivation, Nutritional Value, Medicinal Effect and Environmental Impact. Boca Raton, Florida: CRC Press. ISBN 0-8493-1043-1.
- Alcamo IE, Pommerville J (2004). Alcamo's Fundamentals of Microbiology. Boston, Massachusetts: Jones and Bartlett. p. 590. ISBN 0-7637-0067-3.
- Donoghue MJ, Cracraft J (2004). Assembling the Tree of Life. Oxford (Oxfordshire), UK: Oxford University Press. p. 187. ISBN 0-19-517234-5.
- Silar P (2016). "Protistes Eucaryotes: Origine, Evolution et Biologie des Microbes Eucaryotes". HAL: 462. ISBN 978-2-9555841-0-1.
- Blackwell M, Spatafora JW (2004). "Fungi and their allies". In Bills GF, Mueller GM, Foster MS. Biodiversity of Fungi: Inventory and Monitoring Methods. Amsterdam: Elsevier Academic Press. pp. 18–20. ISBN 0-12-509551-1.
- Brodo IM, Sharnoff SD (2001). Lichens of North America. New Haven, Connecticut: Yale University Press. ISBN 0-300-08249-5.
- Raven PH, Evert RF, Eichhorn, SE (2005). "14—Fungi". Biology of Plants (7 ed.). W. H. Freeman. p. 290. ISBN 978-0-7167-1007-3.
- Purvis W (2000). Lichens. Washington, D.C.: Smithsonian Institution Press in association with the Natural History Museum, London. pp. 49–75. ISBN 1-56098-879-7.
- Nielsen K, Heitman J (2007). "Sex and virulence of human pathogenic fungi". Advances in Genetics. Advances in Genetics. 57: 143–73. doi:10.1016/S0065-2660(06)57004-X. ISBN 978-0-12-017657-1. PMID 17352904.
- Cook GC, Zumla AI (2008). Manson's Tropical Diseases: Expert Consult. Edinburgh, Scotland: Saunders Ltd. p. 347. ISBN 1-4160-4470-1.
- Halpern GM, Miller A (2002). Medicinal Mushrooms: Ancient Remedies for Modern Ailments. New York, New York: M. Evans and Co. p. 116. ISBN 0-87131-981-0.
- Stamets P (2000). Growing Gourmet and Medicinal Mushrooms = [Shokuyō oyobi yakuyō kinoko no saibai]. Berkeley, California: Ten Speed Press. pp. 233–248. ISBN 1-58008-175-4.
- Orr DB, Orr RT (1979). Mushrooms of Western North America. Berkeley, California: University of California Press. p. 17. ISBN 0-520-03656-5.
- Ammirati JF, McKenny M, Stuntz DE (1987). The New Savory Wild Mushroom. Seattle, Washington: University of Washington Press. pp. xii–xiii. ISBN 0-295-96480-4.
- Datta A, Ganesan K, Natarajan K (1989). "Current trends in Candida albicans research". Advances in Microbial Physiology. Advances in Microbial Physiology. 30: 53–88. doi:10.1016/S0065-2911(08)60110-1. ISBN 978-0-12-027730-8. PMID 2700541.
- Gordon Wasson, The Road To Eleusis: Unveiling The Secret of The Mysteries (New York: Harcourt Brace Jovanovich, 1977) ISBN 0151778728

- Isikhuemhen, O.S.; LeBauer, D.S. (2004). "Growing Pleurotus tuber-regium". Oyster Mushroom Cultivation (PDF). Seoul (Korea): Mushworld. pp. 270–281. ISBN 1-883956-01-3.
- John L. Ingraham (15 February 2010). March of the Microbes: Sighting the Unseen. Harvard University Press. p. 201. ISBN 978-0-674-03582-9.
- Pegler DN. (2000). "Taxonomy, nomenclature and description of Armillaria". In Fox RTV. Armillaria Root Rot: Biology and Control of Honey Fungus. Intercept. pp. 81–93. ISBN 1-898298-64-5.

5

Bacteria: A Biotic Stressor

Bacteria can be found in a number of shapes and sizes. They are usually a few micrometers in length. It is the oldest habitant of this world and is very important for the process of recycling nutrients. Some of the bacteria discussed in the content are beet vascular necrosis, phytoplasma, Rhodococcus fascians, Agrobacterium tumefaciens, xanthomonas etc. The chapter provides the reader with an in-depth understanding on bacteria.

Bacteria

Bacteria constitute a large domain of prokaryotic microorganisms. Typically a few micrometres in length, bacteria have a number of shapes, ranging from spheres to rods and spirals. Bacteria were among the first life forms to appear on Earth, and are present in most of its habitats. Bacteria inhabit soil, water, acidic hot springs, radioactive waste, and the deep portions of Earth's crust. Bacteria also live in symbiotic and parasitic relationships with plants and animals.

There are typically 40 million bacterial cells in a gram of soil and a million bacterial cells in a millilitre of fresh water. There are approximately 5×10^{30} bacteria on Earth, forming a biomass which exceeds that of all plants and animals. Bacteria are vital in recycling nutrients, with many of the stages in nutrient cycles dependent on these organisms, such as the fixation of nitrogen from the atmosphere and putrefaction. In the biological communities surrounding hydrothermal vents and cold seeps, bacteria provide the nutrients needed to sustain life by converting dissolved compounds, such as hydrogen sulphide and methane, to energy. On 17 March 2013, researchers reported data that suggested bacterial life forms thrive in the Mariana Trench, which with a depth of up to 11 kilometres is the deepest part of the Earth's oceans. Other researchers reported related studies that microbes thrive inside rocks up to 580 metres below the sea floor under 2.6 kilometres of ocean off the coast of the northwestern United States. According to one of the researchers, "You can find microbes everywhere — they're extremely adaptable to conditions, and survive wherever they are."

Most bacteria have not been characterised, and only about half of the bacterial phyla have species that can be grown in the laboratory. The study of bacteria is known as bacteriology, a branch of microbiology.

There are approximately ten times as many bacterial cells in the human flora as there are human cells in the body, with the largest number of the human flora being in the gut flora, and a large number on the skin. The vast majority of the bacteria in the body are rendered harmless by the protective effects of the immune system, and some are beneficial. However, several species of bacteria are pathogenic and cause infectious diseases, including cholera, syphilis, anthrax, leprosy, and bubonic plague. The most common fatal bacterial diseases are respiratory infections, with tuberculosis alone killing about 2 million people per year, mostly in sub-Saharan Africa. In developed countries, antibiotics are used to treat bacterial infections and are also used in farming, making antibiotic resistance a growing problem. In industry, bacteria are important in sewage treatment and the breakdown of oil spills, the production of cheese and yogurt through fermentation, and the recovery of gold, palladium, copper and other metals in the mining sector, as well as in biotechnology, and the manufacture of antibiotics and other chemicals.

Once regarded as plants constituting the class *Schizomycetes*, bacteria are now classified as prokaryotes. Unlike cells of animals and other eukaryotes, bacterial cells do not contain a nucleus and rarely harbour membrane-bound organelles. Although the term *bacteria* traditionally included all prokaryotes, the scientific classification changed after the discovery in the 1990s that prokaryotes consist of two very different groups of organisms that evolved from an ancient common ancestor. These evolutionary domains are called *Bacteria* and *Archaea*.

Origin and Early Evolution

The ancestors of modern bacteria were unicellular microorganisms that were the first forms of life to appear on Earth, about 4 billion years ago. For about 3 billion years, most organisms were microscopic, and bacteria and archaea were the dominant forms of life. In 2008, fossils of macroorganisms were discovered and named as the Francevillian biota. Although bacterial fossils exist, such as stromatolites, their lack of distinctive morphology prevents them from being used to examine the history of bacterial evolution, or to date the time of origin of a particular bacterial species. However, gene sequences can be used to reconstruct the bacterial phylogeny, and these studies indicate that bacteria diverged first from the archaeal/eukaryotic lineage. Bacteria were also involved in the second great evolutionary divergence, that of the archaea and eukaryotes. Here, eukaryotes resulted from the entering of ancient bacteria into endosymbiotic associations with the ancestors of eukaryotic cells, which were themselves possibly related to the Archaea. This involved the engulfment by proto-eukaryotic cells of alphaproteobacterial symbionts to form either mitochondria or hydrogenosomes, which are still found in all known Eukarya (sometimes in highly reduced form, e.g. in ancient "amitochondrial" protozoa). Later on, some eukaryotes that already contained mitochondria also engulfed cyanobacterial-like organisms. This led to the formation of chloroplasts in algae and plants. There are also some algae that originated from even later endosymbiotic events. Here, eukaryotes engulfed a eukaryotic algae that developed into a "second-generation" plastid. This is known as secondary endosymbiosis.

Morphology

Bacteria display a wide diversity of shapes and sizes, called *morphologies*. Bacterial cells are about one-tenth the size of eukaryotic cells and are typically 0.5–5.0 micrometres in length. However, a few species are visible to the unaided eye — for example, *Thiomargarita namibiensis* is up to half a millimetre long and *Epulopiscium fishelsoni* reaches 0.7 mm. Among the smallest bacteria are members of the genus *Mycoplasma*, which measure only 0.3 micrometres, as small as the largest viruses. Some bacteria may be even smaller, but these ultramicrobacteria are not well-studied.

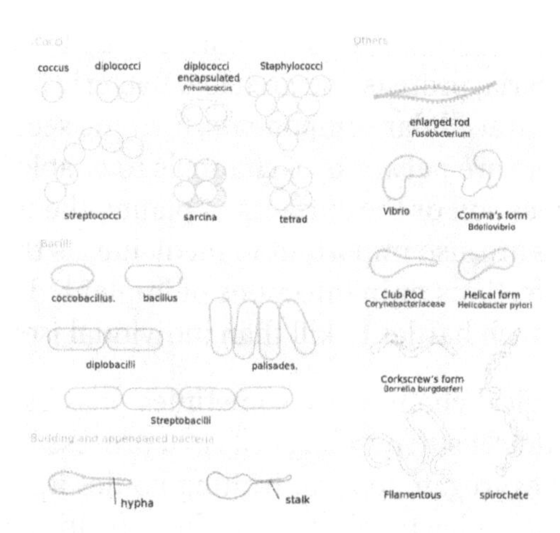

Bacteria display many cell morphologies and arrangements

Most bacterial species are either spherical, called *cocci* (*sing*. coccus, from Greek *kókkos*, grain, seed), or rod-shaped, called *bacilli* (*sing*. bacillus, from Latin *baculus*, stick). Elongation is associated with swimming. Some bacteria, called *vibrio*, are shaped like slightly curved rods or comma-shaped; others can be spiral-shaped, called *spirilla*, or tightly coiled, called *spirochaetes*. A small number of species even have tetrahedral or cuboidal shapes. More recently, some bacteria were discovered deep under Earth's crust that grow as branching filamentous types with a star-shaped cross-section. The large surface area to volume ratio of this morphology may give these bacteria an advantage in nutrient-poor environments. This wide variety of shapes is determined by the bacterial cell wall and cytoskeleton, and is important because it can influence the ability of bacteria to acquire nutrients, attach to surfaces, swim through liquids and escape predators.

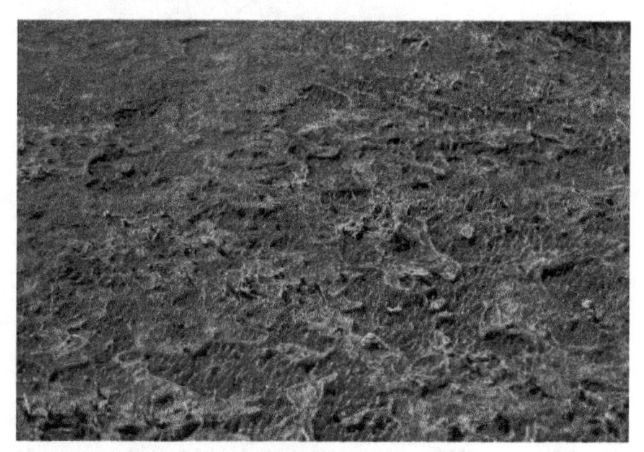

A biofilm of thermophilic bacteria in the outflow of Mickey Hot Springs, Oregon, approximately 20 mm thick.

Many bacterial species exist simply as single cells, others associate in characteristic patterns: *Neisseria* form diploids (pairs), *Streptococcus* form chains, and *Staphylococcus* group together in "bunch of grapes" clusters. Bacteria can also be elongated to form filaments, for example the Actinobacteria. Filamentous bacteria are often surrounded by a sheath that contains many individual cells. Certain types, such as species of the genus *Nocardia*, even form complex, branched filaments, similar in appearance to fungal mycelia.

Bacteria often attach to surfaces and form dense aggregations called *biofilms* or bacterial mats.

These films can range from a few micrometres in thickness to up to half a metre in depth, and may contain multiple species of bacteria, protists and archaea. Bacteria living in biofilms display a complex arrangement of cells and extracellular components, forming secondary structures, such as microcolonies, through which there are networks of channels to enable better diffusion of nutrients. In natural environments, such as soil or the surfaces of plants, the majority of bacteria are bound to surfaces in biofilms. Biofilms are also important in medicine, as these structures are often present during chronic bacterial infections or in infections of implanted medical devices, and bacteria protected within biofilms are much harder to kill than individual isolated bacteria.

Even more complex morphological changes are sometimes possible. For example, when starved of amino acids, Myxobacteria detect surrounding cells in a process known as quorum sensing, migrate towards each other, and aggregate to form fruiting bodies up to 500 micrometres long and containing approximately 100,000 bacterial cells. In these fruiting bodies, the bacteria perform separate tasks; this type of cooperation is a simple type of multicellular organisation. For example, about one in 10 cells migrate to the top of these fruiting bodies and differentiate into a specialised dormant state called myxospores, which are more resistant to drying and other adverse environmental conditions than are ordinary cells.

Cellular Structure

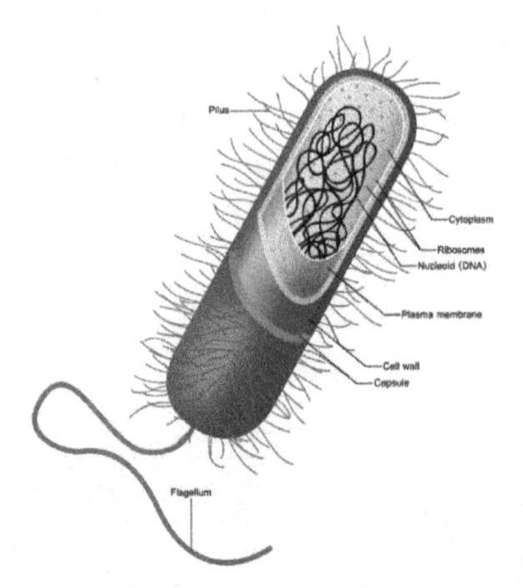

Structure and contents of a typical gram-positive bacterial cell (seen by the fact that only one cell membrane is present).

Intracellular Structures

The bacterial cell is surrounded by a cell membrane (also known as a lipid, cytoplasmic or plasma membrane). This membrane encloses the contents of the cell and acts as a barrier to hold nutrients, proteins and other essential components of the *cytoplasm* within the cell. As they are prokaryotes, bacteria do not usually have membrane-bound organelles in their cytoplasm, and thus contain few large intracellular structures. They lack a true nucleus, mitochondria, chloroplasts and the other organelles present in eukaryotic cells. Bacteria were once seen as simple bags of cytoplasm, but structures such as the *prokaryotic cytoskeleton* and the localisation of proteins to specific locations within the cytoplasm that give bacteria some complexity have been discovered. These subcellular levels of organisation have been called "bacterial hyperstructures".

Bacterial microcompartments, such as carboxysomes, provide a further level of organisation; they are compartments within bacteria that are surrounded by polyhedral protein shells, rather than by lipid membranes. These "polyhedral organelles" localise and compartmentalise bacterial metabolism, a function performed by the membrane-bound organelles in eukaryotes.

Many important biochemical reactions, such as energy generation, use concentration gradients across membranes. The general lack of internal membranes in bacteria means reactions such as electron transport occur across the cell membrane between the cytoplasm and the *periplasmic space*. However, in many photosynthetic bacteria the plasma membrane is highly folded and fills most of the cell with layers of light-gathering membrane. These light-gathering complexes may even form lipid-enclosed structures called chlorosomes in green sulfur bacteria. Other proteins import nutrients across the cell membrane, or expel undesired molecules from the cytoplasm.

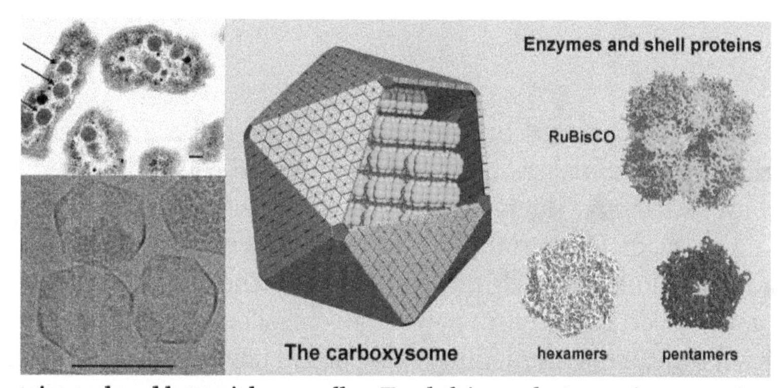

Carboxysomes are protein-enclosed bacterial organelles. Top left is an electron microscope image of carboxysomes in Halothiobacillus neapolitanus, below is an image of purified carboxysomes. On the right is a model of their structure. Scale bars are 100 nm.

Bacteria do not have a membrane-bound nucleus, and their genetic material is typically a single circular DNA chromosome located in the cytoplasm in an irregularly shaped body called the *nucleoid*. The nucleoid contains the chromosome with its associated proteins and RNA. The phylum Planctomycetes and candidate phylum Poribacteria may be exceptions to the general absence of internal membranes in bacteria, because they appear to have a double membrane around their nucleoids and contain other membrane-bound cellular structures. Like all living organisms, bacteria contain *ribosomes*, often grouped in chains called polyribosomes, for the production of proteins, but the structure of the bacterial ribosome is different from that of eukaryotes and Archaea. Bacterial ribosomes have a sedimentation rate of 70S (measured in Svedberg units): their subunits have rates of 30S and 50S. Some antibiotics bind specifically to 70S ribosomes and inhibit bacterial protein synthesis. Those antibiotics kill bacteria without affecting the larger 80S ribosomes of eukaryotic cells and without harming the host.

Some bacteria produce intracellular nutrient storage granules for later use, such as glycogen, polyphosphate, sulfur or polyhydroxyalkanoates. Certain bacterial species, such as the photosynthetic Cyanobacteria, produce internal gas vesicles, which they use to regulate their buoyancy – allowing them to move up or down into water layers with different light intensities and nutrient levels. *Intracellular membranes* called *chromatophores* are also found in membranes of phototrophic bacteria. Used primarily for photosynthesis, they contain bacteriochlorophyll pigments and carotenoids. An early idea was that bacteria might contain membrane folds termed mesosomes, but these were later shown to be artefacts produced by the chemicals used to prepare the cells for electron microscopy. *Inclusions* are considered to be nonliving components of the cell that do not possess metabolic activity and are not bounded by membranes. The most common inclusions are glycogen, lipid droplets, crystals, and pigments. *Volutin granules* are cytoplasmic inclusions of complexed inorganic polyphosphate. These granules are called *metachromatic granules* due to their displaying the metachromatic effect; they appear red or blue when stained with the blue dyes methylene blue or toluidine blue. *Gas vacuoles*, which are freely permeable to gas, are membrane-bound vesicles present in some species of *Cyanobacteria*. They allow the bacteria to control their buoyancy. *Microcompartments* are widespread, membrane-bound organelles that are made of a protein shell that surrounds and encloses various enzymes. *Carboxysomes* are bacterial microcompartments that contain enzymes involved in carbon fixation. *Magnetosomes* are bacterial microcompartments, present in magnetotactic bacteria, that contain magnetic crystals.

Extracellular Structures

In most bacteria, a *cell wall* is present on the outside of the cell membrane. The cell membrane and cell wall comprise the *cell envelope*. A common bacterial cell wall material is *peptidoglycan* (called "murein" in older sources), which is made from polysaccharide chains cross-linked by peptides containing D-amino acids. Bacterial cell walls are different from the cell walls of plants and fungi, which are made of cellulose and chitin, respectively. The cell wall of bacteria is also distinct from that of Archaea, which do not contain peptidoglycan. The cell wall is essential to the survival of many bacteria, and the antibiotic penicillin is able to kill bacteria by inhibiting a step in the synthesis of peptidoglycan.

There are broadly speaking two different types of cell wall in bacteria, a thick one in the gram-positives and a thinner one in the gram-negatives. The names originate from the reaction of cells to the Gram stain, a long-standing test for the classification of bacterial species.

Gram-positive bacteria possess a thick cell wall containing many layers of peptidoglycan and *teichoic acids*. In contrast, *gram-negative bacteria* have a relatively thin cell wall consisting of a few layers of peptidoglycan surrounded by a second lipid membrane containing *lipopolysaccharides* and lipoproteins. Lipopolysaccharides, also called *endotoxins*, are composed of polysaccharides and *lipid A* that is responsible for much of the toxicity of gram-negative bacteria. Most bacteria have the gram-negative cell wall, and only the Firmicutes and Actinobacteria have the alternative gram-positive arrangement. These two groups were previously known as the low G+C and high G+C gram-positive bacteria, respectively. These differences in structure can produce differences in antibiotic susceptibility; for instance, vancomycin can kill only gram-positive bacteria and is ineffective against gram-negative pathogens, such as *Haemophilus influenzae* or *Pseudomonas aeruginosa*. If the bacterial cell wall is entirely removed, it is called a *protoplast*, whereas if it is partially removed, it is called a *spheroplast*. *β*-Lactam antibiotics, such as penicillin, inhibit the formation of peptidoglycan cross-links in the bacterial cell wall. The enzyme lysozyme, found in human tears, also digests the cell wall of bacteria and is the body's main defence against eye infections.

Acid-fast bacteria, such as *Mycobacteria*, are resistant to decolorisation by acids during staining procedures. The high mycolic acid content of *Mycobacteria*, is responsible for the staining pattern of poor absorption followed by high retention. The most common staining technique used to identify acid-fast bacteria is the Ziehl-Neelsen stain or acid-fast stain, in which the acid-fast bacilli are stained bright-red and stand out clearly against a blue background. *L-form bacteria* are strains of bacteria that lack cell walls. The main pathogenic bacteria in this class is *Mycoplasma*

In many bacteria, an *S-layer* of rigidly arrayed protein molecules covers the outside of the cell. This layer provides chemical and physical protection for the cell surface and can act as a macromolecular diffusion barrier. S-layers have diverse but mostly poorly understood functions, but are known to act as virulence factors in *Campylobacter* and contain surface enzymes in *Bacillus stearothermophilus.*

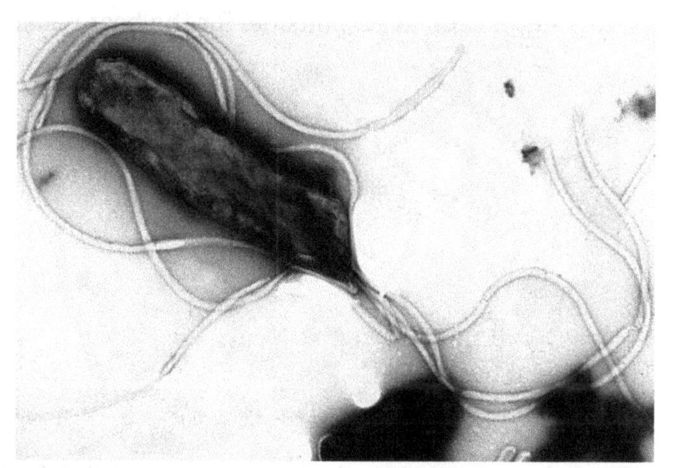

Helicobacter pylori electron micrograph, showing multiple flagella on the cell surface

Flagella are rigid protein structures, about 20 nanometres in diameter and up to 20 micrometres in length, that are used for motility. Flagella are driven by the energy released by the transfer of ions down an electrochemical gradient across the cell membrane.

Fimbriae (sometimes called "attachment pili") are fine filaments of protein, usually 2–10 nanometres in diameter and up to several micrometres in length. They are distributed over the surface of the cell, and resemble fine hairs when seen under the electron microscope. Fimbriae are believed to be involved in attachment to solid surfaces or to other cells, and are essential for the virulence of some bacterial pathogens. *Pili* (*sing.* pilus) are cellular appendages, slightly larger than fimbriae, that can transfer genetic material between bacterial cells in a process called conjugation where they are called *conjugation pili* or "sex pili". They can also generate movement where they are called *type IV pili.*

Glycocalyx are produced by many bacteria to surround their cells, and vary in structural complexity: ranging from a disorganised *slime layer* of extra-cellular polymer to a highly structured *capsule.* These structures can protect cells from engulfment by eukaryotic cells such as macrophages (part of the human immune system). They can also act as antigens and be involved in cell recognition, as well as aiding attachment to surfaces and the formation of biofilms.

The assembly of these extracellular structures is dependent on bacterial secretion systems. These transfer proteins from the cytoplasm into the periplasm or into the environment around the cell. Many types of secretion systems are known and these structures are often essential for the virulence of pathogens, so are intensively studied.

Endospores

Certain genera of gram-positive bacteria, such as *Bacillus, Clostridium, Sporohalobacter, Anaerobacter*, and *Heliobacterium*, can form highly resistant, dormant structures called *endospores*. In almost all cases, one endospore is formed and this is not a reproductive process, although *Anaerobacter* can make up to seven endospores in a single cell. Endospores have a central core of cytoplasm containing DNA and ribosomes surrounded by a cortex layer and protected by an impermeable and rigid coat. Dipicolinic acid is a chemical compound that composes 5% to 15% of the dry weight of bacterial spores. It is implicated as responsible for the heat resistance of the endospore.

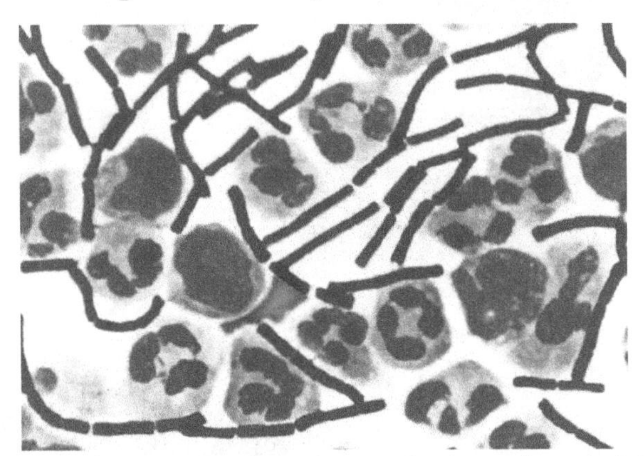

Bacillus anthracis (stained purple) growing in cerebrospinal fluid

Endospores show no detectable metabolism and can survive extreme physical and chemical stresses, such as high levels of UV light, gamma radiation, detergents, disinfectants, heat, freezing, pressure, and desiccation. In this dormant state, these organisms may remain viable for millions of years, and endospores even allow bacteria to survive exposure to the vacuum and radiation in space. According to scientist Dr. Steinn Sigurdsson, "There are viable bacterial spores that have been found that are 40 million years old on Earth — and we know they're very hardened to radiation." Endospore-forming bacteria can also cause disease: for example, anthrax can be contracted by the inhalation of *Bacillus anthracis* endospores, and contamination of deep puncture wounds with *Clostridium tetani* endospores causes tetanus.

Metabolism

Bacteria exhibit an extremely wide variety of metabolic types. The distribution of metabolic traits within a group of bacteria has traditionally been used to define their taxonomy, but these traits often do not correspond with modern genetic classifications. Bacterial metabolism is classified into nutritional groups on the basis of three major criteria: the kind of energy used for growth, the source of carbon, and the electron donors used for growth. An additional criterion of respiratory microorganisms are the electron acceptors used for aerobic or anaerobic respiration.

Nutritional types in bacterial metabolism

Nutritional type	Source of energy	Source of carbon	Examples
Phototrophs	Sunlight	Organic compounds (photoheterotrophs) or carbon fixation (photoautotrophs)	Cyanobacteria, Green sulfur bacteria, Chloroflexi, or Purple bacteria

Lithotrophs	Inorganic compounds	Organic compounds (lithoheterotrophs) or carbon fixation (lithoautotrophs)	Thermodesulfobacteria, *Hydrogenophilaceae*, or Nitrospirae
Organotrophs	Organic compounds	Organic compounds (chemoheterotrophs) or carbon fixation (chemoautotrophs)	*Bacillus*, *Clostridium* or *Enterobacteriaceae*

Carbon metabolism in bacteria is either *heterotrophic*, where organic carbon compounds are used as carbon sources, or *autotrophic*, meaning that cellular carbon is obtained by fixing carbon dioxide. Heterotrophic bacteria include parasitic types. Typical autotrophic bacteria are phototrophic cyanobacteria, green sulfur-bacteria and some purple bacteria, but also many chemolithotrophic species, such as nitrifying or sulfur-oxidising bacteria. Energy metabolism of bacteria is either based on *phototrophy*, the use of light through photosynthesis, or based on *chemotrophy*, the use of chemical substances for energy, which are mostly oxidised at the expense of oxygen or alternative electron acceptors (aerobic/anaerobic respiration).

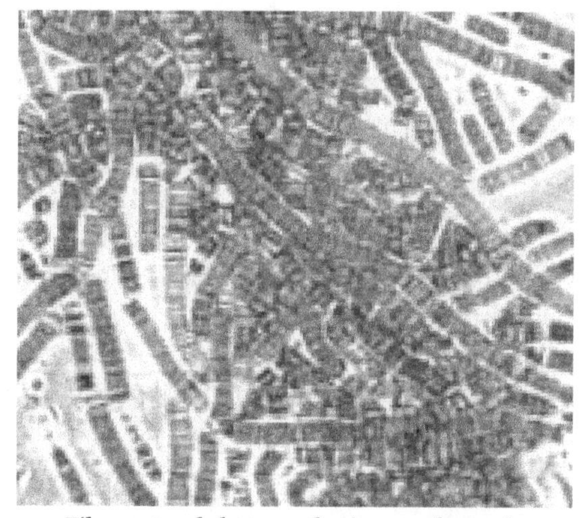

Filaments of photosynthetic cyanobacteria

Bacteria are further divided into *lithotrophs* that use inorganic electron donors and *organotrophs* that use organic compounds as electron donors. Chemotrophic organisms use the respective electron donors for energy conservation (by aerobic/anaerobic respiration or fermentation) and biosynthetic reactions (e.g., carbon dioxide fixation), whereas phototrophic organisms use them only for biosynthetic purposes. Respiratory organisms use chemical compounds as a source of energy by taking electrons from the reduced substrate and transferring them to a terminal electron acceptor in a redox reaction. This reaction releases energy that can be used to synthesise ATP and drive metabolism. In *aerobic organisms*, oxygen is used as the electron acceptor. In *anaerobic organisms* other inorganic compounds, such as nitrate, sulfate or carbon dioxide are used as electron acceptors. This leads to the ecologically important processes of denitrification, sulfate reduction, and acetogenesis, respectively.

Another way of life of chemotrophs in the absence of possible electron acceptors is fermentation, wherein the electrons taken from the reduced substrates are transferred to oxidised intermediates to generate reduced fermentation products (e.g., lactate, ethanol, hydrogen, butyric acid). Fermentation is possible, because the energy content of the substrates is higher than that of the products, which allows the organisms to synthesise ATP and drive their metabolism.

These processes are also important in biological responses to pollution; for example, sulfate-reducing bacteria are largely responsible for the production of the highly toxic forms of mercury (methyl- and dimethylmercury) in the environment. Non-respiratory anaerobes use fermentation to generate energy and reducing power, secreting metabolic by-products (such as ethanol in brewing) as waste. Facultative anaerobes can switch between fermentation and different terminal electron acceptors depending on the environmental conditions in which they find themselves.

Lithotrophic bacteria can use inorganic compounds as a source of energy. Common inorganic electron donors are hydrogen, carbon monoxide, ammonia (leading to nitrification), ferrous iron and other reduced metal ions, and several reduced sulfur compounds. In unusual circumstances, the gas methane can be used by methanotrophic bacteria as both a source of electrons and a substrate for carbon anabolism. In both aerobic phototrophy and chemolithotrophy, oxygen is used as a terminal electron acceptor, whereas under anaerobic conditions inorganic compounds are used instead. Most lithotrophic organisms are autotrophic, whereas organotrophic organisms are heterotrophic.

In addition to fixing carbon dioxide in photosynthesis, some bacteria also fix nitrogen gas (nitrogen fixation) using the enzyme nitrogenase. This environmentally important trait can be found in bacteria of nearly all the metabolic types listed above, but is not universal.

Regardless of the type of metabolic process they employ, the majority of bacteria are able to take in raw materials only in the form of relatively small molecules, which enter the cell by diffusion or through molecular channels in cell membranes. The Planctomycetes are the exception (as they are in possessing membranes around their nuclear material). It has recently been shown that *Gemmata obscuriglobus* is able to take in large molecules via a process that in some ways resembles endocytosis, the process used by eukaryotic cells to engulf external items.

Growth and Reproduction

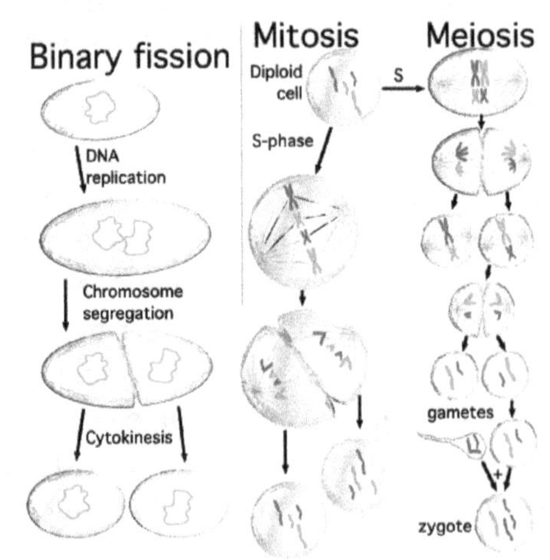

Many bacteria reproduce through binary fission, which is compared to mitosis and meiosis in this image.

Unlike in multicellular organisms, increases in cell size (cell growth) and reproduction by cell division are tightly linked in unicellular organisms. Bacteria grow to a fixed size and then reproduce through *binary fission*, a form of asexual reproduction. Under optimal conditions, bacteria can grow and divide extremely rapidly, and bacterial populations can double as quickly as every 9.8 minutes. In cell division, two identical clone daughter cells are produced. Some bacteria, while still reproducing asexually, form more complex reproductive structures that help disperse the newly formed daughter cells. Examples include fruiting body formation by *Myxobacteria* and aerial hyphae formation by *Streptomyces*, or budding. Budding involves a cell forming a protrusion that breaks away and produces a daughter cell.

A colony of *Escherichia coli*

In the laboratory, bacteria are usually grown using solid or liquid media. Solid *growth media*, such as agar plates, are used to isolate pure cultures of a bacterial strain. However, liquid growth media are used when measurement of growth or large volumes of cells are required. Growth in stirred liquid media occurs as an even cell suspension, making the cultures easy to divide and transfer, although isolating single bacteria from liquid media is difficult. The use of selective media (media with specific nutrients added or deficient, or with antibiotics added) can help identify specific organisms.

Most laboratory techniques for growing bacteria use high levels of nutrients to produce large amounts of cells cheaply and quickly. However, in natural environments, nutrients are limited, meaning that bacteria cannot continue to reproduce indefinitely. This nutrient limitation has led the evolution of different growth strategies. Some organisms can grow extremely rapidly when nutrients become available, such as the formation of algal (and cyanobacterial) blooms that often occur in lakes during the summer. Other organisms have adaptations to harsh environments, such as the production of multiple antibiotics by *Streptomyces* that inhibit the growth of competing microorganisms. In nature, many organisms live in communities (e.g., biofilms) that may allow for increased supply of nutrients and protection from environmental stresses. These relationships can be essential for growth of a particular organism or group of organisms (syntrophy).

Bacterial growth follows four phases. When a population of bacteria first enter a high-nutrient environment that allows growth, the cells need to adapt to their new environment. The first phase of growth is the *lag phase*, a period of slow growth when the cells are adapting to the high-nutrient environment and preparing for fast growth. The lag phase has high biosynthesis rates, as proteins necessary for rapid growth are produced. The second phase of growth is the *log phase*, also known as the *logarithmic or exponential phase*. The log phase is marked by rapid exponential growth. The rate at which cells grow during this phase is known as the *growth rate* (*k*), and the time it takes the cells to double is known as the *generation time* (*g*). During log phase, nutrients are metabolised at maximum speed until one of the nutrients is depleted and starts limiting growth.

The third phase of growth is the *stationary phase* and is caused by depleted nutrients. The cells reduce their metabolic activity and consume non-essential cellular proteins. The stationary phase is a transition from rapid growth to a stress response state and there is increased expression of genes involved in DNA repair, antioxidant metabolism and nutrient transport. The final phase is the *death phase* where the bacteria run out of nutrients and die.

Genomes

The genomes of thousands of bacterial species have been sequenced, with at least 9,000 sequences completed and more than 42,000 left as "permanent" drafts (as of Sep 2016).

Most bacteria have a single circular chromosome that can range in size from only 160,000 base pairs in the endosymbiotic bacteria *Candidatus Carsonella ruddii*, to 12,200,000 base pairs in the soil-dwelling bacteria *Sorangium cellulosum*. The genes in bacterial genomes are usually a single continuous stretch of DNA and although several different types of introns do exist in bacteria, these are much rarer than in eukaryotes. Some bacteria, including the Spirochaetes of the genus *Borrelia* are a notable exception to this arrangement. *Borrelia burgdorferi*, the cause of Lyme disease, contains a single linear chromosome and several linear and circular plasmids.

Plasmids are small extra-chromosomal DNAs that may contain genes for antibiotic resistance or virulence factors. Plasmids replicate independently of chromosomes, so it is possible that plasmids could be lost in bacterial cell division. Against this possibility is the fact that a single bacterium can contain hundreds of copies of a single plasmid.

Genetics

Bacteria, as asexual organisms, inherit identical copies of their parent's genes (i.e., they are clonal). However, all bacteria can evolve by selection on changes to their genetic material DNA caused by genetic recombination or mutations. Mutations come from errors made during the replication of DNA or from exposure to mutagens. Mutation rates vary widely among different species of bacteria and even among different clones of a single species of bacteria. Genetic changes in bacterial genomes come from either random mutation during replication or "stress-directed mutation", where genes involved in a particular growth-limiting process have an increased mutation rate.

DNA Transfer

Some bacteria also transfer genetic material between cells. This can occur in three main ways. First, bacteria can take up exogenous DNA from their environment, in a process called *transformation*. Genes can also be transferred by the process of *transduction*, when the integration of a bacteriophage introduces foreign DNA into the chromosome. The third method of gene transfer is *conjugation*, whereby DNA is transferred through direct cell contact.

Transduction of bacterial genes by bacteriophage appears to be a consequence of infrequent errors during intracellular assembly of virus particles, rather than a bacterial adaptation. Conjugation, in the much-studied E. coli system is determined by plasmid genes, and is an adaptation for transferring copies of the plasmid from one bacterial host to another. It is seldom that a conjugative plasmid integrates into the host bacterial chromosome, and subsequently transfers part of the host bacterial DNA to another bacterium. Plasmid-mediated transfer of host bacterial DNA also appears to be an accidental process rather than a bacterial adaptation.

Transformation, unlike transduction or conjugation, depends on numerous bacterial gene products that specifically interact to perform this complex process, and thus transformation is clearly a bacterial adaptation for DNA transfer. In order for a bacterium to bind, take up and recombine donor DNA into its own chromosome, it must first enter a special physiological state termed competence. In *Bacillus subtilis*, about 40 genes are required for the development of competence. The length of DNA transferred during *B. subtilis* transformation can be between a third of a chromosome up to the whole chromosome. Transformation appears to be common among bacterial species, and thus far at least 60 species are known to have the natural ability to become competent for transformation. The development of competence in nature is usually associated with stressful environmental conditions, and seems to be an adaptation for facilitating repair of DNA damage in recipient cells.

In ordinary circumstances, transduction, conjugation, and transformation involve transfer of DNA between individual bacteria of the same species, but occasionally transfer may occur between individuals of different bacterial species and this may have significant consequences, such as the transfer of antibiotic resistance. In such cases, gene acquisition from other bacteria or the environment is called *horizontal gene transfer* and may be common under natural conditions. Gene transfer is particularly important in antibiotic resistance as it allows the rapid transfer of resistance genes between different pathogens.

Bacteriophages

Bacteriophages are viruses that infect bacteria. Many types of bacteriophage exist, some simply infect and lyse their host bacteria, while others insert into the bacterial chromosome. A bacteriophage can contain genes that contribute to its host's phenotype: for example, in the evolution of *Escherichia coli* O157:H7 and *Clostridium botulinum*, the toxin genes in an integrated phage converted a harmless ancestral bacterium into a lethal pathogen. Bacteria resist phage infection through restriction modification systems that degrade foreign DNA, and a system that uses CRISPR sequences to retain fragments of the genomes of phage that the bacteria have come into contact with in the past, which allows them to block virus replication through a form of RNA interference. This CRISPR system provides bacteria with acquired immunity to infection.

Behaviour

Secretion

Bacteria frequently secrete chemicals into their environment in order to modify it favourably. The secretions are often proteins and may act as enzymes that digest some form of food in the environment.

Bioluminescence

A few bacteria have chemical systems that generate light. This bioluminescence often occurs in bacteria that live in association with fish, and the light probably serves to attract fish or other large animals.

Multicellularity

Bacteria often function as multicellular aggregates known as biofilms, exchanging a variety of molecular signals for inter-cell communication, and engaging in coordinated multicellular behaviour.

The communal benefits of multicellular cooperation include a cellular division of labour, accessing resources that cannot effectively be used by single cells, collectively defending against antagonists, and optimising population survival by differentiating into distinct cell types. For example, bacteria in biofilms can have more than 500 times increased resistance to antibacterial agents than individual "planktonic" bacteria of the same species.

One type of inter-cellular communication by a molecular signal is called quorum sensing, which serves the purpose of determining whether there is a local population density that is sufficiently high that it is productive to invest in processes that are only successful if large numbers of similar organisms behave similarly, as in excreting digestive enzymes or emitting light.

Quorum sensing allows bacteria to coordinate gene expression, and enables them to produce, release and detect autoinducers or pheromones which accumulate with the growth in cell population.

Movement

Many bacteria can move using a variety of mechanisms: flagella are used for swimming through fluids; bacterial gliding and twitching motility move bacteria across surfaces; and changes of buoyancy allow vertical motion.

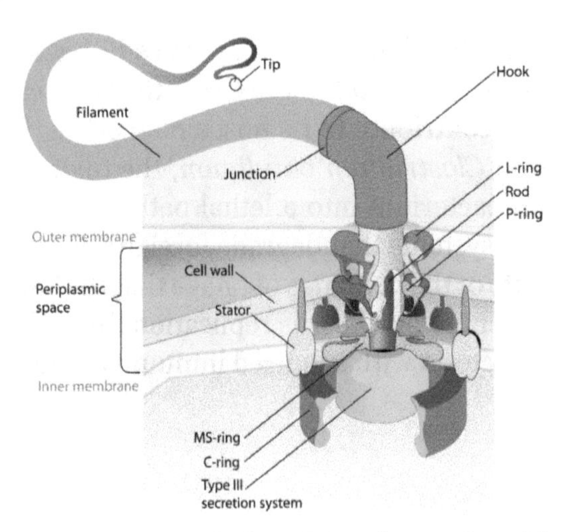

Flagellum of gram-negative bacteria. The base drives the rotation of the hook and filament.

Swimming bacteria frequently move near 10 body lengths per second and a few as fast as 100. This makes them at least as fast as fish, on a relative scale.

In bacterial gliding and twitching motility, bacteria use their *type IV pili* as a grappling hook, repeatedly extending it, anchoring it and then retracting it with remarkable force (>80 pN).

"Our observations redefine twitching motility as a rapid, highly organized mechanism of bacterial translocation by which Pseudomonas aeruginosa can disperse itself over large areas to colonize new

territories. It is also now clear, both morphologically and genetically, that twitching motility and social gliding motility, such as occurs in Myxococcus xanthus, are essentially the same process."

— "A re-examination of twitching motility in Pseudomonas aeruginosa" – Semmler, Whitchurch & Mattick (1999)

Flagella are semi-rigid cylindrical structures that are rotated and function much like the propeller on a ship. Objects as small as bacteria operate a low Reynolds number and cylindrical forms are more efficient than the flat, paddle-like, forms appropriate at human-size scale.

Bacterial species differ in the number and arrangement of flagella on their surface; some have a single flagellum (*monotrichous*), a flagellum at each end (*amphitrichous*), clusters of flagella at the poles of the cell (*lophotrichous*), while others have flagella distributed over the entire surface of the cell (*peritrichous*). The bacterial flagella is the best-understood motility structure in any organism and is made of about 20 proteins, with approximately another 30 proteins required for its regulation and assembly. The flagellum is a rotating structure driven by a reversible motor at the base that uses the electrochemical gradient across the membrane for power. This motor drives the motion of the filament, which acts as a propeller.

Many bacteria (such as *E. coli*) have two distinct modes of movement: forward movement (swimming) and tumbling. The tumbling allows them to reorient and makes their movement a three-dimensional random walk. The flagella of a unique group of bacteria, the spirochaetes, are found between two membranes in the periplasmic space. They have a distinctive helical body that twists about as it moves.

Motile bacteria are attracted or repelled by certain stimuli in behaviours called taxes: these include chemotaxis, phototaxis, energy taxis, and magnetotaxis. In one peculiar group, the myxobacteria, individual bacteria move together to form waves of cells that then differentiate to form fruiting bodies containing spores. The myxobacteria move only when on solid surfaces, unlike *E. coli*, which is motile in liquid or solid media.

Several *Listeria* and *Shigella* species move inside host cells by usurping the cytoskeleton, which is normally used to move organelles inside the cell. By promoting actin polymerisation at one pole of their cells, they can form a kind of tail that pushes them through the host cell's cytoplasm.

Classification and Identification

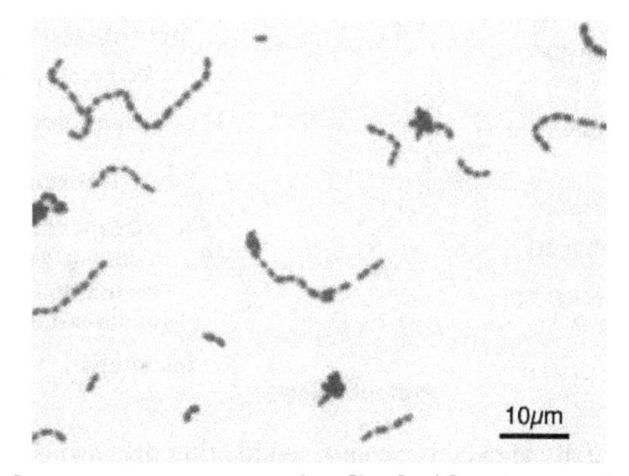

Streptococcus mutans visualised with a Gram stain

Classification seeks to describe the diversity of bacterial species by naming and grouping organisms based on similarities. Bacteria can be classified on the basis of cell structure, cellular metabolism or on differences in cell components, such as DNA, fatty acids, pigments, antigens and quinones. While these schemes allowed the identification and classification of bacterial strains, it was unclear whether these differences represented variation between distinct species or between strains of the same species. This uncertainty was due to the lack of distinctive structures in most bacteria, as well as lateral gene transfer between unrelated species. Due to lateral gene transfer, some closely related bacteria can have very different morphologies and metabolisms. To overcome this uncertainty, modern bacterial classification emphasises molecular systematics, using genetic techniques such as guanine cytosine ratio determination, genome-genome hybridisation, as well as sequencing genes that have not undergone extensive lateral gene transfer, such as the rRNA gene. Classification of bacteria is determined by publication in the International Journal of Systematic Bacteriology, and Bergey's Manual of Systematic Bacteriology. The International Committee on Systematic Bacteriology (ICSB) maintains international rules for the naming of bacteria and taxonomic categories and for the ranking of them in the International Code of Nomenclature of Bacteria.

The term "bacteria" was traditionally applied to all microscopic, single-cell prokaryotes. However, molecular systematics showed prokaryotic life to consist of two separate domains, originally called *Eubacteria* and *Archaebacteria*, but now called *Bacteria* and *Archaea* that evolved independently from an ancient common ancestor. The archaea and eukaryotes are more closely related to each other than either is to the bacteria. These two domains, along with Eukarya, are the basis of the three-domain system, which is currently the most widely used classification system in microbiolology. However, due to the relatively recent introduction of molecular systematics and a rapid increase in the number of genome sequences that are available, bacterial classification remains a changing and expanding field. For example, a few biologists argue that the Archaea and Eukaryotes evolved from gram-positive bacteria.

The identification of bacteria in the laboratory is particularly relevant in medicine, where the correct treatment is determined by the bacterial species causing an infection. Consequently, the need to identify human pathogens was a major impetus for the development of techniques to identify bacteria.

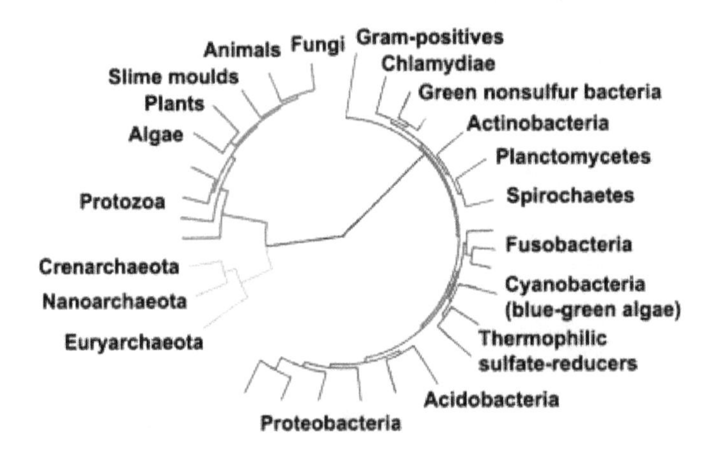

Phylogenetic tree showing the diversity of bacteria, compared to other organisms. Eukaryotes are coloured red, archaea green and bacteria blue.

The *Gram stain*, developed in 1884 by Hans Christian Gram, characterises bacteria based on the structural characteristics of their cell walls. The thick layers of peptidoglycan in the "gram-positive" cell wall stain purple, while the thin "gram-negative" cell wall appears pink. By combining morphology and Gram-staining, most bacteria can be classified as belonging to one of four groups (gram-positive cocci, gram-positive bacilli, gram-negative cocci and gram-negative bacilli). Some organisms are best identified by stains other than the Gram stain, particularly mycobacteria or *Nocardia*, which show acid-fastness on Ziehl–Neelsen or similar stains. Other organisms may need to be identified by their growth in special media, or by other techniques, such as serology.

Culture techniques are designed to promote the growth and identify particular bacteria, while restricting the growth of the other bacteria in the sample. Often these techniques are designed for specific specimens; for example, a sputum sample will be treated to identify organisms that cause pneumonia, while stool specimens are cultured on selective media to identify organisms that cause diarrhoea, while preventing growth of non-pathogenic bacteria. Specimens that are normally sterile, such as blood, urine or spinal fluid, are cultured under conditions designed to grow all possible organisms. Once a pathogenic organism has been isolated, it can be further characterised by its morphology, growth patterns (such as aerobic or anaerobic growth), patterns of hemolysis, and staining.

As with bacterial classification, identification of bacteria is increasingly using molecular methods. Diagnostics using DNA-based tools, such as polymerase chain reaction, are increasingly popular due to their specificity and speed, compared to culture-based methods. These methods also allow the detection and identification of "viable but nonculturable" cells that are metabolically active but non-dividing. However, even using these improved methods, the total number of bacterial species is not known and cannot even be estimated with any certainty. Following present classification, there are a little less than 9,300 known species of prokaryotes, which includes bacteria and archaea; but attempts to estimate the true number of bacterial diversity have ranged from 10^7 to 10^9 total species – and even these diverse estimates may be off by many orders of magnitude.

Interactions with Other Organisms

Despite their apparent simplicity, bacteria can form complex associations with other organisms. These symbiotic associations can be divided into parasitism, mutualism and commensalism. Due to their small size, commensal bacteria are ubiquitous and grow on animals and plants exactly as they will grow on any other surface. However, their growth can be increased by warmth and sweat, and large populations of these organisms in humans are the cause of body odour.

Predators

Some species of bacteria kill and then consume other microorganisms, these species are called *predatory bacteria*. These include organisms such as *Myxococcus xanthus*, which forms swarms of cells that kill and digest any bacteria they encounter. Other bacterial predators either attach to their prey in order to digest them and absorb nutrients, such as *Vampirovibrio chlorellavorus*, or invade another cell and multiply inside the cytosol, such as *Daptobacter*. These predatory bacteria are thought to have evolved from saprophages that consumed dead microorganisms, through adaptations that allowed them to entrap and kill other organisms.

Mutualists

Certain bacteria form close spatial associations that are essential for their survival. One such mutualistic association, called interspecies hydrogen transfer, occurs between clusters of anaerobic bacteria that consume organic acids, such as butyric acid or propionic acid, and produce hydrogen, and methanogenic Archaea that consume hydrogen. The bacteria in this association are unable to consume the organic acids as this reaction produces hydrogen that accumulates in their surroundings. Only the intimate association with the hydrogen-consuming Archaea keeps the hydrogen concentration low enough to allow the bacteria to grow.

In soil, microorganisms that reside in the rhizosphere (a zone that includes the root surface and the soil that adheres to the root after gentle shaking) carry out nitrogen fixation, converting nitrogen gas to nitrogenous compounds. This serves to provide an easily absorbable form of nitrogen for many plants, which cannot fix nitrogen themselves. Many other bacteria are found as symbionts in humans and other organisms. For example, the presence of over 1,000 bacterial species in the normal human gut flora of the intestines can contribute to gut immunity, synthesise vitamins, such as folic acid, vitamin K and biotin, convert sugars to lactic acid, as well as fermenting complex undigestible carbohydrates. The presence of this gut flora also inhibits the growth of potentially pathogenic bacteria (usually through competitive exclusion) and these beneficial bacteria are consequently sold as probiotic dietary supplements.

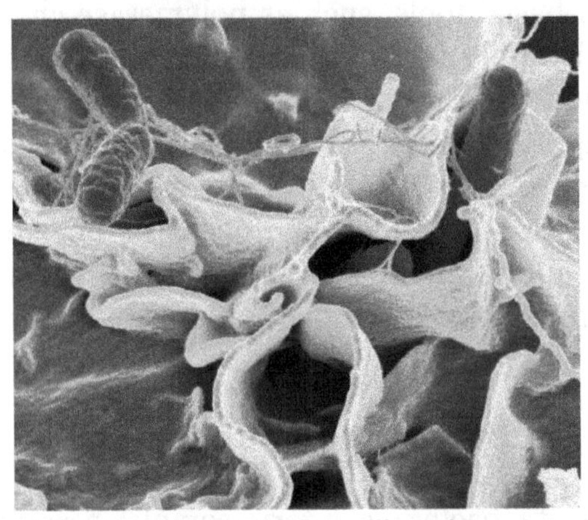

Colour-enhanced scanning electron micrograph showing Salmonella typhimurium (red) invading cultured human cells

Pathogens

If bacteria form a parasitic association with other organisms, they are classed as pathogens. Pathogenic bacteria are a major cause of human death and disease and cause infections such as tetanus, typhoid fever, diphtheria, syphilis, cholera, foodborne illness, leprosy and tuberculosis. A pathogenic cause for a known medical disease may only be discovered many years after, as was the case with *Helicobacter pylori* and peptic ulcer disease. Bacterial diseases are also important in agriculture, with bacteria causing leaf spot, fire blight and wilts in plants, as well as Johne's disease, mastitis, salmonella and anthrax in farm animals.

Each species of pathogen has a characteristic spectrum of interactions with its human hosts. Some organisms, such as *Staphylococcus* or *Streptococcus*, can cause skin infections, pneumonia, meningitis and even overwhelming sepsis, a systemic inflammatory response producing shock, massive vasodilation and death. Yet these organisms are also part of the normal human flora and usually exist on the skin or in the nose without causing any disease at all. Other organisms invariably cause disease in humans, such as the Rickettsia, which are obligate intracellular parasites able to grow and reproduce only within the cells of other organisms. One species of Rickettsia causes typhus, while another causes Rocky Mountain spotted fever. *Chlamydia*, another phylum of obligate intracellular parasites, contains species that can cause pneumonia, or urinary tract infection and may be involved in coronary heart disease. Finally, some species, such as *Pseudomonas aeruginosa*, *Burkholderia cenocepacia*, and *Mycobacterium avium*, are opportunistic pathogens and cause disease mainly in people suffering from immunosuppression or cystic fibrosis.

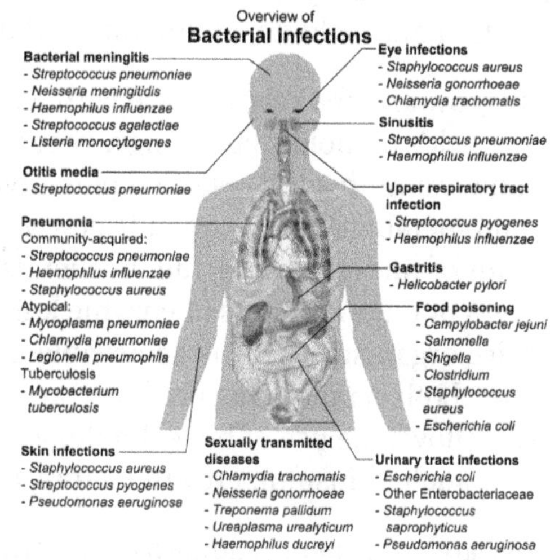

Overview of bacterial infections and main species involved.

Bacterial infections may be treated with antibiotics, which are classified as bacteriocidal if they kill bacteria, or bacteriostatic if they just prevent bacterial growth. There are many types of antibiotics and each class inhibits a process that is different in the pathogen from that found in the host. An example of how antibiotics produce selective toxicity are chloramphenicol and puromycin, which inhibit the bacterial ribosome, but not the structurally different eukaryotic ribosome. Antibiotics are used both in treating human disease and in intensive farming to promote animal growth, where they may be contributing to the rapid development of antibiotic resistance in bacterial populations. Infections can be prevented by antiseptic measures such as sterilising the skin prior to piercing it with the needle of a syringe, and by proper care of indwelling catheters. Surgical and dental instruments are also sterilised to prevent contamination by bacteria. Disinfectants such as bleach are used to kill bacteria or other pathogens on surfaces to prevent contamination and further reduce the risk of infection.

Significance in Technology and Industry

Bacteria, often lactic acid bacteria, such as *Lactobacillus* and *Lactococcus*, in combination with yeasts and moulds, have been used for thousands of years in the preparation of fermented foods, such as cheese, pickles, soy sauce, sauerkraut, vinegar, wine and yogurt.

The ability of bacteria to degrade a variety of organic compounds is remarkable and has been used in waste processing and bioremediation. Bacteria capable of digesting the hydrocarbons in petroleum are often used to clean up oil spills. Fertiliser was added to some of the beaches in Prince

William Sound in an attempt to promote the growth of these naturally occurring bacteria after the 1989 *Exxon Valdez* oil spill. These efforts were effective on beaches that were not too thickly covered in oil. Bacteria are also used for the bioremediation of industrial toxic wastes. In the chemical industry, bacteria are most important in the production of enantiomerically pure chemicals for use as pharmaceuticals or agrichemicals.

Bacteria can also be used in the place of pesticides in the biological pest control. This commonly involves *Bacillus thuringiensis* (also called BT), a gram-positive, soil dwelling bacterium. Subspecies of this bacteria are used as a Lepidopteran-specific insecticides under trade names such as Dipel and Thuricide. Because of their specificity, these pesticides are regarded as environmentally friendly, with little or no effect on humans, wildlife, pollinators and most other beneficial insects.

Because of their ability to quickly grow and the relative ease with which they can be manipulated, bacteria are the workhorses for the fields of molecular biology, genetics and biochemistry. By making mutations in bacterial DNA and examining the resulting phenotypes, scientists can determine the function of genes, enzymes and metabolic pathways in bacteria, then apply this knowledge to more complex organisms. This aim of understanding the biochemistry of a cell reaches its most complex expression in the synthesis of huge amounts of enzyme kinetic and gene expression data into mathematical models of entire organisms. This is achievable in some well-studied bacteria, with models of *Escherichia coli* metabolism now being produced and tested. This understanding of bacterial metabolism and genetics allows the use of biotechnology to bioengineer bacteria for the production of therapeutic proteins, such as insulin, growth factors, or antibodies.

Because of their importance for research in general, samples of bacterial strains are isolated and preserved in Biological Resource Centers. This ensures the availability of the strain to scientists worldwide.

History of Bacteriology

Antonie van Leeuwenhoek, the first microbiologist and the first person to observe bacteria using a microscope.

Bacteria were first observed by the Dutch microscopist Antonie van Leeuwenhoek in 1676, using a single-lens microscope of his own design. He then published his observations in a series of letters to the Royal Society of London. Bacteria were Leeuwenhoek's most remarkable microscopic discovery. They were just at the limit of what his simple lenses could make out and, in one of the most striking hiatuses in the history of science, no one else would see them again for over a century.

Only then were his by-then-largely-forgotten observations of bacteria — as opposed to his famous "animalcules" (spermatozoa) — taken seriously.

Christian Gottfried Ehrenberg introduced the word "bacterium" in 1828. In fact, his *Bacterium* was a genus that contained non-spore-forming rod-shaped bacteria, as opposed to *Bacillus*, a genus of spore-forming rod-shaped bacteria defined by Ehrenberg in 1835.

Louis Pasteur demonstrated in 1859 that the growth of microorganisms causes the fermentation process, and that this growth is not due to spontaneous generation. (Yeasts and moulds, commonly associated with fermentation, are not bacteria, but rather fungi.) Along with his contemporary Robert Koch, Pasteur was an early advocate of the germ theory of disease.

Robert Koch, a pioneer in medical microbiology, worked on cholera, anthrax and tuberculosis. In his research into tuberculosis Koch finally proved the germ theory, for which he received a Nobel Prize in 1905. In *Koch's postulates*, he set out criteria to test if an organism is the cause of a disease, and these postulates are still used today.

Though it was known in the nineteenth century that bacteria are the cause of many diseases, no effective antibacterial treatments were available. In 1910, Paul Ehrlich developed the first antibiotic, by changing dyes that selectively stained *Treponema pallidum* — the spirochaete that causes syphilis — into compounds that selectively killed the pathogen. Ehrlich had been awarded a 1908 Nobel Prize for his work on immunology, and pioneered the use of stains to detect and identify bacteria, with his work being the basis of the Gram stain and the Ziehl–Neelsen stain.

A major step forward in the study of bacteria came in 1977 when Carl Woese recognised that archaea have a separate line of evolutionary descent from bacteria. This new phylogenetic taxonomy depended on the sequencing of 16S ribosomal RNA, and divided prokaryotes into two evolutionary domains, as part of the three-domain system.

Beet Vascular Necrosis

A table beet infected with Pectobacterium carotovorum subsp. betavasculorum. Note the rings of black vascular tissue colonized by the rotting bacteria.

Beet vascular necrosis and rot is a soft rot disease caused by the bacterium Pectobacterium carotovorum subsp. betavasculorum, which has also been known as *Pectobacterium betavasculorum* and *Erwinia carotovora* subsp. *betavasculorum*. It was classified in the genus *Erwinia* until genetic evidence suggested that it belongs to its own group; however, the name Erwinia is still in use. As such, the disease is sometimes called Erwinia rot today. It is a very destructive disease that has been reported across the United States as well as in Egypt. Symptoms include wilting and black streaks on the leaves and petioles. It is usually not fatal to the plant, but in severe cases the beets will become hollowed and unmarketable. The bacteria is a generalist species which rots beets and other plants by secreting digestive enzymes that break down the cell wall and parenchyma tissues. The bacteria thrive in warm and wet conditions, but cannot survive long in fallow soil. However, it is able to persist for long periods of time in the rhizosphere of weeds and non-host crops. While it is difficult to eradicate, there are cultural practices that can be used to control the spread of the disease, such as avoiding injury to the plants and reducing or eliminating application of nitrogen fertilizer.

Hosts

Fodder beets, sugar beets and fodder-sugar crosses are all susceptible to infection by *Pectobacterium carotovorum* subsp. *betavasculorum*. Today most beet cultivars are resistant to the pathogen, however, isolates vary geographically, and some cultivars of beets are only resistant to specific isolates of bacteria. For example, the cultivar USH11 demonstrates resistance to both Montana and California isolates, whereas Beta 4430 is highly susceptible to the Montana isolates but resistant to the California isolate. Other cultivars resistant to California isolates of *Pectobacterium caratovorum* subsp. *betavasculorum* include Beta 4776R, Beta 4430R and Beta 4035R, but HH50 has been found to be susceptible.

Breeding for resistance to other diseases such as beet yellows virus without also selecting for vascular necrosis resistance can leave cultivars susceptible to the pathogen. For example, the use of USH9A and H9B in California's San Joaquin valley is thought to have led to an epiphytotic (severe) outbreak of disease in the early 1970s. This was likely because of the limited gene pool used when selecting strongly for resistance to beet yellows virus.

In addition to beets, *Pectobacterium carotovara* subsp. *betavasculorum* can also infect tomato, potato, carrots, sweet potato, radish, sunflower, artichokes, squash, cucumber and chrysanthemum. Other subspecies of *Pectobacterium carotovora* can also be pathogenic to beets. *Erwinia carotovara* subsp. *atroseptica* is a bacterial soft rot pathogen that is responsible for the disease Blackleg of Potato (*Solanum tuberosum*), and variants of this bacterium can cause root rot in sugarbeets,. This subspecies also has a wide host-range. *Erwinia carotovora* var. *atroseptica* has been detected in the rhizosphere of native vegetation and on weed species such as *Lupinus blumerii* and *Amaranthus palmeri* (pigweed). It is thought that the source of inoculums survives on these non-host plants in areas in which it is endemic as well as in the rhizosphere of other crops such as wheat and corn.

Symptoms

Symptoms can be found on both beet roots and foliage, although foliar symptoms are not always present. If present, foliar symptoms include dark streaking along petioles and viscous froth deposits on the crown which are a by-product of bacterial metabolism. Petioles can also become necrotic and demonstrate vascular necrosis. When roots become severely affected, wilting also occurs. Below ground symptoms include both soft and dry root rot. Affected vascular bundles in roots become necrotic and brown, and tissue adjacent to necrosis becomes pink upon air contact. The plants that do not die completely may have rotted-out, cavernous roots.

Table beet stem infected with Pectobacterium carotovorum subsp. betavasculorum. Note entry through a wound.

Various pathogens can cause root rot in beets; however the black streaking on petioles and necrotic vascular bundles in roots and adjacent pink tissue help to distinguish this disease from others such as Fusarium Yellows. Additionally, sampling from the rhizosphere of infected plants and serological tests can confirm the presence of *Erwinia caratovora* subs.

Disease Cycle

Pectobacterium carotovorum subsp. *betavasculorum* is a gram negative, rod bacteria with peritichous flagella. For it to enter sugar beet, and thus cause infection, it is essential that there is an injury to the leaves, petioles or crown. Infection will often start at the crown and then move down into the root, and can occur at any point in the growing season if environmental conditions are favorable. Once the bacteria enters the plant, it will invade the vascular tissue and cause symptoms by producing plant cell wall degrading enzymes, like pectinases, polygalactronases, and celluases. This results in discolored or necrotic vascular tissue in the root, and the tissue bordering the vascular bundles will turn reddish upon contact with air. Following the infection of the vascular tissue, the bacteria reproduce as long as food resources are available, and the root begins to rot. There is significant variability in the type of rot – it can range from a dry rot to soft and wet rot – because of the multitude of additional microorganisms that may colonize the damaged tissue

Upon death of the sugar beet, or harvest of the field, the pathogen appears to survive in select living plant tissue like beet roots, or volunteer beets. However, it does not appear to survive in sugar beet seeds, or live in the soil after harvest. It is also possible for the pathogen to infect injured carrots, potato, sweet potato, tomato, radish, sunflower, artichokes, squash, cucumber and chrysanthemums; however, since those are often planted in the same season as sugar beets, they are not likely to be overwintering hosts.

Environment

Injury to the leaves, petioles or crown is mandatory for the pathogen to gain entry to the host tissue. Accordingly, hail damage is correlated with a higher degree of disease outbreak. Young plants (less than eight weeks old) are also considered to be more prone to infection

Temperature and availability of moisture are key factors in determining the rate of disease development. Warm temperatures, 25-30 °C, promote rapid disease development., and can result in acute symptoms. Symptoms are also reported to appear at temperatures as low as 18 °C, but disease development is slowed; below that temperature, infections do not develop. Excessive water also promotes disease development by providing a more optimal environment for the pathogen, and has been shown to be a key factor in augmenting disease outbreak in fields with sprinkler irrigation

Agricultural

The degree of nitrogen fertilization is highly correlated to robust disease development: it has been shown that sugar beets supplied with excessive or adequate nitrogen are more diseased than sugarbeets with sub-optimal nitrogen levels. This is a paradox for farmers because, while increased nitrogen fertilization does increase sugar yield in non-infected sugarbeets, it also increases the severity of the disease if infection takes place. Thus, depending on the severity of infection, yield may go down with increased fertilizer use.

The spacing between plants also impacts the degree of infection: greater in-row spacing results in more diseased roots. This may be due to the fact that greater spacing promotes faster growth, and hence greater probability of cracks in the crown, or because of the increased amount of nitrogen available per plant.

Since the pathogen has multiple hosts, it is important for farmers to be wary of other plants in the surrounding area. It is possible for the pathogen to survive in weedy hosts, and can infect injured carrots, potato, sweet potato, tomato, radish, squash, and cucumber. Hence, the presence of these plants may increase the supply of inoculum.

Laboratory

If the pathogen is cultured in a lab, it can grow on Miller and Schroth media, can use sucrose to make reducing sugars, and can use either lactose, methyl alpha-glucoside, inulin or raffinose to make acids. It is also capable of surviving in culture medium sodium levels of up to 7–9%, and in temperatures as high as 39 °C.

Management

Since the bacteria cannot survive in seeds, the best way to prevent the disease is to ensure that vegetatively propagated plant material are clean of infection, such that the bacterium does not enter the soil. However, if the bacteria is already present, there are some methods that can be used to lessen the infection.

Cultural Practices

Because the bacteria readily enter the plant through wounds, management practices that decrease injury to the plants are important to control the spread of the disease. Cultivation is not recommended, as the machinery can become contaminated and physically spread the bacteria around the soil. Accidental leaf tearing or root scarring can also occur depending on the size of the crop, allowing the bacteria to enter more individual plants. If hilling the beets, great care must be taken to avoid getting soil into the crown, because the pathogen is soil-borne and this could expose the plant to more bacteria, thus increasing the risk of infection.

While most bacteria are motile and can swim, they cannot move very far due to their small size. However, they can be carried along by water, and a significant movement of *Pectobacterium* can be attributed to being carried downstream from irrigation and rainwater. To control the spread of the disease, limiting irrigation is another strategy. The bacteria also flourishes in wet conditions, so limiting excess water can control both the spread and severity of the disease.

Increased in-row spacing also causes more severe disease. In an infected field, yield decreased linearly when spacing was greater than 15 cm (6 in), so a spacing of 6 inches or less is recommended.

The bacteria can also utilize nitrogen fertilizer to accelerate their growth, thus limiting or eliminating the amount of nitrogen fertilizer applied will lessen the disease severity. For example, when fertilizer was applied to an infected field the infection rate per root increased from 11% (with no added nitrogen) to 36% (with 336 kg nitrogen/hectare), and sugar yields decreased.

Cultivar	Resistance
H9	No
H10	No
C17	No
546 H3	Moderate
C13	No
E540	No
E538	No
E534	Moderate
E502	Moderate
E506	Yes
E536	Yes
C930-35	Moderate
C927-4	Moderate
C930-19	Yes
C929-62	Yes

Resistance

The bacteria can survive in the rhizosphere of other crops such as tomato, carrots, sweet potato, radish, and squash as well as weed plants like lupin and pigweed, so it is very hard to get rid of it completely. When it is known that the bacterium is present in the soil, planting resistant varieties can be the best defense against the disease. Many available beet cultivars are resistant to *Pectobacterium carotovorum* subsp. *betavasculorum*, and some examples are provided in the corresponding table. A comprehensive list is maintained by the USDA on the Germplasm Resources Information Network. Even though some genes associated with root defense response have been identified, the specific mechanism of resistance is unknown, and it is currently being researched.

Biological Control

Some bacteriophages, viruses that infect bacteria, have been used as effective controls of bacterial diseases in laboratory experiments. This relatively new technology is a promising control method that is currently being researched. Bacteriophages are extremely host-specific, which makes them environmentally sound as they will not destroy other, beneficial soil microorganisms. Some bacteriophages identified as effective controls of *Pectobacterium carotovorum* subsp. *betavasculorum* are the strains ΦEcc2 ΦEcc3 ΦEcc9 ΦEcc14. When mixed with a fertilizer and applied to inoculated calla lily bulbs in a greenhouse, they reduced diseased tissue by 40 to 70%. ΦEcc3 appeared to be the most effective, reducing the percent of diseased plants from 30 to 5% in one trial, to 50 to 15% in a second trial. They have also been used successfully to reduce rotting in lettuce caused by *Pectobacterium carotovorum* subsp. *carotovorum*, a different bacterial species closely related to the one that causes beet vascular necrosis.

While it is more difficult to apply bacteriophages in a field setting, it is not impossible, and laboratory and greenhouse trials are showing bacteriophages to potentially be a very effective control mechanism. However, there are a few obstacles to surmount before field trials can begin. A large problem is that they are damaged by UV light, so applying the phage mixture during the evening will help promote its viability. Also, providing the phages with susceptible non-pathogenic bacteria to replicate with can ensure there is adequate persistence until the bacteriophages can spread to the targeted bacteria. The bacteriophages are unable to kill all the bacteria, because they need a dense population of bacteria in order to effectively infect and spread, so while the phages were able to decrease the number of diseased plants by up to 35%, around 2,000 Colony Forming Units per milliliter (an estimate of living bacteria cells) were able to survive the treatment. Lastly, the use of these bacteriophages places strong selection on the host bacteria, which causes a high probability of developing resistance to the attacking bacteriophage. Thus it is recommended that multiple strains of the bacteriophage be used in each application so the bacteria do not have a chance to develop resistance to any one strain.

Importance

The disease was first identified in the western states of, California, Washington, Texas, Arizona and Idaho in the 1970s and initially led to substantial yield losses in those areas. *Erwinia caratovara* subsp *betavascularum* was not discovered in Montana until 1998. When it first appeared, beet vascular necrosis caused individual farm yield loss ranging from 5–70% in Montana's Bighorn Valley. Today, yield losses from the disease are generally infrequent and patchy as most producers plant resistant varieties. Infection rate is generally low if resistant cultivars are chosen; however, warmer and wetter conditions can lead to higher than normal instance of disease

If infection does occur, bacterial root rots can not only cause economic losses in the field, but also can in storage and processing as well. In processing plants, rotten roots complicate slicing and the bacterially-produced slime can clog filters. This is especially problematic with late-infected beets which are generally harvested and processed along with healthy beets. The disease can also lower sugar-content which greatly reduces the quality.

Phytoplasma

Phytoplasmas are specialised bacteria that are obligate parasites of plant phloem tissue and transmitting insects (vectors). They were discovered by scientists in 1967 and were named mycoplasma-like organisms or MLOs. Since their discovery, phytoplasmas have resisted all attempts to be cultured in vitro in any cell free media, hence routine cultivation in artificial media is still to be established. Nevertheless, still under trial stage, phytoplasma growth in specific artificial media has recently been shown. They are characterised by their lack of a cell wall, a pleiomorphic or filamentous shape, normally with a diameter less than 1 μm, and their very small genomes.

Phytoplasmas are pathogens of agriculturally important plants, including coconut, sugarcane, and sandalwood, causing a wide variety of symptoms that range from mild yellowing to death of infected plants. They are most prevalent in tropical and subtropical regions of the world. They require a vector to be transmitted from plant to plant, and this normally takes the form of sap-sucking insects such as leaf hoppers, in which they are also able to survive and replicate.

History

References to diseases now known to be caused by phytoplasmas occurred as far back as 1603 for mulberry dwarf disease in Japan. Such diseases were originally thought to be caused by viruses, which, like phytoplasmas, require insect vectors, cannot be cultured, and have some symptom similarity. In 1967, phytoplasmas were discovered in ultrathin sections of plant phloem tissue and named mycoplasma-like organisms (MLOs), because they physically resembled mycoplasmas The organisms were renamed phytoplasmas in 1994, at the 10th Congress of The International Organization for Mycoplasmology.

Morphology

Being Mollicutes, a phytoplasma lacks a cell wall and instead is bound by a triple-layered membrane. The cell membranes of all phytoplasmas studied so far usually contain a single immunodominant protein (of unknown function) that makes up the majority of the protein content of the cell membrane. The typical phytoplasma exhibits a pleiomorphic or filamentous shape and is less than 1 μm in diameter. As prokaryotes, phytoplasmas' DNA is found throughout the cytoplasm, rather than being concentrated in a nucleus.

Symptoms

A common symptom caused by phytoplasma infection is phyllody, the production of leaf-like structures in place of flowers. Evidence suggests the phytoplasma downregulates a gene involved in petal formation (*AP3* and its orthologues) and genes involved in the maintenance of the apical meristem (*Wus* and *CLV1*). Other symptoms, such as the yellowing of leaves, are thought to be caused by the phytoplasma's presence in the phloem, affecting its function and changing the transport of carbohydrates.

Phytoplasma-infected plants may also suffer from virescence, the development of green flowers due to the loss of pigment in the petal cells. Phytoplasma-harboring plants which are able to flower may nevertheless be sterile. A phytoplasma effector protein (SAP54) has been identified as inducing symptoms of virescence and phyllody when expressed in plants.

Many plants infected by phytoplasmas gain a bushy or "witches' broom" appearance due to changes in their normal growth patterns. Most plants show apical dominance, but phytoplasma infection can cause the proliferation of auxiliary (side) shoots and an increase in size of the internodes. Such symptoms are actually useful in the commercial production of poinsettias. The infection produces more axillary shoots, which enables production of poinsettia plants that have more than one flower.

Phytoplasmas may cause many other symptoms that are induced because of the stress placed on the plant by infection rather than specific pathogenicity of the phytoplasma. Photosynthesis, especially photosystem II, is inhibited in many phytoplasma-infected plants. Phytoplasma-infected plants often show yellowing which is caused by the breakdown of chlorophyll, the biosynthesis of which is also inhibited.

Effector (Virulence) Proteins

Many plant pathogens produce virulence factors (or effectors) that modulate or interfere with normal host processes in a way that is beneficial to the pathogen. TCP transcription factors normally regulate plant development and control the expression of lipoxygenase (*LOX*) genes that are required for the biosynthesis of jasmonate. In infected *Arabidopsis* plants (and plants that express SAP11 transgenically), jasmonate levels are decreased. The downregulation of jasmonate production is beneficial to the phytoplasma because jasmonate is involved in plant defence against herbivorous insects such as leafhoppers, and leafhoppers have been shown to lay more eggs on AY-WB-infected plants at least in part because of SAP11. For example, the leafhopper *Macrosteles quadrilineatus* lays 30% more eggs on plants that express SAP11 transgenically, and 60% more eggs on plants infected with AY-WB. Phytoplasmas cannot survive in the external environment and are dependent upon insects such as leafhoppers for transmission to new (healthy) plants. Thus, by interfering with jasmonate production, SAP11 'encourages' leafhoppers to lay more eggs on phytoplasma-infected plants, thereby ensuring that newly hatching leafhopper nymphs feed upon infected plants and become vectors for the bacteria.

Transmission

Movement between Plants

Phytoplasmas are mainly spread by insects of the families Cicadellidea (leafhoppers), Fulgoridea (planthoppers), and Psyllidae (jumping plant lice) , which feed on the phloem tissues of infected plants, picking up the phytoplasmas and transmitting them to the next plant on which they feed. So, the host range of phytoplasmas is strongly dependent upon its insect vector. Phytoplasmas contain a major antigenic protein that makes up the majority of their cell surface proteins. This protein has been shown to interact with insect microfilament complexes and is believed to be the determining factor in insect-phytoplasma interaction. Phytoplasmas may overwinter in insect vectors or perennial plants. Phytoplasmas can have varying effects on their insect hosts; examples of both reduced and increased fitness have been seen.

Phytoplasmas enter the insect's body through the stylet, move through the intestine, and are then absorbed into the haemolymph. From there they proceed to colonise the salivary glands, a process that can take up to three weeks. Once established, phytoplasmas are found in most major organs of an infected insect host. The time between being taken up by the insect and reaching an infectious titre in the salivary glands is called the latency period.

Phytoplasmas can also be spread via dodders (Cuscuta) or vegetative propagation such as the grafting of a piece of infected plant onto a healthy plant.

Movement within Plants

Phytoplasmas are able to move within the phloem from source to sink, and they are able to pass through sieve tube elements. But since they spread more slowly than solutes, for this and other reasons, movement by passive translocation is not supported.

Detection and Diagnosis

Before molecular techniques were developed, the diagnosis of phytoplasma diseases was difficult because they could not be cultured. Thus, classical diagnostic techniques, such as observation of symptoms, were used. Ultrathin sections of the phloem tissue from suspected phytoplasma-infected plants would also be examined for their presence. Treating infected plants with antibiotics such as tetracycline to see if this cured the plant was another diagnostic technique employed.

Molecular diagnostic techniques for the detection of phytoplasma began to emerge in the 1980s and included ELISA-based methods. In the early 1990s, polymerase chain reaction-based methods were developed that were far more sensitive than those that used ELISA, and RFLP analysis allowed the accurate identification of different strains and species of phytoplasma.

More recently, techniques have been developed that allow for assessment of the level of infection. Both quantitative PCR and bioimaging have been shown to be effective methods of quantifying the titre of phytoplasmas within the plant.

Control

Phytoplasmas are normally controlled by the breeding and planting of disease resistant varieties of crops (believed to the most economically viable option) and by the control of the insect vector.

Tissue culture can be used to produce clones of phytoplasma-infected plants that are healthy. The chances of gaining healthy plants in this manner can be enhanced by the use of cryotherapy, freezing the plant samples in liquid nitrogen, before using them for tissue culture.

Work has also been carried out investigating the effectiveness of plantibodies targeted against phytoplasmas.

Tetracyclines are bacteriostatic to phytoplasmas. However, without continuous use of the antibiotic, disease symptoms reappear. Thus, tetracycline is not a viable control agent in agriculture, but it is used to protect ornamental coconut trees.

Genetics

The genomes of three phytoplasmas have been sequenced: aster yellows witches broom, onion yellows (*Ca.* Phytoplasma asteris) and *Ca.* Phytoplasma australiense Phytoplasmas have very small genomes, which also have extremely low levels of the nucleotides G and C, sometimes as little as 23%, which is thought to be the threshold for a viable genome. In fact Bermuda grass white leaf phytoplasma has a genome size of just 530 kb, one of the smallest known genomes of living organisms. Larger phytoplasma genomes are around 1350 kb. The small genome size associated with phytoplasmas is due to their being the product of reductive evolution from *Bacillus*/*Clostridium* ancestors. They have lost 75% or more of their original genes, so can no longer survive outside of insects or plant phloem. Some phytoplasmas contain extrachromosomal DNA such as plasmids.

Despite their very small genomes, many predicted genes are present in multiple copies. Phytoplasmas lack many genes for standard metabolic functions and have no functioning homologous recombination pathways, but do have a *sec* transport pathway. Many phytoplasmas contain two rRNA operons. Unlike the rest of the Mollicutes, the triplet code of UGA is used as a stop codon in phytoplasmas.

Phytoplasma genomes contain large numbers of transposon genes and insertion sequences. They also contain a unique family of repetitive extragenic palindromes called PhREPS whose role is unknown though it is theorised that the stem loop structures the PhREPS are capable of forming may play a role in transcription termination or genome stability.

Taxonomy

Phytoplasmas belong to the monophyletic order Acholeplasmatales. In 1992, the Subcommittee on the Taxonomy of Mollicutes proposed the use of the name *Phytoplasma* in place of the use of the term MLO (mycoplasma-like organism) "for reference to the phytopathogenic mollicutes". In 2004, the genus name *Phytoplasma* was adopted and is currently at Candidatus status which is used for bacteria that cannot be cultured. Its taxonomy is complicated because it can not be cultured, thus methods normally used for classification of prokaryotes are not possible. Phytoplasma taxonomic groups are based on differences in the fragment sizes produced by the restriction digest of the 16S rRNA gene sequence (RFLP) or by comparison of DNA sequences from the 16s/23s spacer regions. There is some disagreement over how many taxonomic groups the phytoplasmas fall into, recent work involving computer simulated restriction digests of the 16Sr gene suggest there may be up to 28 groups whereas other papers argue for less groups, but more subgroups. Each group includes at least one *Ca.* Phytoplasma species, characterized by distinctive biological, phytopathological, and genetic properties.

Rhodococcus Fascians

Rhodococcus fascians (known as Corynebacterium fascians until 1984) is a Gram positive bacterial phytopathogen that causes leafy gall disease. R. fascians is the only phytopathogenic member of the Rhodococcus genus; its host range includes both dicotyledonous and monocotyledonous hosts. Because it commonly afflicts tobacco (Nicotiana) plants, it is an agriculturally significant pathogen.

Physiology and Morphology

R. fascians is an aerobic, pleiomorphic, actinomycete that is nonmotile and does not form spores. When grown on the surface of an agar plate, colonies are orange in color and appear both smooth or rough.

Virulence

R. fascians can be a pathogen of plants, both angiosperm or gymnosperm. Infected plants show typical symptoms, such as leaf deformation, witches broom and leaf gall, which development depends on the plant's cultivar, plant's age, and the bacterial strain.

Leaf deformation consists of widening of parenchyma and growth of vascular system, resulting in wrinkling of laminae and widening of veins. Leafy gall is a gall originated from a bud which would not develop under normal conditions. All effects coming from the infection of *R. fascians* do not depend on plant cells transformation (as for Agrobacterium tumefaciens or Agrobacterium rhizogenes), but on expression of virulence-related genes of bacterium and on the production of compounds that can interfere with normal plant growth and development. During the infection, *R. fascians* usually stays outside vegetal tissues, near a junction or cavity of a plant's cell walls, maybe to avoid environmental stresses. Presence of *R. fascians* was also observed in intercellular spaces inside tissues (in leaf or galls) and even inside cell walls. Presence of *R. fascians* on the infected plant is necessary, not only for the initiation of infection, but also for its maintenance.

Genes that Control Virulence

Virulence in *R. fascians* is controlled by genes on a plasmid (strains lacking that plasmid are not virulent) and on the chromosome. Using deletion mutations, it was possible to identify three loci on the plasmid: *fas*, *att*, and *hyp*, and one locus on the chromosome, *vic*.

The *fas* is an operon made of six genes (orf 1-6) and a regulatory gene, *fasR*. Because deletions of some *fas* genes give a non-virulent phenotype, for fas a main role in virulence was proposed . Gene *fasR* is an araC-like transcriptional regulator. Its transcription can be induced *in vitro* in cultures containing certain carbon sources (such as glucose, sucrose, arabinose, glycerol, pyruvate, mannitol, mannose) or nitrogen sources (such as histidine), and is influenced by culture pH and optical density. Also, fasR can be induced by gall extract created by virulent strain. The operon codifies for genes involved in cytokinin synthesis and degradation (orf 4,5,6), in particular for an isopentenyl transferase, a cytokinin oxidase and a glutation-s transferase. The orf1,2,3 transcribe for a cytochrome 450, a ferridoxine containing also a pyruvate dehydrogenase alfa-like domain and a pyruvate dehydrogenase beta subunit. It was supposed that the first three genes supply energy for the synthesis and degradation of cytokinin, performed by the last three genes of the operon: *R. fascians* can actually produce and degrade zeatin and isopentenil adenine. The compound cytokin oxidase(orf4) can also create adenine with a reactive nitrogen in position 6, which can react with other lateral chains, to form cytokininn-like compounds, more efficient in inducing plant tissue growth.

The *att* is an operon composed of nine genes: *attR*, a transcriptional regulator, *attX*, a gene including domains for transmembrane localization (perhaps needed for exportation of compounds made by other *att* genes), and several genes *attA-attH*. Many point and Δ*att* mutants show an attenuated virulence.

Gene *attR* is a transcriptional factor including a helix-turn-helix motif. Its transcription is regulated by the same factors that regulate *fasR* transcription, but with a higher intensity, suggesting, with the attenuation of virulence in *att* mutants, that *att* may regulate fas transcription. Transcription of *att* operon is regulated with a quorum-sensing mechanism: indeed, density of cultures can influence transcription of *attR*, and leafy gall extracts coming from galls made by *att* mutant strains are less effective on transcription of *attR*.

Genes *attA-attH* may be involved in synthesis of compounds needed for transcription of *attR* and *attX*. In fact, *attA*, *attD* & *attH* are involved in betalactamase synthesis, but no traces of those compounds were found in culture supernatants.

The *hyp* codifies for an RNA-helicase; mutants for this gene are hypervirulent. Also, *hyp* is involved in post transcriptional control of virulence-related genes, maybe on *fas* products.

Operon *vic* is an operon made of five genes, located on the bacterial chromosome. The only known gene is *vicA*, the fourth gene in the operon, whose product is a Mas homologue, a protein needed for the switch from citric acid cycle to glyoxylate cycle, both for metabolic reasons and to avoid glyoxylate accumulation, which is toxic for the bacteria. Mutations in *vicA* reduce virulence due to incapacity of *R. fascians* to resist glyoxylate accumulation.

Induction of Transcription in Infected Plant

In tobacco, infection of *R. fascians* leads to hyperexpression of a cytochrome P450, homologue to a gene involved in inactivation of abscisic acid in *Arabidopsis thaliana*, of a gibberellic acid oxidase, which inactivates this hormone and its precursors, a proline dehydrogenase, which has its transcription induced by cytokinin and turns proline into glutamic acid, and a factor involved in molybdenum cofactor, needed for sulfur, carbon and nitrogen metabolism control and for abscisic acid synthesis.

Role of Phytohormones During Infection

All the effects of *R. fascians* infection can be attributed to hormone hyperdosage. In particular, most of the effects are connected to auxin and cytokinin, such as: formation of green islands on leaves, wrinkling of laminae, bud proliferation, delay of senescence, and inhibition of lateral roots. In fact, *R. fascians* can produce itself cytokinin, or cytokinin-like compounds: using *orf4* and *orf5* in the *fas* operon, it can stimulate infected plants to produce cytokinin, and it can produce indole-3-acetic acid itself, using a pathway starting from tryptophan and passing through production of 3-indol-piruvic acid and 3-indol-acetaldeid. *R.fascians* can also degrade cytokinin to influence the cytokinin/auxin ratio.

Beside cytokinin and auxin, *R. fascians* acts on other hormones: in particular, it can block abscisic acid and gibberellic acid synthesis in infected plants. Abscisic acid represses growth, so a block of production is needed to allow proliferation of cells in leafy galls. Gibberellic acid controls cellular differentiation, so its block is needed for maintenance of meristematic cells and for their proliferation.

Plant Diseases

R. fascians causes diseases in several host plants including tobacco, small fruits (caneberries, strawberries) and ornamental plants (butterfly flowers, *Primula*, kalanchoes, *Impatiens*, geraniums, carnations)

Agrobacterium Tumefaciens

Agrobacterium tumefaciens (updated scientific name Rhizobium radiobacter, synonym Agrobacterium radiobacter) is the causal agent of crown gall disease (the formation of tumours) in over 140 species of eudicots. It is a rod-shaped, Gram-negative soil bacterium. Symptoms are caused by the insertion of a small segment of DNA (known as the T-DNA, for 'transfer DNA', not to be confused with tRNA that transfers amino acids during protein synthesis, confusingly also called transfer RNA), from a plasmid, into the plant cell, which is incorporated at a semi-random location into the plant genome.

A. tumefaciens is an alphaproteobacterium of the family Rhizobiaceae, which includes the nitrogen-fixing legume symbionts. Unlike the nitrogen-fixing symbionts, tumor-producing *Agrobacterium* species are pathogenic and do not benefit the plant. The wide variety of plants affected by *Agrobacterium* makes it of great concern to the agriculture industry.

Economically, *A. tumefaciens* is a serious pathogen of walnuts, grape vines, stone fruits, nut trees, sugar beets, horse radish, and rhubarb.

Conjugation

To be virulent, the bacterium must contain a tumour-inducing plasmid (Ti plasmid or pTi), of 200 kb, which contains the T-DNA and all the genes necessary to transfer it to the plant cell. Many strains of *A. tumefaciens* do not contain a pTi.

Since the Ti plasmid is essential to cause disease, prepenetration events in the rhizosphere occur to promote bacterial conjugation - exchange of plasmids amongst bacteria. In the presence of opines, *A. tumefaciens* produces a diffusible conjugation signal called 3oC8HSL or the *Agrobacterium* autoinducer. This activates the transcription factor TraR, positively regulating the transcription of genes required for conjugation.

Method of Infection

A. tumefaciens infects the plant through its Ti plasmid. The Ti plasmid integrates a segment of its DNA, known as T-DNA, into the chromosomal DNA of its host plant cells. *A. tumefaciens* has flagella that allow it to swim through the soil towards photoassimilates that accumulate in the rhizosphere around roots. Some strains may chemotactically move towards chemical exudates from plants, such as acetosyringone and sugars. The former is recognised by the VirA protein, a transmembrane protein encoded in the virA gene on the Ti plasmid. Sugars are recognised by the chvE protein, a chromosomal gene-encoded protein located in the periplasmic space.

At least 25 vir genes on the Ti plasmid are necessary for tumor induction. In addition to their perception role, virA and chvE induce other vir genes. The virA protein has autokinase activity: it phosphorylates itself on a histidine residue. Then the virA protein phosphorylates the virG protein on its aspartate residue. The virG protein is a cytoplasmic protein produced from the virG Ti plasmid gene. It is a transcription factor, inducing the transcription of the vir operons. The chvE protein regulates the second mechanism of the vir genes' activation. It increases VirA protein sensitivity to phenolic compounds.

Attachment is a two-step process. Following an initial weak and reversible attachment, the bacteria synthesize cellulose fibrils that anchor them to the wounded plant cell to which they were attracted. Four main genes are involved in this process: *chvA, chvB, pscA*, and *att*. The products of the first three genes apparently are involved in the actual synthesis of the cellulose fibrils. These fibrils also anchor the bacteria to each other, helping to form a microcolony.

VirC, the most important virulent gene, is a necessary step in the recombination of illegitimate recolonization. It selects the section of the DNA in the host plant that will be replaced and it cuts into this strand of DNA.

After production of cellulose fibrils, a calcium-dependent outer membrane protein called rhicadhesin is produced, which also aids in sticking the bacteria to the cell wall. Homologues of this protein can be found in other rhizobia.

Possible plant compounds that initiate *Agrobacterium* to infect plant cells:

- Acetosyringone and other phenolic compounds
- alpha-Hydroxyacetosyringone
- Catechol
- Ferulic acid
- Gallic acid
- p-Hydroxybenzoic acid
- Protocatechuic acid
- Pyrogallic acid
- Resorcylic acid
- Sinapinic acid
- Syringic acid
- Vanillin

Formation of the T-pilus

To transfer the T-DNA into the plant cell, *A. tumefaciens* uses a type IV secretion mechanism, involving the production of a T-pilus. When acetosyringone and other substances are detected, a signal transduction event activates the expression of 11 genes within the VirB operon which are responsible for the formation of the T-pilus.

The pro-pilin is formed first. This is a polypeptide of 121 amino acids which requires processing by the removal of 47 residues to form a T-pilus subunit. The subunit is circularized by the formation of a peptide bond between the two ends of the polypeptide.

Products of the other VirB genes are used to transfer the subunits across the plasma membrane. Yeast two-hybrid studies provide evidence that VirB6, VirB7, VirB8, VirB9 and VirB10 may all encode components of the transporter. An ATPase for the active transport of the subunits would also be required.

Transfer of T-DNA into the Plant Cell

The T-DNA must be cut out of the circular plasmid. A VirD1/D2 complex nicks the DNA at the left and right border sequences. The VirD2 protein is covalently attached to the 5' end. VirD2 contains a motif that leads to the nucleoprotein complex being targeted to the type IV secretion system (T4SS).

In the cytoplasm of the recipient cell, the T-DNA complex becomes coated with VirE2 proteins, which are exported through the T4SS independently from the T-DNA complex. Nuclear localization signals, or NLSs, located on the VirE2 and VirD2, are recognised by the importin alpha protein, which then associates with importin beta and the nuclear pore complex to transfer the T-DNA into the nucleus. VIP1 also appears to be an important protein in the process, possibly acting as an adapter to bring the VirE2 to the importin. Once inside the nucleus, VIP2 may target the T-DNA to areas of chromatin that are being actively transcribed, so that the T-DNA can integrate into the host genome.

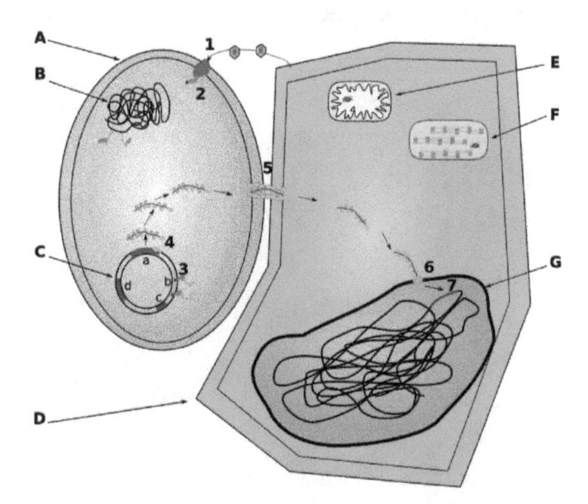

A: *Agrobacterium tumefaciens*
B: Agrobacterium genome
C: Ti Plasmid : a: T-DNA , b: Vir genes , c: Replication origin , d: Opines catabolism genes
D: Plant cell
E: Mitochondria
F: Chloroplast
G: Nucleus

Genes in the T-DNA

Hormones

To cause gall formation, the T-DNA encodes genes for the production of auxin or indole-3-acetic acid via the IAM pathway. This biosynthetic pathway is not used in many plants for the production of auxin, so it means the plant has no molecular means of regulating it and auxin will be produced constitutively. Genes for the production of cytokinins are also expressed. This stimulates cell proliferation and gall formation.

Opines

The T-DNA contains genes for encoding enzymes that cause the plant to create specialized amino acids which the bacteria can metabolize, called opines. Opines are a class of chemicals that serve as a source of nitrogen for *A. tumefaciens*, but not for most other organisms. The specific type of opine produced by *A. tumefaciens* C58 infected plants is nopaline (Escobar *et al.*, 2003).

Two nopaline type Ti plasmids, pTi-SAKURA and pTiC58, were fully sequenced. *A. tumefaciens* C58, the first fully sequenced pathovar, was first isolated from a cherry tree crown gall. The genome was simultaneously sequenced by Goodner *et al.* and Wood *et al.* in 2001. The genome of *A. tumefaciens* C58 consists of a circular chromosome, two plasmids, and a linear chromosome. The presence of a covalently bonded circular chromosome is common to Bacteria, with few exceptions. However, the presence of both a single circular chromosome and single linear chromosome is unique to a group in this genus. The two plasmids are pTiC58, responsible for the processes involved in virulence, and pAtC58, dubbed the "cryptic" plasmid.

The pAtC58 plasmid has been shown to be involved in the metabolism of opines and to conjugate with other bacteria in the absence of the pTiC58 plasmid. If the pTi plasmid is removed, the tumor growth that is the means of classifying this species of bacteria does not occur.

Biotechnological Uses

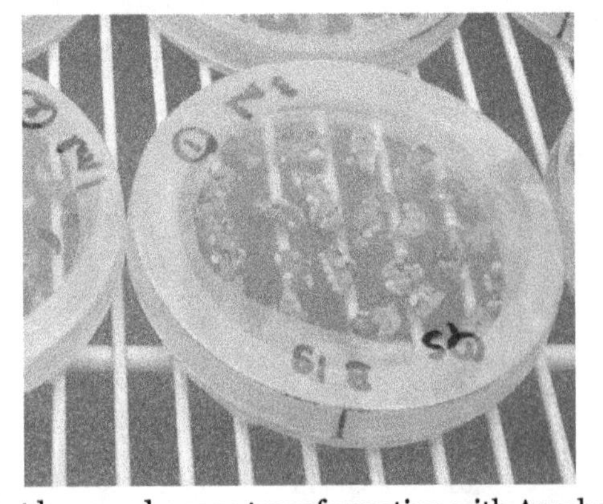

Plants that have undergone transformation with Agrobacterium

The DNA transmission capabilities of *Agrobacterium* have been vastly explored in biotechnology as a means of inserting foreign genes into plants. Marc Van Montagu and Jeff Schell, (University of Ghent and Plant Genetic Systems, Belgium) discovered the gene transfer mechanism between *Agrobacterium* and plants, which resulted in the development of methods to alter the bacterium into an efficient delivery system for genetic engineering in plants. The plasmid T-DNA that is transferred to the plant is an ideal vehicle for genetic engineering. This is done by cloning a desired gene sequence into the T-DNA that will be inserted into the host DNA. This process has been performed using firefly luciferase gene to produce glowing plants. This luminescence has been a useful device in the study of plant 'chloroplast' function and as a reporter gene. It is also possible to transform *Arabidopsis thaliana* by dipping flowers into a broth of *Agrobacterium*: the seed produced will be transgenic. Under laboratory conditions, the T-DNA has also been transferred to human cells, demonstrating the diversity of insertion application.

The mechanism by which *Agrobacterium* inserts materials into the host cell by a type IV secretion system is very similar to mechanisms used by pathogens to insert materials (usually proteins) into human cells by type III secretion. It also employs a type of signaling conserved in many Gram-negative bacteria called quorum sensing. This makes *Agrobacterium* an important topic of medical research, as well.

Natural Genetic Transformation

Natural genetic transformation in bacteria is a sexual process involving the transfer of DNA from one cell to another through the intervening medium, and the integration of the donor sequence into the recipient genome by homologous recombination. *A. tumefaciens* can undergo natural transformation in soil without any specific physical or chemical treatment.

Xylella Fastidiosa

Xylella fastidiosa, a bacterium in the class Gammaproteobacteria, is an important plant pathogen that causes phoney peach disease in the southern United States, bacterial leaf scorch, oleander leaf scorch, and Pierce's disease (in grapevines), and citrus variegated chlorosis disease (CVC) in Brazil. In Europe it has attacked olive trees in the Salento area of Southern Italy causing the Olive Quick Decline Syndrome (OQDS).

Pierce's Disease (Grapevines)

Pierce's disease (PD) was discovered in 1892 by Newton B. Pierce (1856–1916; California's first professional plant pathologist) on grapes in California near Anaheim, where it was known as "Anaheim disease." The disease is endemic in northern California, being spread by the blue-green sharpshooter, which only attacks grapevines that are adjacent to riparian habitats. It became a real threat to California's wine industry when the glassy-winged sharpshooter (GWSS), native to the southeast United States, was discovered in the Temecula Valley in California in 1996. The GWSS spreads PD much more extensively than other vectors do. It triggered a unique effort from growers, administrators, policy makers and researchers to work together in finding a solution for this immense threat. No cure has yet been found, but the understanding of *Xylella fastidiosa* and glassy-winged sharpshooter biology has increased much since 2000, when the California Department of Food and Agriculture, in collaboration with different universities, such as University of California, Davis (UC Davis); University of California, Berkeley, and University of California, Riverside, and the University of Houston–Downtown started to focus their research on this pest. The research explores the different aspects of the disease propagation from the vector to the host plant and within the host plant, to the impact of the disease on California's economy. All researchers working on Pierce's disease meet annually in San Diego in mid-December to discuss the progress in their field. All proceedings from this symposium can be found on the Pierce's disease website, developed and managed by the Public Intellectual Property Resource for Agriculture (PIPRA).

There are no resistant *Vitis vinifera* varieties, and Chardonnay and Pinot noir are especially sensitive, but muscadine grapes (*Vitis rotundifolia*) have a natural resistance. Pierce's disease is found in the southeastern United States and Mexico. Also it was reported by Luis G. Jiménez-Arias in Costa Rica, and Venezuela, and possibly in other parts of Central and South America. There are isolated hot spots of the disease near creeks in Napa and Sonoma in northern California. Work is underway at UC Davis to breed PD resistance from *Vitis rotundifolia* into *Vitis vinifera*. The first generation was 50% high quality vinifera genes, the next 75%, the third 87% and the fourth 94%. In the spring of 2007, seedlings that are 94% vinifera were planted.

When a vine becomes infected, the bacterium causes a gel to form in the xylem tissue of the vine, preventing water from being drawn through the vine. Leaves on vines with Pierce's disease will turn yellow and brown, and eventually drop off the vine. Shoots will also die. After one to five years, the vine itself will die. The proximity of vineyards to citrus orchards compounds the threat, because citrus is not only a host for the sharpshooter eggs, but it is also a popular overwintering site for the insect. Likewise, oleander, a common landscaping plant in California, serves as a reservoir for *Xylella*.

Nerium oleander infected with Xylella fastidiosa in Phoenix, Arizona

Oleander Leaf Scorch

Oleander leaf scorch is a disease of landscape oleanders (*Nerium oleander*) caused by a strain of *X. fastidiosa* which has become prevalent in California and Arizona, USA starting in the mid 1990s. This disease is transmitted by a type of leafhopper (insect) called the Glassy-winged sharpshooter (*Homalodisca coagulata*).

Olive Tree Decline

In October 2013 the bacterium was found to be infecting olive trees in the region of Apulia in southern Italy. The disease was causing a rapid decline in olive plantations and by April 2015 it was affecting the whole Province of Lecce and other zones of Apulia. The bacterium had never previously been confirmed in Europe. Almond and oleander plants in the region have also tested positive for the pathogen.

The disease has been called **o**live quick decline syndrome (OQDS; in Italian: *complesso del disseccamento rapido dell'olivo*). The disease causes withering and desiccation of terminal shoots, distributed randomly at first but which then expands to the rest of the canopy. This results in the collapse and death of the trees. In the affected groves, all of the plants show symptoms. By the beginning of 2015 it had infected up to a million trees in the southern region of Apulia.

By July 2015, *Xylella fastidiosa* had reached Corsica; by October 2015, it had reached Mainland France, near Nice. On 18th August 2016 in Corsica have been detected 279 focuses of the infection, concentrated mostly in the south and the west of the island.

In August 2016, the bacterium has been detected in Germany in an oleander plant.

Genome Sequencing

The genome sequencing of *X. fastidiosa* was realized by a pool of over 30 research labs in the State of São Paulo, Brazil, and funded by this State's Science Foundation (FAPESP).

Pseudomonas Syringae

Pseudomonas syringae is a rod-shaped, Gram-negative bacterium with polar flagella. As a plant pathogen, it can infect a wide range of species, and exists as over 50 different pathovars, all of which are available to researchers from international culture collections such as the NCPPB, ICMP, and others. Whether these pathovars represent a single species is unclear.

P. syringae is a member of the *Pseudomonas* genus, and based on 16S rRNA analysis, it has been placed in the *P. syringae* group. It is named after the lilac tree (*Syringa vulgaris*), from which it was first isolated.

Tomato plant leaf infected with bacterial speck

P. syringae tests negative for arginine dihydrolase and oxidase activity, and forms the polymer levan on sucrose nutrient agar. Many, but not all, strains secrete the lipodepsinonapeptide plant toxin syringomycin, and it owes its yellow fluorescent appearance when cultured *in vitro* on King's B medium to production of the siderophore pyoverdin.

P. syringae also produces ice nucleation active (INA) proteins which cause water (in plants) to freeze at fairly high temperatures (-4 to -2°C), resulting in injury. Since the 1970s, *P. syringae* has been implicated as an atmospheric "biological ice nucleator", with airborne bacteria serving as cloud condensation nuclei. Recent evidence has suggested the species plays a larger role than previously thought in producing rain and snow. They have also been found in the cores of hailstones, aiding in bioprecipitation. These INA proteins are also used in making artificial snow.

P. syringae pathogenesis is dependent on effector proteins secreted into the plant cell by the bacterial type III secretion system. Nearly 60 different type III effector families encoded by *hop* genes have been identified in *P. syringae*. Type III effectors contribute to pathogenesis chiefly through their role in suppressing plant defense. Owing to early availability of the genome sequence for three *P. syringae* strains and the ability of selected strains to cause disease on well-characterized host plants, including *Arabidopsis thaliana*, *Nicotiana benthamiana*, and the tomato, *P. syringae* has come to represent an important model system for experimental characterization of the molecular dynamics of plant-pathogen interactions.

Bacterial speck on tomato in Upstate New York

History

In 1961, Paul Hoppe of the U.S. Department of Agriculture studied a corn fungus by grinding up infected leaves each season, then applying the powder to test corn for the following season to track the disease. A surprise frost occurred that year, leaving peculiar results. Only plants infected with the diseased powder incurred frost damage, leaving healthy plants unfrozen. This phenomenon baffled scientists until graduate student Stephen Lindow of the University of Wisconsin–Madison with D.C. Arny and C. Upper found a bacterium in the dried leaf powder in the early 1970s. Dr. Lindow, now a plant pathologist at the University of California-Berkeley, found that when this particular bacterium was introduced to plants where it is originally absent, the plants became very vulnerable to frost damage. He went on to identify the bacterium as *P. syringae*, investigate *P. syringae*'s role in ice nucleation and in 1977, discover the mutant ice-minus strain. He was later successful at developing the ice-minus strain of *P. syringae* through recombinant DNA technology, as well.

Epidemiology

Disease by *P. syringae* tends to be favoured by wet, cool conditions—optimum temperatures for disease tend to be around 12–25°C, although this can vary according to the pathovar involved. The bacteria tend to be seed-borne, and are dispersed between plants by rain splash.

Although it is a plant pathogen, it can also live as a saprotroph in the phyllosphere when conditions are not favourable for disease. Some saprotrophic strains of *P. syringae* have been used as biocontrol agents against postharvest rots.

Mechanisms of Pathogenicity

The mechanisms of *P. syringae* pathogenicity can be separated into several categories: ability to invade a plant, ability to overcome host resistance, biofilm formation, and production of proteins with ice-nucleating properties.

Ability to Invade Plants

Planktonic *P. syringae* is able to enter plants using its flagella and pilli to swim towards a target host. It enters the plant via wounds of natural opening sites, as it is not able to breach the plant cell wall. The role of taxis in *P. syringae* has not been well-studied, but the bacteria are thought to use chemical signals released by the plant to find their host and cause infection.

Overcoming Host Resistance

P. syringae isolates carry a range of virulence factors called type III secretion system (T3SS) effector proteins. These proteins primarily function to cause disease symptoms and manipulate the host's immune response to facilitate infection. The major family of T3SS effectors in *P. syringae* is the *hrp* gene cluster, coding for the Hrp secretion apparatus.

The pathogens also produce phytotoxins which injure the plant and can suppress the host immune system. One such phytotoxin is coronatine, found in pathovars *Pto* and *Pgl*.

Biofilm Formation

P. syringae produces polysaccharides which allow it to adhere to the surface of plant cells. It also releases quorum sensing molecules, which allows it to sense the presence of other bacterial cells nearby. If these molecules pass a threshold level, the bacteria change their pattern of gene expression to form a biofilm and begin expression virulence-related genes. The bacteria secrete highly viscous compounds such as polysaccharides and DNA to create a protective environment in which to grow.

Ice-nucleating Properties

P. syringae—more than any mineral or other organism—is responsible for the surface frost damage in plants exposed to the environment. For plants without antifreeze proteins, frost damage usually occurs between -4 and -12°C as the water in plant tissue can remain in a supercooled liquid state. *P. syringae* can cause water to freeze at temperatures as high as −1.8°C (28.8°F), but strains causing ice nucleation at lower temperatures (down to −8 °C) are more common. The freezing causes injuries in the epithelia and makes the nutrients in the underlying plant tissues available to the bacteria.

P. syringae has *ina* (ice nucleation-active) genes that make INA proteins which translocate to the outer bacterial membrane on the surface of the bacteria, where the proteins act as nuclei for ice formation. Artificial strains of *P. syringae* known as ice-minus bacteria have been created to reduce frost damage.

P. syringae has been found in the center of hailstones, suggesting the bacterium may play a role in Earth's hydrological cycle.

Pathovars

Following ribotypical analysis, incorporation of several pathovars of *P. syringae* into other species was proposed (*P. amygdali*, *'P. tomato'*, *P. coronafaciens*, *P. avellanae*, *'P. helianthi'*, *P. tremae*, *P. cannabina*, and *P. viridiflava*). According to this schema, the remaining pathovars are:

- *P. s.* pv. *aceris* attacks maple *Acer* species.
- *P. s.* pv. *actinidiae* attacks kiwifruit *Actinidia deliciosa.*
- *P. s.* pv. *aesculi* attacks horse chestnut *Aesculus hippocastanum*, causing bleeding canker.
- *P. s.* pv. *aptata* attacks beets *Beta vulgaris.*
- *P. s.* pv. *atrofaciens* attacks wheat *Triticum aestivum.*
- *P. s.* pv. *dysoxylis* attacks the kohekohe tree *Dysoxylum spectabile.*
- *P. s.* pv. *japonica* attacks barley *Hordeum vulgare.*
- *P. s.* pv. *lapsa* attacks wheat *Triticum aestivum.*
- *P. s.* pv. *panici* attacks *Panicum* grass species.
- *P. s.* pv. *papulans* attacks crabapple *Malus sylvestris* species.
- *P. s.* pv. *phaseolicola* causes halo blight of beans.
- *P. s.* pv. *pisi* attacks peas *Pisum sativum.*
- *P. s.* pv. *syringae* attacks *Syringa*, *Prunus*, and *Phaseolus* species.
- *P. s.* pv. *glycinea attacks soybean* Glycine max, *causing bacterial blight of soybean.*

However, many of the strains for which new species groupings were proposed continue to be referred to in the scientific literature as pathovars of *P. syringae*, including pathovars tomato, phaseolicola, and maculicola. *Pseudomonas savastanoi* was once considered a pathovar or subspecies of *P. syringae*, and in many places continues to be referred to as *P. s.* pv. *savastanoi*, although as a result of DNA-relatedness studies, it has been instated as a new species. It has three host-specific pathovars: *P. s. fraxini* (which causes ash canker), *P. s. nerii* (which attacks oleander), and *P. s. oleae* (which causes olive knot).

Determinants of Host Specificity

A combination of the pathogen's effector genes and the plant's resistance genes is thought to determine which species a particular pathovar can infect. Plants can develop resistance to a pathovar by recognising pathogen-associated molecular patterns (PAMPs) and launching an immune response. These PAMPs are necessary for the microbe to function, so cannot be lost, but the pathogen may find ways to suppress this immune response, leading to an evolutionary arms race between the pathogen and the host.

P. Syringae as a Model System

Owing to early availability of genome sequences for *P syringae* pv, tomato strain DC3000, *P. syringae* pv. *syringae* strain B728a, and *P. syringae* pv. *phaseolicola* strain 1448A, together with the ability of selected strains to cause disease on well-characterized host plants such as *Arabidopsis thaliana, Nicotiana benthamiana*, and tomato, *P. syringae* has come to represent an important model system for experimental characterization of the molecular dynamics of plant-pathogen interactions. The *P. syringae* experimental system has been a source of pioneering evidence for the important role of pathogen gene products in suppressing plant defense. The nomenclature system developed for *P. syringae* effectors has been adopted by researchers characterizing effector repertoires in other bacteria, and methods used for bioinformatic effector identification have been adapted for other organisms. In addition, researchers working with *P. syringae* have played an integral role in the Plant-Associated Microbe Gene Ontology working group, aimed at developing gene ontology terms that capture biological processes occurring during the interactions between organisms, and using the terms for annotation of gene products.

P. Syringae Pv. Tomato Strain DC3000 and Arabidopsis Thaliana

As mentioned above, the genome of *P. syringae* pv. tomato DC3000 has been sequenced, and approximately 40 Hop (Hrp Outer Protein) effectors, pathogenic proteins that attenuate the host cell, have been identified. These 40 effectors are not recognized by *A. thaliana* thus making *P. syringae* pv. tomato DC3000 virulent, that is, *P. syringae* pv. tomato DC3000 is able to infect *A. thaliana* which is susceptible to this pathogen.

Many gene-for-gene relationships have been identified using the two model organisms, *P. syringae* pv. tomato strain DC3000 and *Arabidopsis*. The gene-for-gene relationship describes the recognition of pathogenic avirulence (*avr*) genes by host resistance genes (R-genes). *P. syringae* pv. tomato DC3000 is a useful tool for studying *avr*: R-gene interactions in *A. thaliana* because it can be transformed with *avr* genes from other bacterial pathogens, and furthermore, because none of the endogenous *hops* genes is recognized by *A. thaliana*, any observed aver recognition identified using this model can be attributed to recognition of the introduced *avr* by *A. thaliana*. The transformation of *P. syringae* pv tomato DC3000 with effectors from other pathogens have led to the identification of many R-genes in *Arabidopsis* to further advance knowledge of plant pathogen interactions.

Xanthomonas

Xanthomonas is a genus of Proteobacteria, many of which cause plant diseases.

Taxonomy

The *Xanthomonas* genus has been subject of numerous taxonomic and phylogenetic studies and was first described as *Bacterium vesicatorium* as a pathogen of pepper and tomato in 1921. Dowson later reclassified the bacterium as *Xanthomonas campestris* and proposed the genus *Xanthomonas.Xanthomonas* was first described as a monotypic genus and further research resulted in the division into two groups, A and B. Later work using DNA:DNA hybridization has served as a framework for the general *Xanthomonas* species classification. Other tools, including multilocus sequence analysis and amplified fragment-length polymorphism, have been used for classification within clades. While previous research has illustrated the complexity of the *Xanthomonas* genus, recent research appears to have resulted in a clearer picture. More recently, genome-wide analysis of multiple *Xanthomonas* strains mostly supports the previous phylogenies.

Morphology and Growth

Individual cell characteristics include:

- Cell type – straight rods
- Size – 0.4 – 1.0 μm wide by 1.2 – 3.0 μm long
- Motility – motile by a single polar flagellum

Colony growth characteristics include:

- Mucoid, convex, and yellow colonies on YDC medium
- Yellow pigment from xanthomonadin, which contains bromine
- Most produce large amounts of extracellular polysaccharide
- Temperature range – 4 to 37 °C

Biochemical and physiological test results are:

- Gram stain – negative
- Catalase positive
- Oxidase negative

Xanthomonas Plant Pathogens

Xanthomonas species can cause bacterial spots and blights of leaves, stems, and fruits on a wide variety of plant species. Pathogenic species show high degrees of specificity and some are split into multiple pathovars, a species designation based on host specificity.

Bacterial blight of cotton, caused by *Xanthomonas citri* subsp. *malvacearum* is the most important bacterial disease on cotton which infects all aerial parts of the host. Loss due to this disease was estimated for about 10 to 30% on different cultivars and can be found in Asia, Africa and southern America. Citrus canker, caused by *Xanthomonas citri* subsp. *citri* is an economically important disease of many citrus species (lime, orange, lemon, pamelo, etc.)

Bacterial leaf spot has caused significant crop losses over the years. Causes of this disease include *Xanthomonas euvesicatoria* and *Xanthomonas perforans* = [*Xanthomonas axonopodis* (syn. *campestris*) pv. *vesicatoria*], *Xanthomonas vesicatoria*, and *Xanthomonas gardneri*. In some areas where infection begins soon after transplanting, the total crop can be lost as a result of this disease.

Bacterial blight of rice, caused by *Xanthomonas oryzae* pv. *oryzae*, is a disease found worldwide and particularly destructive in the rice-producing regions in Asia.

Plant Pathogenesis and Disease Control

Xanthomonas species can be easily spread in water, movement of infected material such as seed or propagation plants, and by mechanical means such as infected pruning tools. Upon contact with a susceptible host, bacteria enter through wounds or natural plant openings as a means to infect. They inject a number of effector proteins, including TAL effectors, into the plant by their secretion systems (i.e., type III secretion system).

To prevent infections, limiting the introduction of the bacteria is key. Some resistant cultivars of certain plant species are available as this may be the most economical means for controlling this disease. For chemical control, preventative applications are best to reduce the potential for bacterial development. Copper-containing products offer some protection along with field-grade antibiotics such as oxytetracycline, which is labeled for use on some food crops in the United States. Curative applications of chemical pesticides may slow or reduce the spread of the bacterium, but will not cure already diseased plants. It is important to consult chemical pesticide labels when attempting to control bacterial diseases, as different *Xanthomonas* species can have different responses to these applications. Over-reliance on chemical control methods can also result in the selection of resistant isolates, so these applications should be considered a last resort.

Industrial Use

Xanthomonas species produce an extrapolysaccharide called xanthan gum that has a wide range of industrial uses, including foods, petroleum products, and cosmetics.

Xanthomonas Resources

Isolates of most species of *Xanthomonas* are available from the National Collection of Plant Pathogenic Bacteria in the United Kingdom and other international culture collections such as ICMP in New Zealand, CFBP in France, and VKM in Russia. It also can be taken out from MTCC India.

Multiple genomes of *Xanthomonas* have been sequenced and additional data sets/tools are available at The *Xanthomonas* Resource.

References

- Dusenbery, David B. (2009). Living at Micro Scale, pp. 20–25. Harvard University Press, Cambridge, Mass. ISBN 978-0-674-03116-6.
- Hecker M, Völker U (2001). "General stress response of Bacillus subtilis and other bacteria". Adv Microb Physiol. Advances in Microbial Physiology. 44: 35–91. doi:10.1016/S0065-2911(01)44011-2. ISBN 978-0-12-027744-5. PMID 11407115.
- Dusenbery, David B. (2009). Living at Micro Scale, p. 136. Harvard University Press, Cambridge, Mass. ISBN 978-0-674-03116-6.
- Dusenbery, David B. (2009). Living at Micro Scale, Chapter 13. Harvard University Press, Cambridge, Mass. ISBN 978-0-674-03116-6.
- Fisher B, Harvey RP, Champe PC (2007). Lippincott's Illustrated Reviews: Microbiology (Lippincott's Illustrated Reviews Series). Hagerstwon, MD: Lippincott Williams & Wilkins. pp. Chapter 33, pages 367–392. ISBN 0-7817-8215-5.
- "Taxonomy browser (Agrobacterium radiobacter K84)". National Center for Biotechnology Information. Retrieved 7 December 2015.
- Bennett, J. Michael; Rhetoric, Emeritus; Hicks, Dale R.; Naeve, Seth L.; Bennett, Nancy Bush (2014). The Minnesota Soybean Field Book (PDF). St Paul, MN: University of Minnesota Extension. p. 84. Retrieved 21 February 2016.
- The University of Waikato (March 25, 2014). "Bacterial DNA – the role of plasmids". Themes — Bacteria in biotech. Biotechnology Learning Hub. Retrieved 2014-09-03.
- Perombelon, Michel CM; Kelman, Arthur (1980). "Ecology of the soft rot erwinias". Annual Review of Phytopathology. 18 (1): 361–387. doi:10.1146/annurev.py.18.090180.002045. Retrieved 17 October 2013.
- Thomson, S.V.; et al. (1977). "Beet Vascular Necrosis and Rot of Sugarbeet: General Description and Etiology" (PDF). Phytopathology. 67 (10): 1183–1189. doi:10.1094/phyto-67-1183. Retrieved 17 October 2013.
- "Sugar Beet Production Guide, Chapter 11: Disease Management, pg 138-139" (PDF). University of Nebraska – Lincoln Extension, 2013. Retrieved 17 October 2013.
- Zidack, Nina; Barry Jacobsen (2001). "First Report and Virulence Evaluation of Erwinia caratovora subs. Betavasculorum on Sugarbeet in Montana". Plant Health Progress. Retrieved 18 October 2013.
- Strausbaugh, Carl A.; Anne M. Gillen (2008). "Bacterial and yeast associated with sugar beet root rot at harvest in the Intermountain West". Plant Disease. 92: 357–363. doi:10.1094/pdis-92-3-0357. Retrieved 15 October 2013.
- "Sugar Beet (Beta vulgaris)-Bacterial Vascular Necrosis and Rot {Erwinia Root Rot}". pacific northwest plant disease management handbook. Retrieved 17 October 2013.
- Stanghellini, M.E.; et al. (1977). "Serological and Physiological Differences of Erwinia carotovora between Potato and Sugar Beet" (PDF). Phytopathology. 67 (10): 1178–1182. doi:10.1094/phyto-67-1178. Retrieved 17 October 2013.
- Gallian, John J. "Management of Sugarbeet Root Rots" (PDF). Pacific Northwest Extension. Retrieved 17 October 2013.
- "UC Pest Management Guidelines". University of California Agriculture and Natural Resources. Retrieved 17 October 2013.
- Lewellen, R. T.; E. D. Whitney; C. K. Goulas (1978). "Inheritance of resistance to Erwinia root rot in sugarbeet" (PDF). Phytopathology. 68: 947–950. doi:10.1094/phyto-68-947. Retrieved 17 October 2013.
- "Germplasm Resources Information Network". United States Department of Agriculture. Retrieved 28 September 2013.

- Smigocki, A C. "Molecular Approaches To Pest And Pathogen Resistance in Sugar Beet". united states department of agriculture agricultural research service. Retrieved 17 October 2013.
- Duffy, B (2006). "Biological control of bacterial diseases in field crops". Symposium on Biological Control of Bacterial Plant Diseases: 93–98. Retrieved 17 October 2013.
- Ichinose, Yuki; Taguchi, Fumiko; Mukaihara, Takafumi (2013). "Pathogenicity and virulence factors of Pseudomonas syringae". J Gen Plant Pathol. 79: 285–296. doi:10.1007/s10327-013-0452-8.

6

Understanding Abiotic Stress

Abiotic stress is the negative stress caused by non-living factors. Abiotic stress causes the most damage to the growth and the productivity of crops across the globe. In agriculture, abiotic stress causes stress by natural environment factors such as high winds, droughts and floods. This chapter elucidates all the factors related to abiotic stress.

Abiotic Stress

Abiotic stress is defined as the negative impact of non-living factors on the living organisms in a specific environment. The non-living variable must influence the environment beyond its normal range of variation to adversely affect the population performance or individual physiology of the organism in a significant way.

Whereas a biotic stress would include such living disturbances as fungi or harmful insects, abiotic stress factors, or stressors, are naturally occurring, often intangible, factors such as intense sunlight or wind that may cause harm to the plants and animals in the area affected. Abiotic stress is essentially unavoidable. Abiotic stress affects animals, but plants are especially dependent on environmental factors, so it is particularly constraining. Abiotic stress is the most harmful factor concerning the growth and productivity of crops worldwide. Research has also shown that abiotic stressors are at their most harmful when they occur together, in combinations of abiotic stress factors.

Examples

Abiotic stress comes in many forms. The most common of the stressors are the easiest for people to identify, but there are many other, less recognizable abiotic stress factors which affect environments constantly.

The most basic stressors include:

- High winds
- Extreme temperatures
- Drought
- Flood
- Other natural disasters, such as tornadoes and wildfires.

Lesser-known stressors generally occur on a smaller scale. They include: poor edaphic conditions like rock content and pH levels, high radiation, compaction, contamination, and other, highly specific conditions like rapid rehydration during seed germination.

Effects

Abiotic stress, as a natural part of every ecosystem, will affect organisms in a variety of ways. Although these effects may be either beneficial or detrimental, the location of the area is crucial in determining the extent of the impact that abiotic stress will have. The higher the latitude of the area affected, the greater the impact of abiotic stress will be on that area. So, a taiga or boreal forest is at the mercy of whatever abiotic stress factors may come along, while tropical zones are much less susceptible to such stressors.

Benefits

One example of a situation where abiotic stress plays a constructive role in an ecosystem is in natural wildfires. While they can be a human safety hazard, it is productive for these ecosystems to burn out every once in a while so that new organisms can begin to grow and thrive. Even though it is healthy for an ecosystem, a wildfire can still be considered an abiotic stressor, because it puts an obvious stress on individual organisms within the area. Every tree that is scorched and each bird nest that is devoured is a sign of the abiotic stress. On the larger scale, though, natural wildfires are positive manifestations of abiotic stress.

What also needs to be taken into account when looking for benefits of abiotic stress, is that one phenomenon may not affect an entire ecosystem in the same way. While a flood will kill most plants living low on the ground in a certain area, if there is rice there, it will thrive in the wet conditions. Another example of this is in phytoplankton and zooplankton. The same types of conditions are usually considered stressful for these two types of organisms. They act very similarly when exposed to ultraviolet light and most toxins, but at elevated temperatures the phytoplankton reacts negatively, while the thermophilic zooplankton reacts positively to the increase in temperature. The two may be living in the same environment, but an increase in temperature of the area would prove stressful only for one of the organisms.

Lastly, abiotic stress has enabled species to grow, develop, and evolve, furthering natural selection as it picks out the weakest of a group of organisms. Both plants and animals have evolved mechanisms allowing them to survive extremes.

Detriments

The most obvious detriment concerning abiotic stress involves farming. It has been claimed by one study that abiotic stress causes the most crop loss of any other factor and that most major crops are reduced in their yield by more than 50% from their potential yield.

Because abiotic stress is widely considered a detrimental effect, the research on this branch of the issue is extensive. For more information on the harmful effects of abiotic stress, see the sections below on plants and animals.

In Plants

A plant's first line of defense against abiotic stress is in its roots. If the soil holding the plant is healthy and biologically diverse, the plant will have a higher chance of surviving stressful conditions.

Facilitation, or the positive interactions between different species of plants, is an intricate web of association in a natural environment. It is how plants work together. In areas of high stress, the level of facilitation is especially high as well. This could possibly be because the plants need a stronger network to survive in a harsher environment, so their interactions between species, such as cross-pollination or mutualistic actions, become more common to cope with the severity of their habitat.

Plants also adapt very differently from one another, even from a plant living in the same area. When a group of different plant species was prompted by a variety of different stress signals, such as drought or cold, each plant responded uniquely. Hardly any of the responses were similar, even though the plants had become accustomed to exactly the same home environment.

Rice (*Oryza sativa*) is a classic example. Rice is a staple food throughout the world, especially in China and India. Rice plants experience different types of abiotic stresses, like drought and high salinity. These stress conditions have a negative impact on rice production. Genetic diversity has been studied among several rice varieties with different genotypes using molecular markers.

Serpentine soils (media with low concentrations of nutrients and high concentrations of heavy metals) can be a source of abiotic stress. Initially, the absorption of toxic metal ions is limited by cell membrane exclusion. Ions that are absorbed into tissues are sequestered in cell vacuoles. This sequestration mechanism is facilitated by proteins on the vacuole membrane.

Chemical priming has been proposed to increase tolerance to abiotic stresses in crop plants. In this method, which is analogous to vaccination, stress-inducing chemical agents are introduced to the plant in brief doses so that the plant begins preparing defense mechanisms. Thus, when the abiotic stress occurs, the plant has already prepared defense mechanisms that can be activated faster and increase tolerance.

Phosphate Starvation in Plants

Phosphorus (P) is an essential macronutrient required for plant growth and development, but most of the world's soil is limited in this important plant nutrient. Plants can utilize P mainly in the form if soluble inorganic phosphate (Pi) but are subjected to abiotic stress of P-limitation when there is not sufficient soluble PO4 available in the soil. Phosphorus forms insoluble complexes with Ca and Mg in basic soils and Al and Fe in acidic soils that makes it unavailable for plant roots. When there is limited bioavailable P in the soil, plants show extensive abiotic stress phenotype such as short primary roots and more lateral roots and root hairs to make more surface available for Pi absorption, exudation of organic acids and phosphatase to release Pi from complex P containing molecules and make it available for growing plants organs. It has been shown that PHR1, a MYB - related transcription factor is a master regulator of P-starvation response in plants. PHR1 also has been shown to regulate extensive remodeling of lipids and metabolites during phosphorus limitation stress

In Animals

For animals, the most stressful of all the abiotic stressors is heat. This is because many species are unable to regulate their internal body temperature. Even in the species that are able to regulate their own temperature, it is not always a completely accurate system. Temperature determines metabolic rates, heart rates, and other very important factors within the bodies of animals, so an extreme temperature change can easily distress the animal's body. Animals can respond to extreme heat, for example, through natural heat acclimation or by burrowing into the ground to find a cooler space.

It is also possible to see in animals that a high genetic diversity is beneficial in providing resiliency against harsh abiotic stressors. This acts as a sort of stock room when a species is plagued by the perils of natural selection. A variety of galling insects are among the most specialized and diverse herbivores on the planet, and their extensive protections against abiotic stress factors have helped the insect in gaining that position of honor.

In Endangered Species

Biodiversity is determined by many things, and one of them is abiotic stress. If an environment is highly stressful, biodiversity tends to be low. If abiotic stress does not have a strong presence in an area, the biodiversity will be much higher.

This idea leads into the understanding of how abiotic stress and endangered species are related. It has been observed through a variety of environments that as the level of abiotic stress increases, the number of species decreases. This means that species are more likely to become population threatened, endangered, and even extinct, when and where abiotic stress is especially harsh.

Natural Stress

In regard to agriculture, Abiotic stress is stress produced by natural environment factors such as extreme temperatures, wind, drought, and salinity. Humankind doesn't have much control over abiotic stresses. It is very important for humans to understand how stress factors affect plants and other living things so that we can take some preventative measures.

Preventative measures are the only way that humans can protect themselves and their possessions from abiotic stress. There are many different types of abiotic stressors, and several methods that humans can use to reduce the negative effects of stress on living things.

Cold

One of the types of Abiotic Stress is cold. This has a huge impact on farmers. Cold impacts crop growers all over the world in every single country. Yields suffer and farmers also suffer huge losses because the weather is just too cold to produce crops (Xiong & Zhu, 2001).

Humans have planned the planting of our crops around the seasons. Even though the seasons are fairly predictable, there are always unexpected storms, heat waves, or cold snaps that can ruin our growing seasons.(Suzuki & Mittler, 2006)

ROS stands for reactive oxygen species. ROS plays a large role in mediating events through transduction. Cold stress was shown to enhance the transcript, protein, and activity of different

ROS-scavenging enzymes. Low temperature stress has also been shown to increase the H_2O_2 accumulation in cells.(Suzuki & Mittler, 2006)

Plants can be acclimated to low or even freezing temperatures. If a plant can go through a mild cold spell this activates the cold-responsive genes in the plant. Then if the temperature drops again, the genes will have conditioned the plant to cope with the low temperature. Even below freezing temperatures can be survived if the proper genes are activated (Suzuki & Mittler, 2006).

Heat

Heat stress has been shown to cause problems in mitochondrial functions and can result in oxidative damage. Activators of heat stress receptors and defenses are thought to be related to ROS. Heat is another thing that plants can deal with if they have the proper pretreatment. This means that if the temperature gradually warms up the plants are going to be better able to cope with the change. A sudden long temperature increase could cause damage to the plant because their cells and receptors haven't had enough time to prepare for a major temperature change.

Heat stress can also have a detrimental effect on plant reproduction. Temperatures 10 degrees Celsius or more above normal growing temperatures can have a bad effect on several plant reproductive functions. Pollen meiosis, pollen germination, ovule development, ovule viability, development of the embryo, and seedling growth are all aspects of plant reproduction that are affected by heat.(Cross, McKay, McHughen, & Bonham-Smith, 2003)

There have been many studies on the effects of heat on plant reproduction. One study on plants was conducted on Canola plants at 28 degrees Celsius, the result was decreased plant size, but the plants were still fertile. Another experiment was conducted on Canola plants at 32 degrees Celsius, this resulted in the production of sterile plants. Plants seem to be more easily damaged by extreme temperatures during the late flower to early seed development stage (Cross, McKay, McHughen, & Bonham-Smith, 2003).

Wind

Wind is a huge part of abiotic stress. There is simply no way to stop the wind from blowing. This is definitely a bigger problem in some parts of the world than in others. Barren areas such as deserts are very susceptible to natural wind erosion. These types of areas don't have any vegetation to hold the soil particles in place. Once the wind starts to blow the soil around, there is nothing to stop the process. The only chance for the soil to stay in place is if the wind doesn't blow. This is usually not an option.

Plant growth in windblown areas is very limited. Because the soil is constantly moving, there is no opportunity for plants to develop a root system. Soil that blows a lot usually is very dry also. This leaves little nutrients to promote plant growth.

Farmland is typically very susceptible to wind erosion. Most farmers do not plant cover crops during the seasons when their main crops are not in the fields. They simply leave the ground open and uncovered. When the soil is dry, the top layer becomes similar to powder. When the wind blows, the powdery top layer of the farmland is picked up and carried for miles. This is the exact scenario that occurred during the "Dust Bowl" in the 30's. The combination of drought and poor farming practices allowed the wind to moves thousands of tons of dirt from one area to the next.

Wind is one of the factors that humans can really have some control over. Simply practice good farming practices. Don't leave ground bare and without any type of vegetation. During dry seasons it is especially important to have the land covered because dry soil moves much easier than wet soil in the wind.

When soil is not blowing due to the wind, conditions are much better for plant growth. Plants cannot grow in a soil that is constantly blowing. Their root systems do not have time to be established. Also, when soil particles are blowing they wear away at the plants that they run into. Plants are essentially "sand blasted."

Drought

Drought is very detrimental to all types of plant growth. When there is no water in the soil there are not very many nutrients to support plant growth. Drought also enhances the effects of wind. When drought occurs the soil becomes very dry and light. The wind picks up this dry dirt and carries it away. This action severely degrades the soil and creates a poor condition for growing plants.

Adaptation of Plants

Plants have been exposed to the elements for thousands of years. During this time they have evolved in order to lessen the effects of abiotic stress. Signal transduction is the mechanism in plants that is responsible for the adaptation of plants (Xiong & Zhu, 2001). Many signaling transduction networks have been discovered and studied in microbial and animal systems. There is limited knowledge in the plant field because it is very difficult to find exactly which phenotypes in the plant are affected by stressors. These phenotypes are very valuable to the researchers. They need to know the phenotypes so that they can create a method to screen for mutant genes. Mutants are the key to finding signaling pathways in living creatures.

Animals and microbes easier to run tests on because they show a reaction fairly quickly when a stress factor is put on them, this leads to the isolation of the specific gene. There have been decades of research on the effects of temperature, drought, and salinity, but not very many answers.

Receptors

The part of the plants, animal, or microbe that first senses an abiotic stress factor is a receptor. Once a signal is picked up by a receptor, a lot of different things can happen. Signals are transmitted intercellularly and then they activate nuclear transcription to get the effects of a certain set of genes. These genes that are activated allow the plant to respond to the stress that it is experiencing. Even though none of the receptors for cold, drought, salinity or the stress hormone abscisic acid in plants is known for sure, the knowledge that we have today shows that receptor-like protein kinases, two-component histidine kinases, as well as G-protein receptors may be the possible sensors of these different signals.

Receptor like kinases can be found in plants as well as animals. There are many more RLKs in plants than there are in animals. They are also a little bit different. Unlike animal RLKs that usually possess tyrosine signature sequences, plant RLKs have serine or threonine signature sequences (Xiong & Zhu, 2001).

Genetically Modified Plants

Plants are most commonly modified to be resistant to specific herbicides or pathogens, but we have the technology to modify plants in order to make them resistant to specific abiotic stressors. Cold, heat, drought, or salt are all factors that could possibly be defended against by genetically modified plants.

Some plants could have genes added to them from other species of plants that have a resistance to a specific stress. Plants implanted with these genes would then become transgenic plants because they have the genes from another species of plant in them. Scientists first have to isolate the specific gene in a plant that is responsible for its resistance. The gene would then be taken out of the plant and put in to another plant. The plant that is injected with the new resistant gene would have a resistance to an abiotic stressor and be able to tolerate a wider range of conditions (Weil, 2005).

This process of creating transgenic plants could have a huge impact on our nation's economy. If plants could be genetically engineered to be resistant to a wider variety of stress, crop yields would skyrocket. With the expansion of town and cities there is a decreasing number of farm acres. Although the farm acres are being built on, the number of people consuming agriculture products is going up. Ethanol is also responsible for using much more of the corn that is grown here in the U.S. The production of this fuel has put a strain on the corn market. Prices of corn have gone up and this price is having a negative impact on the people who feed animals using corn. The combination of reduced acres of farmland and a higher demand on crops have left producers and consumers in a severe dilemma. The only solution to this problem is to keep getting higher and higher yields from the cropland that we have left.

Genetically modified plants are a good answer to the problem of not enough crops to go around. These plants can be engineered to be resistant to all types of abiotic stress. This would eliminate crop yield loss due to extreme temperatures, drought, wind, or salinity. The consumers of crops would enjoy a little bit lower prices because the demand on them would be a little lower.

The Midwestern U.S. is experiencing a severe drought. Farmers are being limited on how much they can irrigate due to the shortage of water. There is also very little rain during the growing season so the crops do not yield very well. This problem could be solved by genetically modifying plants to become more drought resistant. If plants could use less water and produce yields that are superior or equal to current ones, it would be better for the people and also the environment. People would enjoy an abundance of crops to consume and export for a profit. The environment would be able to have more water in its aquifers and rivers throughout the country.

Another environmental factor that would be improved would be the amount of land left for wildlife. Crops modified to be resistant to abiotic stress and other factors that decrease yields would require less land use. Producers would be able to grow enough crops on less acres if the plants were modified to produce very high yields. This would allow some of the cropland that is in use today to be set aside for wildlife. Instead of farming "fence line to fence line" farmers would be able to create large buffers in their fields. These buffers would provide a great habitat for plants and animals.

A lot of people do not like genetically modified organisms. People opposed to these modified plants often claim that they are not safe for the environment or for human consumption. There are many videos and reports in circulation that discredit the safety of genetically modified organisms. King Corn is one video that claims that corn is bad for humans to consume.

There are strict regulations and protocols that go along with genetically modifying plants. A company that specializes in producing genetically modifying organisms must put their plants through a huge variety of tests to ensure the safety of their product. Each of these tests must be passed by the product in order to produce more of the plant seeds.

When seeds are mass-produced, the fields that they are grown in have to meet specific criteria. They must have no vegetation zones around them to prevent the spread of the modified plants into the native population. The plots must be carefully labeled and marked so that the company knows exactly what is planted in the field. All of these protocols are in place to ensure the safety of the consumers and also of the environment. Because genetically modified plants are given stress resistant genes or high yielding genes they are better for the environment. They only help create more land to be put back into natural habitats for plants and animals.

Conclusion

onAbiotic stress is a naturally occurring factor that cannot be controlled by humans. Some of the stress factors go hand in hand. One example of two stressors that are complimentary to each other is wind and draught. Drought dries out the soil and kills the plants that are growing in the soil. After this occurs the soil is left barren and dry. When the wind picks up then the soil is picked up and carried for miles. Irrigation is a way that humans can try to keep this from happening, but sometimes it is not possible to irrigate some areas.

Genetically modified plants can be implemented to slow down the effects of the abiotic stressors. These plants can be given genes that allow them to survive several types of natural stressors. This allows more crops to be grown on a smaller amount of land. This is important because there is less and less farmland available. Also, less need for farmland allows some of it to be set aside for natural wildlife habitat.

Abiotic stress only poses a problem to people or the environment if they are not prepared for it. There can be steps taken by humans to lessen the effects. Plants and animals have the ability to adapt to abiotic stress over time. This is natures way of taking care of itself and keeping everything in balance.

Abiotic Component

In biology and ecology, abiotic components or abiotic factors are non-living chemical and physical parts of the environment that affect living organisms and the functioning of ecosystems. Abiotic factors and phenomena associated with them underpin all biology.

Abiotic components include physical conditions and non-living resources that affect living organisms in terms of growth, maintenance, and reproduction. Resources are distinguished as substances or objects in the environment required by one organism and consumed or otherwise made unavailable for use by other organisms.

Component degradation of a substance by chemical or physical processes, e.g. hydrolysis. All non-living components of an ecosystem, such as the atmosphere or water, are called abiotic components.

Examples

In biology, abiotic factors can include water, light, radiation, temperature, humidity, atmosphere, and soil. The macroscopic climate often influences each of the above. Pressure and sound waves may also be considered in the context of marine or sub-terrestrial environments.

All of these factors affect different organisms to different extents. If there is little or no sunlight then plants may wither and die from not being able to get enough sunlight to complete the cycle of photosynthesis. Many Archea require very high temperatures, or pressures, or unusual concentrations of chemical substances, such as sulfur, because of their specialization into extreme conditions. Certain fungi have evolved to survive mostly at the temperature, the humidity, and stability of their environment.

For example, there is a significant difference in access to water as well as humidity between temperate rain forests and deserts. This difference in water access causes a diversity in the types of plants and animals that grow in these areas.

References

- Savvides, Andreas (December 15, 2015). "Chemical Priming of Plants Against Multiple Abiotic Stresses: Mission Possible?". Trends in Plant Science. Retrieved March 10, 2016.
- Palm, Brady; Van Volkenburgh (2012). "Serpentine tolerance in Mimuslus guttatus does not rely on exclusion of magnesium". Functional Plant Biology.
- Chapin, F.S. III, H.A. Mooney, M.C. Chapin, and P. Matson. 2011. Principles of terrestrial ecosystem ecology. Springer, New York.
- Hogan, C. Benito (2010). "Abiotic factor". Encyclopedia of Earth. Washington,D.C.: National Council for Science and the Environment.

7

Basic Abiotic Stressors

The basic abiotic stressors are wildfire and drought. Wildfire is the fire that occurs either in forests or in rural areas whereas droughts are the shortage of water supply in a particular area. Droughts can last for months or even years, and it certainly has substantial impact on the ecosystem. The text discusses the major stressors of abiotic stress in critical manner providing key analysis to the subject matter.

Wildfire

A wildfire in California on September 5, 2008

A wildfire or wildland fire is a fire in an area of combustible vegetation that occurs in the countryside or rural area. Depending on the type of vegetation where it occurs, a wildfire can also be classified more specifically as a brush fire, bush fire, desert fire, forest fire, grass fire, hill fire, peat fire, vegetation fire, or veld fire. Fossil charcoal indicates that wildfires began soon after the appearance of terrestrial plants 420 million years ago. Wildfire's occurrence throughout the history of terrestrial life invites conjecture that fire must have had pronounced evolutionary effects on most ecosystems' flora and fauna. Earth is an intrinsically flammable planet owing to its cover of carbon-rich vegetation, seasonally dry climates, atmospheric oxygen, and widespread lightning and volcano ignitions.

Wildfires can be characterized in terms of the cause of ignition, their physical properties, the combustible material present, and the effect of weather on the fire. Wildfires can cause damage to property and human life, but they have many beneficial effects on native vegetation, animals, and ecosystems that have evolved with fire. Many plant species depend on the effects of fire for growth and reproduction. However, wildfire in ecosystems where wildfire is uncommon or where non-native vegetation has encroached may have negative ecological effects. Wildfire behaviour and severity result from the combination of factors such as available fuels, physical setting, and weather. Analyses of historical meteorological data and national fire records in western North America show the primacy of climate in driving large regional fires via wet periods that create substantial fuels or drought and warming that extend conducive fire weather.

Strategies of wildfire prevention, detection, and suppression have varied over the years. One common and inexpensive technique is controlled burning: permitting or even igniting smaller fires to minimize the amount of flammable material available for a potential wildfire. Vegetation may be burned periodically to maintain high species diversity and frequent burning of surface fuels limits fuel accumulation. Wildland fire use is the cheapest and most ecologically appropriate policy for many forests. Fuels may also be removed by logging, but fuels treatments and thinning have no effect on severe fire behavior Wildfire itself is reportedly "the most effective treatment for reducing a fire's rate of spread, fireline intensity, flame length, and heat per unit of area" according to Jan Van Wagtendonk, a biologist at the Yellowstone Field Station. Building codes in fire-prone areas typically require that structures be built of flame-resistant materials and a defensible space be maintained by clearing flammable materials within a prescribed distance from the structure.

Causes

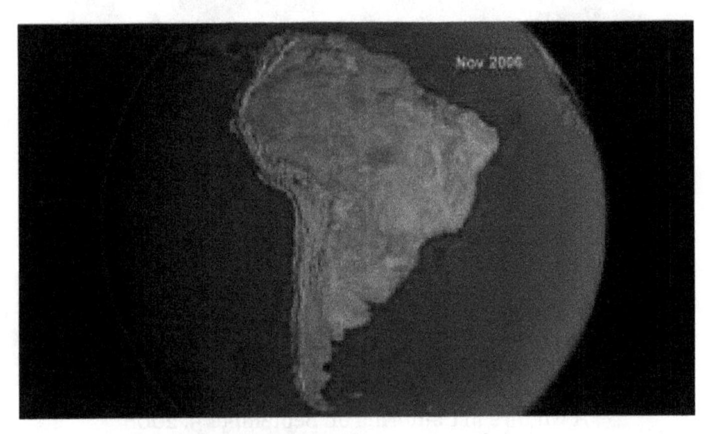

Forecasting South American fires.

UC Irvine scientist James Randerson discusses new research linking ocean temperatures and fire seasons severity

Four major natural causes of wildfire ignitions exist:

- lightning
- sparks from rockfalls
- spontaneous combustion
- volcanic eruption

The most common direct human causes of wildfire ignition include arson, discarded cigarettes, power-line arcs (as detected by arc mapping), and sparks from equipment Ignition of wildland fires via contact with hot rifle-bullet fragments is also possible under the right conditions. Wildfires can also be started in communities experiencing shifting cultivation, where land is cleared quickly and farmed until the soil loses fertility, and slash and burn clearing. Forested areas cleared by logging encourage the dominance of flammable grasses, and abandoned logging roads overgrown by vegetation may act as fire corridors. Annual grassland fires in southern Vietnam stem in part from the destruction of forested areas by US military herbicides, explosives, and mechanical land-clearing and -burning operations during the Vietnam War.

The most common cause of wildfires varies throughout the world. In Canada and northwest China, for example, lightning operates as the major source of ignition. In other parts of the world, human involvement is a major contributor. In Africa, Central America, Fiji, Mexico, New Zealand, South America, and Southeast Asia, wildfires can be attributed to human activities such as agriculture, animal husbandry, and land-conversion burning. In China and in the Mediterranean Basin, human carelessness is a major cause of wildfires. In the United States and Australia, the source of wildfires can be traced both to lightning strikes and to human activities (such as machinery sparks, cast-away cigarette butts, or arson)." Coal seam fires burn in the thousands around the world, such as those in Burning Mountain, New South Wales; Centralia, Pennsylvania; and several coal-sustained fires in China. They can also flare up unexpectedly and ignite nearby flammable material.

Spread

A surface fire in the western desert of Utah, U.S.

Charred landscape following a crown fire in the North Cascades, U.S.

The spread of wildfires varies based on the flammable material present, its vertical arrangement and moisture content, and weather conditions. Fuel arrangement and density is governed in part by topography, as land shape determines factors such as available sunlight and water for plant growth. Overall, fire types can be generally characterized by their fuels as follows:

- Ground fires are fed by subterranean roots, duff and other buried organic matter. This fuel type is especially susceptible to ignition due to spotting. Ground fires typically burn by smoldering, and can burn slowly for days to months, such as peat fires in Kalimantan and Eastern Sumatra, Indonesia, which resulted from a riceland creation project that unintentionally drained and dried the peat.
- Crawling or surface fires are fueled by low-lying vegetation such as leaf and timber litter, debris, grass, and low-lying shrubbery.
- Ladder fires consume material between low-level vegetation and tree canopies, such as small trees, downed logs, and vines. Kudzu, Old World climbing fern, and other invasive plants that scale trees may also encourage ladder fires.
- Crown, canopy, or aerial fires burn suspended material at the canopy level, such as tall trees, vines, and mosses. The ignition of a crown fire, termed *crowning*, is dependent on the density of the suspended material, canopy height, canopy continuity, sufficient surface and ladder fires, vegetation moisture content, and weather conditions during the blaze. Stand-replacing fires lit by humans can spread into the Amazon rain forest, damaging ecosystems not particularly suited for heat or arid conditions.

Physical Properties

Experimental fire in Canada

Wildfires occur when all of the necessary elements of a fire triangle come together in a susceptible area: an ignition source is brought into contact with a combustible material such as vegetation, that is subjected to sufficient heat and has an adequate supply of oxygen from the ambient air. A high moisture content usually prevents ignition and slows propagation, because higher temperatures are required to evaporate any water within the material and heat the material to its fire point. Dense forests usually provide more shade, resulting in lower ambient temperatures and greater humidity, and are therefore less susceptible to wildfires. Less dense material such as grasses and leaves are easier to ignite because they contain less water than denser material such as branches and trunks. Plants continuously lose water by evapotranspiration, but water loss is usually balanced by water absorbed from the soil, humidity, or rain. When this balance is not maintained, plants dry out and are therefore more flammable, often a consequence of droughts.

A wildfire *front* is the portion sustaining continuous flaming combustion, where unburned material meets active flames, or the smoldering transition between unburned and burned material. As the front approaches, the fire heats both the surrounding air and woody material through convection and thermal radiation. First, wood is dried as water is vaporized at a temperature of 100 °C (212 °F). Next, the pyrolysis of wood at 230 °C (450 °F) releases flammable gases. Finally, wood can smoulder at 380 °C (720 °F) or, when heated sufficiently, ignite at 590 °C (1,000 °F). Even before the flames of a wildfire arrive at a particular location, heat transfer from the wildfire front warms the air to 800 °C (1,470 °F), which pre-heats and dries flammable materials, causing materials to ignite faster and allowing the fire to spread faster. High-temperature and long-duration surface wildfires may encourage flashover or *torching*: the drying of tree canopies and their subsequent ignition from below.

Torching of Juniper Tree on the Palisade Wildfire in Nevada

Wildfires have a rapid *forward rate of spread* (FROS) when burning through dense, uninterrupted fuels. They can move as fast as 10.8 kilometres per hour (6.7 mph) in forests and 22 kilometres per hour (14 mph) in grasslands. Wildfires can advance tangential to the main front to form a *flanking* front, or burn in the opposite direction of the main front by *backing*. They may also spread by *jumping* or *spotting* as winds and vertical convection columns carry *firebrands* (hot wood embers) and other burning materials through the air over roads, rivers, and other barriers that may otherwise act as firebreaks. Torching and fires in tree canopies encourage spotting, and dry ground fuels that surround a wildfire are especially vulnerable to ignition from firebrands. Spotting can create *spot fires* as hot embers and firebrands ignite fuels downwind from the fire. In Australian bushfires, spot fires are known to occur as far as 20 kilometres (12 mi) from the fire front.

Especially large wildfires may affect air currents in their immediate vicinities by the stack effect: air rises as it is heated, and large wildfires create powerful updrafts that will draw in new, cooler air from surrounding areas in thermal columns. Great vertical differences in temperature and humidity encourage pyrocumulus clouds, strong winds, and fire whirls with the force of tornadoes at speeds of more than 80 kilometres per hour (50 mph). Rapid rates of spread, prolific crowning or spotting, the presence of fire whirls, and strong convection columns signify extreme conditions.

The thermal heat from wildfire can cause significant weathering of rocks and boulders, heat can rapidly expand a boulder and thermal shock can occur, which may cause an object's structure to fail.

Effect of Weather

Heat waves, droughts, cyclical climate changes such as El Niño, and regional weather patterns such as high-pressure ridges can increase the risk and alter the behavior of wildfires dramatically. Years of precipitation followed by warm periods can encourage more widespread fires and longer fire seasons. Since the mid-1980s, earlier snowmelt and associated warming has also been associated with an increase in length and severity of the wildfire season in the Western United States. Global warming may increase the intensity and frequency of droughts in many areas, creating more intense and frequent wildfires. A 2015 study indicates that the increase in fire risk in California may be attributable to human-induced climate change. A study of alluvial sediment deposits going back over 8,000 years found warmer climate periods experienced severe droughts and stand-replacing fires and concluded climate was such a powerful influence on wildfire that trying to recreate presettlement forest structure is likely impossible in a warmer future.

Lightning-sparked wildfires are frequent occurrences during the dry summer season in Nevada.

Intensity also increases during daytime hours. Burn rates of smoldering logs are up to five times greater during the day due to lower humidity, increased temperatures, and increased wind speeds. Sunlight warms the ground during the day which creates air currents that travel uphill. At night the land cools, creating air currents that travel downhill. Wildfires are fanned by these winds and often follow the air currents over hills and through valleys. Fires in Europe occur frequently during the hours of 12:00 p.m. and 2:00 p.m. Wildfire suppression operations in the United States revolve around a 24-hour *fire day* that begins at 10:00 a.m. due to the predictable increase in intensity resulting from the daytime warmth.

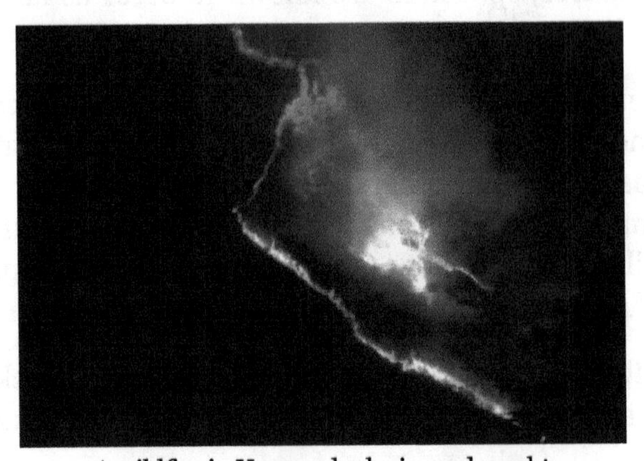

A wildfire in Venezuela during a drought

Ecology

Wildfire's occurrence throughout the history of terrestrial life invites conjecture that fire must have had pronounced evolutionary effects on most ecosystems' flora and fauna. Wildfires are common in climates that are sufficiently moist to allow the growth of vegetation but feature extended dry, hot periods. Such places include the vegetated areas of Australia and Southeast Asia, the veld in southern Africa, the fynbos in the Western Cape of South Africa, the forested areas of the United States and Canada, and the Mediterranean Basin.

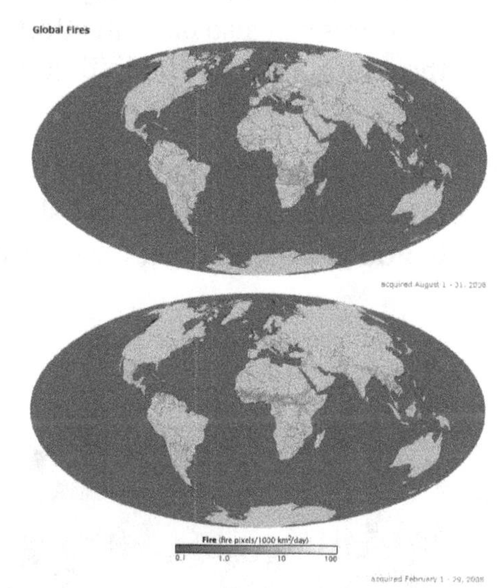

Global fires during the year 2008 for the months of August (top image) and February (bottom image), as detected by the Moderate Resolution Imaging Spectroradiometer (MODIS) on NASA's Terra satellite.

High-severity wildfire creates complex early seral forest habitat (also called "snag forest habitat"), which often has higher species richness and diversity than unburned old forest. Plant and animal species in most types of North American forests evolved with fire, and many of these species depend on wildfires, and particularly high-severity fires, to reproduce and grow. Fire helps to return nutrients from plant matter back to soil, the heat from fire is necessary to the germination of certain types of seeds, and the snags (dead trees) and early successional forests created by high-severity fire create habitat conditions that are beneficial to wildlife. Early successional forests created by high-severity fire support some of the highest levels of native biodiversity found in temperate conifer forests. Post-fire logging has no ecological benefits and many negative impacts; the same is often true for post-fire seeding.

Although some ecosystems rely on naturally occurring fires to regulate growth, some ecosystems suffer from too much fire, such as the chaparral in southern California and lower elevation deserts in the American Southwest. The increased fire frequency in these ordinarily fire-dependent areas has upset natural cycles, damaged native plant communities, and encouraged the growth of non-native weeds. Invasive species, such as *Lygodium microphyllum* and *Bromus tectorum*, can grow rapidly in areas that were damaged by fires. Because they are highly flammable, they can increase the future risk of fire, creating a positive feedback loop that increases fire frequency and further alters native vegetation communities.

In the Amazon Rainforest, drought, logging, cattle ranching practices, and slash-and-burn agriculture damage fire-resistant forests and promote the growth of flammable brush, creating a cycle that encourages more burning. Fires in the rainforest threaten its collection of diverse species and produce large amounts of CO_2. Also, fires in the rainforest, along with drought and human involvement, could damage or destroy more than half of the Amazon rainforest by the year 2030. Wildfires generate ash, destroy available organic nutrients, and cause an increase in water runoff, eroding away other nutrients and creating flash flood conditions. A 2003 wildfire in the North Yorkshire Moors destroyed 2.5 square kilometers (600 acres) of heather and the underlying peat layers. Afterwards, wind erosion stripped the ash and the exposed soil, revealing archaeological remains dating back to 10,000 BC. Wildfires can also have an effect on climate change, increasing the amount of carbon released into the atmosphere and inhibiting vegetation growth, which affects overall carbon uptake by plants.

In tundra there is a natural pattern of accumulation of fuel and wildfire which varies depending on the nature of vegetation and terrain. Research in Alaska has shown fire-event return intervals, (FRIs) that typically vary from 150 to 200 years with dryer lowland areas burning more frequently than wetter upland areas.

Plant Adaptation

Ecological succession after a wildfire in a boreal pine forest next to Hara Bog, Lahemaa National Park, Estonia. The pictures were taken one and two years after the fire.

Plants in wildfire-prone ecosystems often survive through adaptations to their local fire regime. Such adaptations include physical protection against heat, increased growth after a fire event, and flammable materials that encourage fire and may eliminate competition. For example, plants of the genus *Eucalyptus* contain flammable oils that encourage fire and hard sclerophyll leaves to resist heat and drought, ensuring their dominance over less fire-tolerant species. Dense bark, shedding lower branches, and high water content in external structures may also protect trees from rising temperatures. Fire-resistant seeds and reserve shoots that sprout after a fire encourage species preservation, as embodied by pioneer species. Smoke, charred wood, and heat can stimulate the germination of seeds in a process called *serotiny*. Exposure to smoke from burning plants promotes germination in other types of plants by inducing the production of the orange butenolide.

Grasslands in Western Sabah, Malaysian pine forests, and Indonesian *Casuarina* forests are believed to have resulted from previous periods of fire. Chamise deadwood litter is low in water content and flammable, and the shrub quickly sprouts after a fire. Cape lilies lie dormant until flames brush away the covering, then blossom almost overnight. Sequoia rely on periodic fires to reduce competition, release seeds from their cones, and clear the soil and canopy for new growth. Caribbean Pine in Bahamian pineyards have adapted to and rely on low-intensity, surface fires for survival and growth. An optimum fire frequency for growth is every 3 to 10 years. Too frequent fires favor herbaceous plants, and infrequent fires favor species typical of Bahamian dry forests.

Atmospheric Effects

Most of the Earth's weather and air pollution resides in the troposphere, the part of the atmosphere that extends from the surface of the planet to a height of about 10 kilometers (6 mi). The vertical lift of a severe thunderstorm or pyrocumulonimbus can be enhanced in the area of a large wildfire, which can propel smoke, soot, and other particulate matter as high as the lower stratosphere. Previously, prevailing scientific theory held that most particles in the stratosphere came from volcanoes, but smoke and other wildfire emissions have been detected from the lower stratosphere. Pyrocumulus clouds can reach 6,100 meters (20,000 ft) over wildfires. Satellite observation of smoke plumes from wildfires revealed that the plumes could be traced intact for distances exceeding 1,600 kilometers (1,000 mi). Computer-aided models such as CALPUFF may help predict the size and direction of wildfire-generated smoke plumes by using atmospheric dispersion modeling.

A Pyrocumulus cloud produced by a wildfire in Yellowstone National Park

Wildfires can affect local atmospheric pollution, and release carbon in the form of carbon dioxide. Wildfire emissions contain fine particulate matter which can cause cardiovascular and respiratory problems. Increased fire byproducts in the troposphere can increase ozone concentration beyond safe levels. Forest fires in Indonesia in 1997 were estimated to have released between 0.81 and 2.57 gigatonnes (0.89 and 2.83 billion short tons) of CO_2 into the atmosphere, which is between 13%–40% of the annual global carbon dioxide emissions from burning fossil fuels. Atmospheric models suggest that these concentrations of sooty particles could increase absorption of incoming solar radiation during winter months by as much as 15%.

Smoke trail from a fire seen while looking towards Dargo from Swifts Creek, Victoria, Australia, 11 January 2007

History

In the Welsh Borders, the first evidence of wildfire is rhyniophytoid plant fossils preserved as charcoal, dating to the Silurian period (about 420 million years ago). Smoldering surface fires started to occur sometime before the Early Devonian period 405 million years ago. Low atmospheric oxygen during the Middle and Late Devonian was accompanied by a decrease in charcoal abundance. Additional charcoal evidence suggests that fires continued through the Carboniferous period. Later, the overall increase of atmospheric oxygen from 13% in the Late Devonian to 30-31% by the Late Permian was accompanied by a more widespread distribution of wildfires. Later, a decrease in wildfire-related charcoal deposits from the late Permian to the Triassic periods is explained by a decrease in oxygen levels.

Wildfires during the Paleozoic and Mesozoic periods followed patterns similar to fires that occur in modern times. Surface fires driven by dry seasons are evident in Devonian and Carboniferous progymnosperm forests. Lepidodendron forests dating to the Carboniferous period have charred peaks, evidence of crown fires. In Jurassic gymnosperm forests, there is evidence of high frequency, light surface fires. The increase of fire activity in the late Tertiary is possibly due to the increase of C_4-type grasses. As these grasses shifted to more mesic habitats, their high flammability increased fire frequency, promoting grasslands over woodlands. However, fire-prone habitats may have contributed to the prominence of trees such as those of the genera *Eucalyptus*, *Pinus* and *Sequoia*, which have thick bark to withstand fires and employ serotiny.

Human Involvement

Aerial view of deliberate wildfires on the Khun Tan Range, Thailand. These fires are lit by local farmers every year in order to promote the growth of a certain mushroom

The human use of fire for agricultural and hunting purposes during the Paleolithic and Mesolithic ages altered the preexisting landscapes and fire regimes. Woodlands were gradually replaced by smaller vegetation that facilitated travel, hunting, seed-gathering and planting. In recorded human history, minor allusions to wildfires were mentioned in the Bible and by classical writers such as Homer. However, while ancient Hebrew, Greek, and Roman writers were aware of fires, they were not very interested in the uncultivated lands where wildfires occurred. Wildfires were used in battles throughout human history as early thermal weapons. From the Middle ages, accounts were written of occupational burning as well as customs and laws that governed the use of fire. In Germany, regular burning was documented in 1290 in the Odenwald and in 1344 in the Black Forest. In the 14th century Sardinia, firebreaks were used for wildfire protection. In Spain during the 1550s, sheep husbandry was discouraged in certain provinces by Philip II due to the harmful effects of fires used in transhumance. As early as the 17th century, Native Americans were observed using fire for many purposes including cultivation, signaling, and warfare. Scottish botanist David Douglas noted the native use of fire for tobacco cultivation, to encourage deer into smaller areas for hunting purposes, and to improve foraging for honey and grasshoppers. Charcoal found in sedimentary deposits off the Pacific coast of Central America suggests that more burning occurred in the 50 years before the Spanish colonization of the Americas than after the colonization. In the post-World War II Baltic region, socio-economic changes led more stringent air quality standards and bans on fires that eliminated traditional burning practices. In the mid-19th century, explorers from the HMS *Beagle* observed Australian Aborigines using fire for ground clearing, hunting, and regeneration of plant food in a method later named fire-stick farming. Such careful use of fire has been employed for centuries in the lands protected by Kakadu National Park to encourage biodiversity.

Wildfires typically occurred during periods of increased temperature and drought. An increase in fire-related debris flow in alluvial fans of northeastern Yellowstone National Park was linked to the period between AD 1050 and 1200, coinciding with the Medieval Warm Period. However, human influence caused an increase in fire frequency. Dendrochronological fire scar data and charcoal layer data in Finland suggests that, while many fires occurred during severe drought conditions, an increase in the number of fires during 850 BC and 1660 AD can be attributed to human influence. Charcoal evidence from the Americas suggested a general decrease in wildfires between 1 AD and 1750 compared to previous years. However, a period of increased fire frequency between 1750 and 1870 was suggested by charcoal data from North America and Asia, attributed to human population growth and influences such as land clearing practices. This period was followed by an overall decrease in burning in the 20th century, linked to the expansion of agriculture, increased livestock grazing, and fire prevention efforts. A meta-analysis found that 17 times more land burned annually in California before 1800 compared to recent decades (1,800,000 hectares/year compared to 102,000 hectares/year).

Invasive species moved by humans have in some cases increased the intensity of wildfires, such as Eucalyptus in California and gamba grass in Australia.

Prevention

1985 Smokey Bear poster with part of his admonition, "Only you can prevent forest fires".

Wildfire prevention refers to the preemptive methods aimed at reducing the risk of fires as well as lessening its severity and spread. Prevention techniques aim to manage air quality, maintain ecological balances, protect resources, and to affect future fires. North American firefighting policies permit naturally caused fires to burn to maintain their ecological role, so long as the risks of escape into high-value areas are mitigated. However, prevention policies must consider the role that humans play in wildfires, since, for example, 95% of forest fires in Europe are related to human involvement. Sources of human-caused fire may include arson, accidental ignition, or the uncontrolled use of fire in land-clearing and agriculture such as the slash-and-burn farming in Southeast Asia.

In 1937, U.S. President Franklin D. Roosevelt initiated a nationwide fire prevention campaign, highlighting the role of human carelessness in forest fires. Later posters of the program featured Uncle Sam, leaders of the Axis powers of World War II, characters from the Disney movie *Bambi*, and the official mascot of the U.S. Forest Service, Smokey Bear. Reducing human-caused ignitions may be the most effective means of reducing unwanted wildfire. Alteration of fuels is commonly undertaken when attempting to affect future fire risk and behavior. Wildfire prevention programs around the world may employ techniques such as *wildland fire use* and *prescribed or controlled burns*. *Wildland fire use* refers to any fire of natural causes that is monitored but allowed to burn. *Controlled burns* are fires ignited by government agencies under less dangerous weather conditions.

A prescribed burn in a *Pinus nigra* stand in Portugal

Vegetation may be burned periodically to maintain high species diversity and frequent burning of surface fuels limits fuel accumulation. Wildland fire use is the cheapest and most ecologically appropriate policy for many forests. Fuels may also be removed by logging, but fuels treatments and thinning have no effect on severe fire behavior Wildfire models are often used to predict and compare the benefits of different fuel treatments on future wildfire spread, but their accuracy is low.

Wildfire itself is reportedly "the most effective treatment for reducing a fire's rate of spread, fireline intensity, flame length, and heat per unit of area" according to Jan Van Wagtendonk, a biologist at the Yellowstone Field Station.

Building codes in fire-prone areas typically require that structures be built of flame-resistant materials and a defensible space be maintained by clearing flammable materials within a prescribed distance from the structure. Communities in the Philippines also maintain fire lines 5 to 10 meters (16 to 33 ft) wide between the forest and their village, and patrol these lines during summer months or seasons of dry weather. Continued residential development in fire-prone areas and rebuilding structures destroyed by fires has been met with criticism. The ecological benefits of fire are often overridden by the economic and safety benefits of protecting structures and human life.

US Wildfire Policy

Poster for forest fire prevention showing a burning cigarette and a forest fire.

History of Wildfire Policy in the U.S.

Since the turn of the 20th century, various federal and state agencies have been involved in wildland fire management in one form or another. In the early 20th century, for example, the federal government, through the U.S. Army and the U.S. Forest Service, solicited fire suppression as a primary goal of managing the nation's forests. At this time in history fire was viewed as a threat to timber, an economically important natural resource. As such, the decision was made to devote public funds to fire suppression and fire prevention efforts. For example, the Forest Fire Emergency Fund Act of 1908 permitted deficit spending in the case of emergency fire situations. As a result, the U.S. Forest Service was able to acquire a deficit of over $1 million in 1910 due to emergency fire suppression efforts. Following the same tone of timber resource protection, the U.S. Forest Service adopted the "10 AM Policy" in 1935. Through this policy, the agency advocated the control of all fires by 10 o'clock of the morning following the discovery of a wildfire. Fire prevention was also heavily advocated through public education campaigns such as Smokey Bear. Through these and similar public education campaigns the general public was, in a sense, trained to perceive all wildfire as a threat to civilized society and natural resources. The negative sentiment towards wildland fire prevailed and helped to shape wildland fire management objectives throughout most of the 20th century.

Texas was particularly hard-hit by wildfires in 2011, as noted by this placard at the state Forestry Museum in Lufkin.

Beginning in the 1970s public perception of wildland fire management began to shift. Despite strong funding for fire suppression in the first half of the 20th century, massive wildfires continued to be prevalent across the landscape of North America. Ecologists were beginning to recognize the presence and ecological importance of natural, lightning-ignited wildfires across the United States. It was learned that suppression of fire in certain ecosystems may in fact increase the likelihood that a wildfire will occur and may increase the intensity of those wildfires. With the emergence of fire ecology as a science also came an effort to apply fire to ecosystems in a controlled manner; however, suppression is still the main tactic when a fire is set by a human or if it threatens life or property. By the 1980s, in light of this new understanding, funding efforts began to support prescribed burning in order to prevent wildfire events. In 2001, the United States implemented a National Fire Plan, increasing the budget for the reduction of hazardous fuels from $108 million in 2000 to $401 million.

In addition to using prescribed fire to reduce the chance of catastrophic wildfires, mechanical methods have recently been adopted as well. Mechanical methods include the use of chippers and other machinery to remove hazardous fuels and thereby reduce the risk of wildfire events. Today the United States' maintains that, "fire, as a critical natural process, will be integrated into land and resource management plans and activities on a landscape scale, and across agency boundaries. Response to wildfire is based on ecological, social and legal consequences of fire. The circumstance under which a fire occurs, and the likely consequences and public safety and welfare, natural and cultural resources, and values to be protected dictate the appropriate management response to fire" (United States Department of Agriculture Guidance for Implementation of Federal Wildland Fire Management Policy, 13 February 2009). The five federal regulatory agencies managing forest fire response and planning for 676 million acres in the United States are the Department of the Interior, the Bureau of Land Management, the Bureau of Indian Affairs, the National Park Service, the United States Department of Agriculture-Forest Service and the United States Fish and Wildlife Services. Several hundred million U.S. acres of wildfire management are also conducted by state, county, and local fire management organizations. In 2014, legislators proposed The Wildfire Disaster Funding Act to provide $2.7 billion fund appropriated by congress for the USDA and Department of Interior to use in fire suppression. The bill is a reaction to United States Forest Service and Department of Interior costs of Western Wildfire suppression appending that amounted to $3.5 billion in 2013.

Wildland-urban Interface Policy

An aspect of wildfire policy that is gaining attention is the wildland-urban interface (WUI). More and more people are living in "red zones," or areas that are at high risk of wildfires. FEMA and the NFPA develop specific policies to guide homeowners and builders in how to build and maintain structures at the WUI and how protect against property losses. For example, NFPA-1141 is a standard for fire protection infrastructure for land development in wildland, rural and suburban areas and NFPA-1144 is a standard for reducing structure ignition hazards from wildland fire. For a full list of these policies and guidelines, Compensation for losses in the WUI are typically negotiated on an incident-by-incident basis. This is generating discussion about the burden of responsibility for funding and fighting a fire in the WUI, in that, if a resident chooses to live in a known red zone, should he or she retain a higher level of responsibility for funding home protection against wildfires. One initiative aimed at helping U.S. WUI communities live more safely with fire is called fire-adapted communities.

Economics of Fire Management Policy

Similar to that of military operations, fire management is often very expensive in the U.S. and the rest of the world. Today, it is not uncommon for suppression operations for a single wildfire to exceed costs of $1 million in just a few days. The United States Department of Agriculture allotted $2.2 billion for wildfire management in 2012. Although fire suppression purports to benefit society, other options for fire management exist. While these options cannot completely replace fire suppression as a fire management tool, other options can play an important role in overall fire management and can therefore affect the costs of fire suppression.

It is commonly accepted that past fire suppression and climate change has resulted in larger, more intense wildfire events which are seen today. In economic terms, expenditures used for wildfire suppression in the early 20th century have contributed to increased suppression costs which are being realized today.

Detection

Dry Mountain Fire Lookout in the Ochoco National Forest, Oregon, circa 1930

Fast and effective detection is a key factor in wildfire fighting. Early detection efforts were focused on early response, accurate results in both daytime and nighttime, and the ability to prioritize fire danger. Fire lookout towers were used in the United States in the early 20th century and fires were reported using telephones, carrier pigeons, and heliographs. Aerial and land photography using instant cameras were used in the 1950s until infrared scanning was developed for fire detection in the 1960s. However, information analysis and delivery was often delayed by limitations in communication technology. Early satellite-derived fire analyses were hand-drawn on maps at a remote site and sent via overnight mail to the fire manager. During the Yellowstone fires of 1988, a data station was established in West Yellowstone, permitting the delivery of satellite-based fire information in approximately four hours.

Currently, public hotlines, fire lookouts in towers, and ground and aerial patrols can be used as a means of early detection of forest fires. However, accurate human observation may be limited by operator fatigue, time of day, time of year, and geographic location. Electronic systems have gained popularity in recent years as a possible resolution to human operator error. A government report on a recent trial of three automated camera fire detection systems in Australia did, however, conclude "...detection by the camera systems was slower and less reliable than by a trained human observer". These systems may be semi- or fully automated and employ systems based on the risk area and degree of human presence, as suggested by GIS data analyses. An integrated approach of multiple systems can be used to merge satellite data, aerial imagery, and personnel position via Global Positioning System (GPS) into a collective whole for near-realtime use by wireless Incident Command Centers.

A small, high risk area that features thick vegetation, a strong human presence, or is close to a critical urban area can be monitored using a local sensor network. Detection systems may include wireless sensor networks that act as automated weather systems: detecting temperature, humidity, and smoke. These may be battery-powered, solar-powered, or *tree-rechargeable*: able to recharge their battery systems using the small electrical currents in plant material. Larger, medium-risk areas can be monitored by scanning towers that incorporate fixed cameras and sensors to detect smoke or additional factors such as the infrared signature of carbon dioxide produced by fires. Additional capabilities such as night vision, brightness detection, and color change detection may also be incorporated into sensor arrays.

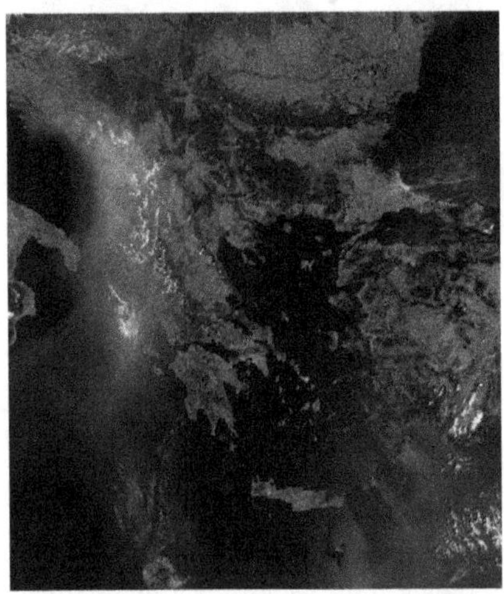

Wildfires across the Balkans in late July 2007 (MODIS image)

Satellite and aerial monitoring through the use of planes, helicopter, or UAVs can provide a wider view and may be sufficient to monitor very large, low risk areas. These more sophisticated systems employ GPS and aircraft-mounted infrared or high-resolution visible cameras to identify and target wildfires. Satellite-mounted sensors such as Envisat's Advanced Along Track Scanning Radiometer and European Remote-Sensing Satellite's Along-Track Scanning Radiometer can measure infrared radiation emitted by fires, identifying hot spots greater than 39 °C (102 °F). The National Oceanic and Atmospheric Administration's Hazard Mapping System combines remote-sensing data from satellite sources such as Geostationary Operational Environmental Satellite (GOES), Moderate-Resolution Imaging Spectroradiometer (MODIS), and Advanced Very High Resolution Radiometer (AVHRR) for detection of fire and smoke plume locations. However, satellite detection is prone to offset errors, anywhere from 2 to 3 kilometers (1 to 2 mi) for MODIS and AVHRR data and up to 12 kilometers (7.5 mi) for GOES data. Satellites in geostationary orbits may become disabled, and satellites in polar orbits are often limited by their short window of observation time. Cloud cover and image resolution and may also limit the effectiveness of satellite imagery.

in 2015 a new fire detection tool is in operation at the U.S. Department of Agriculture (USDA) Forest Service (USFS) which uses data from the Suomi National Polar-orbiting Partnership (NPP) satellite to detect smaller fires in more detail than previous space-based products. The high-resolution data is used with a computer model to predict how a fire will change direction based on weather and land conditions. The active fire detection product using data from Suomi NPP's Visible Infrared Imaging Radiometer Suite (VIIRS) increases the resolution of fire observations to 1,230 feet (375 meters). Previous NASA satellite data products available since the early 2000s observed fires at 3,280 foot (1 kilometer) resolution. The data is one of the intelligence tools used by the USFS and Department of Interior agencies across the United States to guide resource allocation and strategic fire management decisions. The enhanced VIIRS fire product enables detection every 12 hours or less of much smaller fires and provides more detail and consistent tracking of fire lines during long duration wildfires – capabilities critical for early warning systems and support of routine mapping of fire progression. Active fire locations are available to users within minutes from the satellite overpass through data processing facilities at the USFS Remote Sensing Applications Center, which uses technologies developed by the NASA Goddard Space Flight Center Direct Readout Laboratory in Greenbelt, Maryland. The model uses data on weather conditions and the land surrounding an active fire to predict 12–18 hours in advance whether a blaze will shift direction. The state of Colorado decided to incorporate the weather-fire model in its firefighting efforts beginning with the 2016 fire season.

In 2014, an international campaign was organized in South Africa's Kruger National Park to validate fire detection products including the new VIIRS active fire data. In advance of that campaign, the Meraka Institute of the Council for Scientific and Industrial Research in Pretoria, South Africa, an early adopter of the VIIRS 375m fire product, put it to use during several large wildfires in Kruger.

The demand for timely, high-quality fire information has increased in recent years. Wildfires in the United States burn an average of 7 million acres of land each year. For the last 10 years, the USFS and Department of Interior have spent a combined average of about $2–4 billion annually on wildfire suppression.

Suppression

A Russian firefighter extinguishing a wildfire

Wildfire suppression depends on the technologies available in the area in which the wildfire occurs. In less developed nations the techniques used can be as simple as throwing sand or beating the fire with sticks or palm fronds. In more advanced nations, the suppression methods vary due to increased technological capacity. Silver iodide can be used to encourage snow fall, while fire retardants and water can be dropped onto fires by unmanned aerial vehicles, planes, and helicopters. Complete fire suppression is no longer an expectation, but the majority of wildfires are often extinguished before they grow out of control. While more than 99% of the 10,000 new wildfires each year are contained, escaped wildfires under extreme weather conditions are difficult to suppress without a change in the weather. Wildfires in Canada and the US burn an average of 54,500 square kilometers (13,000,000 acres) per year.

Above all, fighting wildfires can become deadly. A wildfire's burning front may also change direction unexpectedly and jump across fire breaks. Intense heat and smoke can lead to disorientation and loss of appreciation of the direction of the fire, which can make fires particularly dangerous. For example, during the 1949 Mann Gulch fire in Montana, USA, thirteen smokejumpers died when they lost their communication links, became disoriented, and were overtaken by the fire. In the Australian February 2009 Victorian bushfires, at least 173 people died and over 2,029 homes and 3,500 structures were lost when they became engulfed by wildfire.

Costs of Wildfire Suppression

In California, the U.S. Forest Service spends about $200 million per year to suppress 98% of wildfires and up to $1 billion to suppress the other 2% of fires that escape initial attack and become large.

Wildland Firefighting Safety

Wildfire fighters cutting down a tree using a chainsaw

Wildland fire fighters face several life-threatening hazards including heat stress, fatigue, smoke and dust, as well as the risk of other injuries such as burns, cuts and scrapes, animal bites, and even rhabdomyolysis.

Especially in hot weather condition, fires present the risk of heat stress, which can entail feeling heat, fatigue, weakness, vertigo, headache, or nausea. Heat stress can progress into heat strain, which entails physiological changes such as increased heart rate and core body temperature. This can lead to heat-related illnesses, such as heat rash, cramps, exhaustion or heat stroke. Various factors can contribute to the risks posed by heat stress, including strenuous work, personal risk factors such as age and fitness, dehydration, sleep deprivation, and burdensome personal protective equipment. Rest, cool water, and occasional breaks are crucial to mitigating the effects of heat stress.

Smoke, ash, and debris can also pose serious respiratory hazards to wildland fire fighters. The smoke and dust from wildfires can contain gases such as carbon monoxide, sulfur dioxide and formaldehyde, as well as particulates such as ash and silica. To reduce smoke exposure, wildfire fighting crews should, whenever possible, rotate firefighters through areas of heavy smoke, avoid downwind firefighting, use equipment rather than people in holding areas, and minimize mop-up. Camps and command posts should also be located upwind of wildfires. Protective clothing and equipment can also help minimize exposure to smoke and ash.

Firefighters are also at risk of cardiac events including strokes and heart attacks. Fire fighters should maintain good physical fitness. Fitness programs, medical screening and examination programs which include stress tests can minimize the risks of firefighting cardiac problems. Other injury hazards wildland fire fighters face include slips, trips and falls, burns, scrapes and cuts from tools and equipment, being struck by trees, vehicles, or other objects, plant hazards such as thorns and poison ivy, snake and animal bites, vehicle crashes, electrocution from power lines or lightning storms, and unstable building structures.

Fire Retardant

Fire retardants are used to help slow wildfires, coat fuels, and lessen oxygen availability as required by various firefighting situations. They are composed of nitrates, ammonia, phosphates and sulfates, as well as other chemicals and thickening agents. The choice of whether to apply retardant depends on the magnitude, location and intensity of the wildfire. Fire retardants are used to reach inaccessible geographical regions where ground firefighting crews are unable to reach a wildfire or in any occasion where human safety and structures are endangered. In certain instances, fire retardant may also be applied ahead of wildfires for protection of structures and vegetation as a precautionary fire defense measure.

The application of aerial fire retardants creates an atypical appearance on land and water surfaces and has the potential to change soil chemistry. Fire retardant can decrease the availability of plant nutrients in the soil by increasing the acidity of the soil. Fire retardant may also affect water quality through leaching, eutrophication, or misapplication. Fire retardant's effects on drinking water remain inconclusive. Dilution factors, including water body size, rainfall, and water flow rates lessen the concentration and potency of fire retardant. Wildfire debris (ash and sediment) clog rivers and reservoirs increasing the risk for floods and erosion that ultimately slow and/or damage water treatment systems. There is continued concern of fire retardant effects on land, water, wildlife habitats, and watershed quality, additional research is needed. However, on the positive side, fire retardant (specifically its nitrogen and phosphorus components) has been shown to have a fertilizing effect on nutrient-deprived soils and thus creates a temporary increase in vegetation.

Current USDA procedure maintains that the aerial application of fire retardant in the United States must clear waterways by a minimum of 300 feet in order to safeguard effects of retardant runoff. Aerial uses of fire retardant are required to avoid application near waterways and endangered species (plant and animal habitats). After any incident of fire retardant misapplication, the U.S. Forest Service requires reporting and assessment impacts be made in order to determine mitigation, remediation, and/or restrictions on future retardant uses in that area.

Modeling

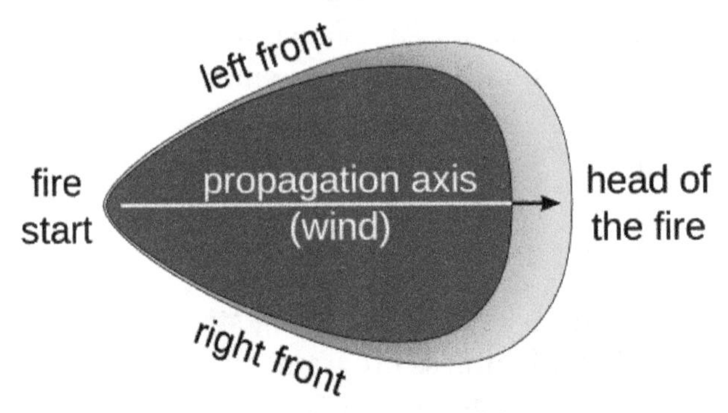

Fire Propagation Model

Wildfire modeling is concerned with numerical simulation of wildfires in order to comprehend and predict fire behavior. Wildfire modeling aims to aid wildfire suppression, increase the safety of firefighters and the public, and minimize damage. Using computational science, wildfire modeling involves the statistical analysis of past fire events to predict spotting risks and front behavior. Various wildfire propagation models have been proposed in the past, including simple ellipses and egg- and fan-shaped models. Early attempts to determine wildfire behavior assumed terrain and vegetation uniformity. However, the exact behavior of a wildfire's front is dependent on a variety of factors, including windspeed and slope steepness. Modern growth models utilize a combination of past ellipsoidal descriptions and Huygens' Principle to simulate fire growth as a continuously expanding polygon. Extreme value theory may also be used to predict the size of large wildfires. However, large fires that exceed suppression capabilities are often regarded as statistical outliers in standard analyses, even though fire policies are more influenced by large wildfires than by small fires.

2003 Canberra firestorm

Human Risk and Exposure

Wildfire risk is the chance that a wildfire will start in or reach a particular area and the potential loss of human values if it does. Risk is dependent on variable factors such as human activities, weather patterns, availability of wildfire fuels, and the availability or lack of resources to suppress a fire. Wildfires have continually been a threat to human populations. However, human induced geographical and climatic changes are exposing populations more frequently to wildfires and increasing wildfire risk. It is speculated that the increase in wildfires arises from a century of wildfire suppression coupled with the rapid expansion of human developments into fire-prone wildlands. Wildfires are naturally occurring events that aid in promoting forest health. Global warming and climate changes are causing an increase in temperatures and more droughts nationwide which contributes to an increase in wildfire risk.

2009 California Wildfires at NASA/JPL – Pasadena, California

Regional Burden of Wildfires in the United States

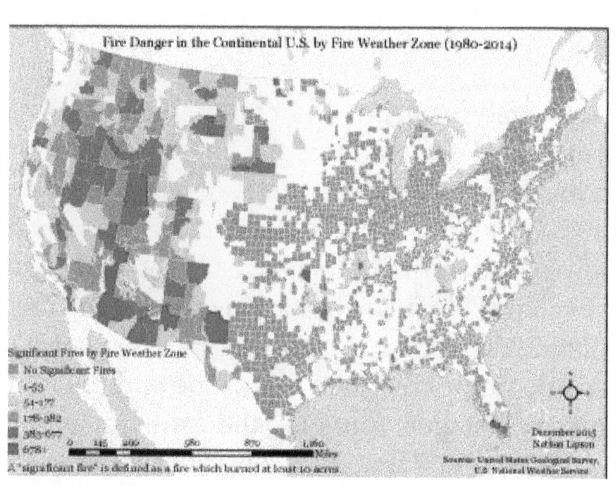

This map depicts regional burden to wildfires in the US from 1980-2014. These are categorized by "fire weather zone" as defined by the National Weather Service.

Nationally, the burden of wildfires is disproportionally heavily distributed in the southern and western regions. The Geographic Area Coordinating Group (GACG) divides the United States and Alaska into 11 geographic areas for the purpose of emergency incident management. One particular area of focus is wildland fires. A national assessment of wildfire risk in the United States based on GACG identified regions (with the slight modification of combining Southern and Northern California, and the West and East Basin); indicate that California (50.22% risk) and the Southern Area (15.53% risk) are the geographic areas with the highest wildfire risk. The western areas of the nation are experiencing an expansion of human development into and beyond what is called the wildland-urban interface (WUI). When wildfires inevitably occur in these fire-prone areas, often communities are threatened due to their proximity to fire-prone forest. The south is one of the fastest growing regions with 88 million acres classified as WUI. The south consistently has the highest number of wildfires per year.

More than 50, 000 communities are estimated to be at high to very high risk of wildfire damage. These statistics are greatly attributable to the South's year-round fire season.

Wild ires' Risk to Human Health

The most noticeable adverse effect of wildfires is the destruction of property. However, the release of hazardous chemicals from the burning of wildland fuels also significantly impacts health in humans.

Wildfire smoke is composed primarily of carbon dioxide and water vapor. Other common smoke components present in lower concentrations are carbon monoxide, formaldehyde, acrolein, polyaromatic hydrocarbons, and benzene. Small particulates suspended in air which come in solid form or in liquid droplets are also present in smoke. 80 -90% of wildfire smoke, by mass, is within the fine particle size class of 2.5 micrometers in diameter or smaller.

Despite carbon dioxide's high concentration in smoke, it poses a low health risk due to its low toxicity. Rather, carbon monoxide and fine particulate matter, particularly 2.5 μm in diameter and smaller, have been identified as the major health threats. Other chemicals are considered to be significant hazards but are found in concentrations that are too low to cause detectable health effects.

The degree of wildfire smoke exposure to an individual is dependent on the length, severity, duration, and proximity of the fire. People are exposed directly to smoke via the respiratory tract though inhalation of air pollutants. Indirectly, communities are exposed to wildfire debris that can contaminate soil and water supplies.

Firefighters are at the greatest risk for acute and chronic health effects resulting from wildfire smoke exposure. Due to firefighters' occupational duties, they are frequently exposed to hazardous chemicals at a close proximity for longer periods of time. A case study on the exposure of wildfire smoke among wildland firefighters shows that firefighters are exposed to significant levels of carbon monoxide and respiratory irritants above OSHA-permissible exposure limits (PEL) and ACGIH threshold limit values (TLV). 5–10% are overexposed. The study obtained exposure concentrations for one wildland firefighter over a 10-hour shift spent holding down a fireline. The firefighter was exposed to a wide range of carbon monoxide and respiratory irritant (combination of particulate matter 3.5 μm and smaller, acrolein, and formaldehype) levels. Carbon monoxide levels reached up to 160ppm and the TLV irritant index value reached a high of 10. In contrast, the OSHA PEL for carbon monoxide is 30ppm and for the TLV respiratory irritant index, the calculated threshold limit value is 1; any value above 1 exceeds exposure limits.

Residents in communities surrounding wildfires are exposed to lower concentrations of chemicals, but they are at a greater risk for indirect exposure through water or soil contamination. Exposure to residents is greatly dependent on individual susceptibility. Vulnerable persons such as children (ages 0–4), the elderly (ages 65 and older), smokers, and pregnant women are at an increased risk due to their already compromised body systems, even when the exposures are present at low chemical concentrations and for relatively short exposure periods.

The U.S. Environmental Protection Agency (EPA) developed the Air Quality Index (AQI), a public resource that provides national air quality standard concentrations for common air pollutants. The public can use this index as a tool to determine their exposure to hazardous air pollutants based on visibility range.

Additionally, there is evidence of an increase in material stress, as documented by researchers M.H. O'Donnell and A.M. Behie, thus affecting birth outcomes. In Australia, studies show that male infants born with drastically higher average birth weights were born in mostly severely fire-affected areas. This is attributed to the fact that maternal signals directly affect fetal growth patterns.

Health Effects

Inhalation of smoke from a wildfire can be a health hazard. Wildfire smoke is composed of carbon dioxide, water vapor, particulate matter, organic chemicals, nitrogen oxides and other compounds. The principal health concern is the inhalation of particulate matter and carbon monoxide.

Particulate matter (PM) is a type of air pollution made up of particles of dust and liquid droplets. They are characterized into two categories based on the diameter of the particle. Coarse particles are between 2.5 micrometers and 10 micrometers and fine particles measure 2.5 micrometers and less. Both sizes can be inhaled. Coarse particles are filtered by the upper airways and can cause eye and sinus irritation as well as sore throat and coughing. The fine particles are more problematic because, when inhaled, they can be deposited deep into the lungs, where they are absorbed into the bloodstream. This is particularly hazardous to the very young, elderly and those with chronic conditions such as asthma, chronic obstructive pulmonary disease (COPD), cystic fibrosis and cardiovascular conditions. The illnesses most commonly with exposure to fine particle from wildfire smoke are bronchitis, exacerbation of asthma or COPD, and pneumonia. Symptoms of these complications include wheezing and shortness of breath and cardiovascular symptoms include chest pain, rapid heart rate and fatigue.

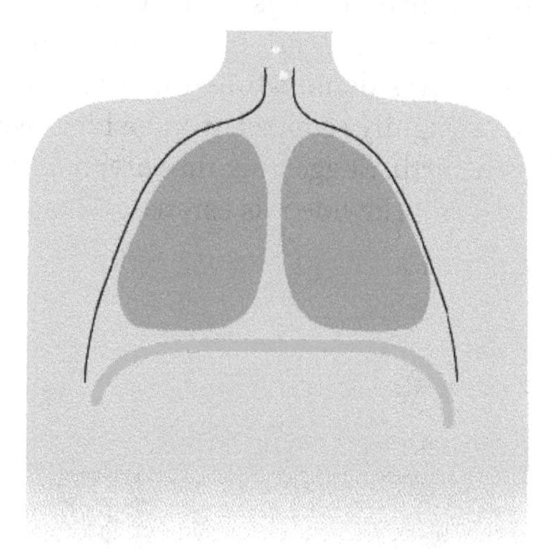

Animation of diaphragmatic breathing with the diaphragm shown in green

Carbon monoxide (CO) is a colorless, odorless gas that can be found at the highest concentration at close proximity to a smoldering fire. For this reason, carbon monoxide inhalation is a serious threat to the health of wildfire firefighters. CO in smoke can be inhaled into the lungs where it is absorbed into the bloodstream and reduces oxygen delivery to the body's vital organs. At high concentrations, it can cause headache, weakness, dizziness, confusion, nausea, disorientation, visual impairment, coma and even death. However, even at lower concentrations, such as those found at wildfires, individuals with cardiovascular disease may experience chest pain and cardiac arrhythmia. A recent study tracking the number and cause of wildfire firefighter deaths from 1990–2006 found that 21.9% of the deaths occurred from heart attacks.

Another important and somewhat less obvious health effect of wildfires is psychiatric diseases and disorders. Both adults and children from countries ranging from the United States and Canada to Greece and Australia who were directly and indirectly affected by wildfires were found by researchers to demonstrate several different mental conditions linked to their experience with the wildfires. These include post-traumatic stress disorder (PTSD), depression, anxiety, and phobias.

In a new twist to wildfire health effects, former uranium mining sites were burned over in the summer of 2012 near North Fork, Idaho. This prompted concern from area residents and Idaho State Department of Environmental Quality officials over the potential spread of radiation in the resultant smoke, since those sites had never been completely cleaned up from radioactive remains.

Epidemiology

The EPA has defined acceptable concentrations of particulate matter in the air, through the National Ambient Air Quality Standards and monitoring of ambient air quality has been mandated. Due to these monitoring programs and the incidence of several large wildfires near populated areas, epidemiological studies have been conducted and demonstrate an association between human health effects and an increase in fine particulate matter due to wildfire smoke.

An increase in PM emitted from the Hayman fire in Colorado in June 2002, was associated with an increase in respiratory symptoms in patients with COPD. Looking at the wildfires in Southern California in October 2003 in a similar manner, investigators have shown an increase in hospital admissions due to asthma during peak concentrations of PM. Children participating in the Children's Health Study were also found to have an increase in eye and respiratory symptoms, medication use and physician visits. Recently, it was demonstrated that mothers who were pregnant during the fires gave birth to babies with a slightly reduced average birth weight compared to those who were not exposed to wildfire during birth. Suggesting that pregnant women may also be at greater risk to adverse effects from wildfire. Worldwide it is estimated that 339,000 people die due to the effects of wildfire smoke each year.

Drought

Contraction/Desiccation cracks in dry earth (Sonoran desert, Mexico).

A drought is a period of below-average precipitation in a given region, resulting in prolonged shortages in its water supply, whether atmospheric, surface water or ground water. A drought can last for months or years, or may be declared after as few as 15 days. It can have a substantial impact on the ecosystem and agriculture of the affected region and harm to the local economy. Annual dry seasons in the tropics significantly increase the chances of a drought developing and subsequent bush fires. Periods of heat can significantly worsen drought conditions by hastening evaporation of water vapour.

Many plant species, such as those in the family Cactaceae (or cacti), have drought tolerance adaptations like reduced leaf area and waxy cuticles to enhance their ability to tolerate drought. Some others survive dry periods as buried seeds. Semi-permanent drought produces arid biomes such as deserts and grasslands. Prolonged droughts have caused mass migrations and humanitarian crises. Most arid ecosystems have inherently low productivity. The most prolonged drought ever in the world in recorded history occurred in the Atacama Desert in Chile (400 Years).

Causes of Drought

Precipitation Deficiency

Mechanisms of producing precipitation include convective, stratiform, and orographic rainfall. Convective processes involve strong vertical motions that can cause the overturning of the atmosphere in that location within an hour and cause heavy precipitation, while stratiform processes involve weaker upward motions and less intense precipitation over a longer duration. Precipitation can be divided into three categories, based on whether it falls as liquid water, liquid water that freezes on contact with the surface, or ice. Drought are mainly course by in low rain areas. If these factors do not support precipitation volumes sufficient to reach the surface over a sufficient time, the result is a drought. Drought can be triggered by a high level of reflected sunlight and above average prevalence of high pressure systems, winds carrying continental, rather than oceanic air masses, and ridges of high pressure areas aloft can prevent or restrict the developing of thunderstorm activity or rainfall over one certain region. Once a region is within drought, feedback mechanisms such as local arid air, hot conditions which can promote warm core ridging, and minimal evapotranspiration can worsen drought conditions.

Ancient Meso-American civilizations may have amplified droughts by deforestation.

Dry Season

Within the tropics, distinct, wet and dry seasons emerge due to the movement of the Intertropical Convergence Zone or Monsoon trough. The dry season greatly increases drought occurrence, and is characterized by its low humidity, with watering holes and rivers drying up. Because of the lack of these watering holes, many grazing animals are forced to migrate due to the lack of water and feed to more fertile spots. Examples of such animals are zebras, elephants, and wildebeest. Because of the lack of water in the plants, bushfires are common. Since water vapor becomes more energetic with increasing temperature, more water vapor is required to increase relative humidity values to 100% at higher temperatures (or to get the temperature to fall to the dew point). Periods of warmth quicken the pace of fruit and vegetable production, increase evaporation and transpiration from plants, and worsen drought conditions.

Sheep on a drought affected paddock near Uranquinty, New South Wales.

El Niño

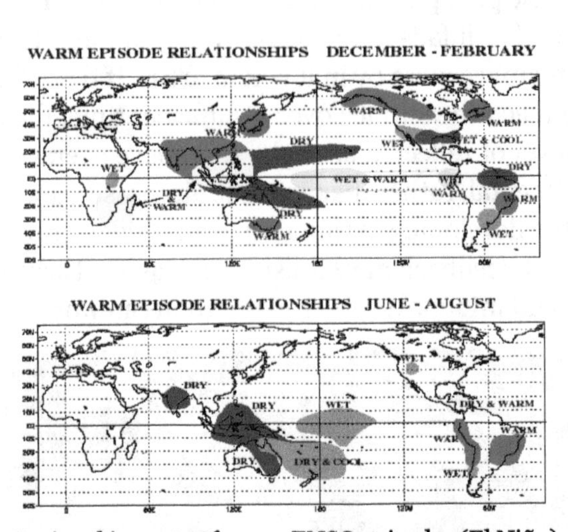

Regional impacts of warm ENSO episodes (El Niño)

Drier and hotter weather occurs in parts of the Amazon River Basin, Colombia, and Central America during El Niño events. Winters during the El Niño are warmer and drier than average conditions in the Northwest, northern Midwest, and northern Mideast United States, so those regions experience reduced snowfalls. Conditions are also drier than normal from December to February in south-central Africa, mainly in Zambia, Zimbabwe, Mozambique, and Botswana. Direct effects of El Niño resulting in drier conditions occur in parts of Southeast Asia and Northern Australia, increasing bush fires, worsening haze, and decreasing air quality dramatically. Drier-than-normal conditions are also in general observed in Queensland, inland Victoria, inland New South Wales, and eastern Tasmania from June to August. As warm water spreads from the west Pacific and the Indian Ocean to the east Pacific, it causes extensive drought in the western Pacific. Singapore experienced the driest February in 2014 since records began in 1869, with only 6.3 mm of rain falling in the month and temperatures hitting as high as 35 °C on 26 February. The years 1968 and 2005 had the next driest Februaries, when 8.4 mm of rain fell.

Erosion and Human Activities

Human activity can directly trigger exacerbating factors such as over farming, excessive irrigation, deforestation, and erosion adversely impact the ability of the land to capture and hold water. In arid climates, the main source of erosion is wind. Erosion can be the result of material movement by the wind. The wind can cause small particles to be lifted and therefore moved to another region (deflation). Suspended particles within the wind may impact on solid objects causing erosion by abrasion (ecological succession). Wind erosion generally occurs in areas with little or no vegetation, often in areas where there is insufficient rainfall to support vegetation.

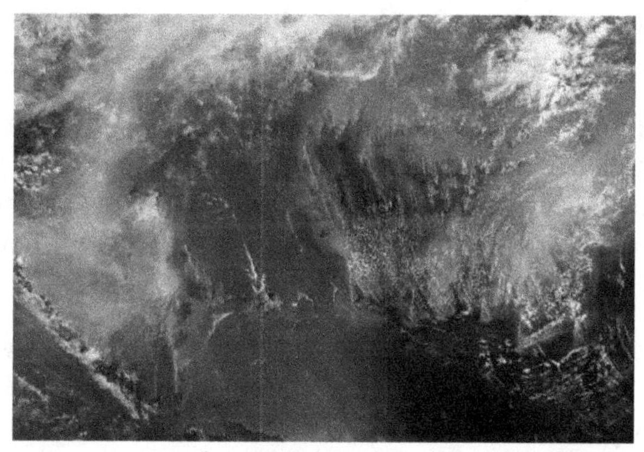

Fires on Borneo and Sumatra, 2006. People use slash-and-burn deforestation to clear land for agriculture.

Loess is a homogeneous, typically nonstratified, porous, friable, slightly coherent, often calcareous, fine-grained, silty, pale yellow or buff, windblown (Aeolian) sediment. It generally occurs as a widespread blanket deposit that covers areas of hundreds of square kilometers and tens of meters thick. Loess often stands in either steep or vertical faces. Loess tends to develop into highly rich soils. Under appropriate climatic conditions, areas with loess are among the most agriculturally productive in the world. Loess deposits are geologically unstable by nature, and will erode very readily. Therefore, windbreaks (such as big trees and bushes) are often planted by farmers to reduce the wind erosion of loess. Wind erosion is much more severe in arid areas and during times of drought. For example, in the Great Plains, it is estimated that soil loss due to wind erosion can be as much as 6100 times greater in drought years than in wet years.

Fields outside Benambra, Victoria, Australia suffering from drought conditions.

Climate Change

Activities resulting in global climate change are expected to trigger droughts with a substantial impact on agriculture throughout the world, and especially in developing nations. Overall, global warming will result in increased world rainfall. Along with drought in some areas, flooding and erosion will increase in others. Paradoxically, some proposed solutions to global warming that focus on more active techniques, solar radiation management through the use of a space sunshade for one, may also carry with them increased chances of drought.

Types

As a drought persists, the conditions surrounding it gradually worsen and its impact on the local population gradually increases. People tend to define droughts in three main ways:

1. Meteorological drought is brought about when there is a prolonged time with less than average precipitation. Meteorological drought usually precedes the other kinds of drought.
2. Agricultural droughts are droughts that affect crop production or the ecology of the range. This condition can also arise independently from any change in precipitation levels when soil conditions and erosion triggered by poorly planned agricultural endeavors cause a shortfall in water available to the crops. However, in a traditional drought, it is caused by an extended period of below average precipitation.
3. Hydrological drought is brought about when the water reserves available in sources such as aquifers, lakes and reservoirs fall below the statistical average. Hydrological drought tends to show up more slowly because it involves stored water that is used but not replenished. Like an agricultural drought, this can be triggered by more than just a loss of rainfall. For instance, Kazakhstan was recently awarded a large amount of money by the World Bank to restore water that had been diverted to other nations from the Aral Sea under Soviet rule. Similar circumstances also place their largest lake, Balkhash, at risk of completely drying out.

Consequences of Drought

A Mongolian gazelle dead due to drought.

The effects of droughts and water shortages can be divided into three groups: environmental, economic and social consequences. In the case of environmental effects: lower surface and subterranean water levels, lower flow levels (with a decrease below the minimum leading to direct danger for amphibian life), increased pollution of surface water, the drying out of wetlands, more and larger fires, higher deflation intensity, losing biodiversity, worse health of trees and the appearance of pests and den droid diseases. Economic losses include lower agricultural, forest, game and fishing output, higher food production costs, lower energy production levels in hydro plants, losses caused by depleted water tourism and transport revenue, problems with water supply for the energy sector and technological processes in metallurgy, mining, the chemical, paper, wood, foodstuff industries etc., disruption of water supplies for municipal economies. Meanwhile, social costs include the negative effect on the health of people directly exposed to this phenomenon (excessive heat waves), possible limitation of water supplies and its increased pollution levels, high food costs, stress caused by failed harvests, etc. This is why droughts and fresh water shortages may be considered as a factor which increases the gap between developed and developing countries.

The effect varies according to vulnerability. For example, subsistence farmers are more likely to migrate during drought because they do not have alternative food sources. Areas with populations that depend on water sources as a major food source are more vulnerable to famine.

Drought can also reduce water quality, because lower water flows reduce dilution of pollutants and increase contamination of remaining water sources. Common consequences of drought include:

- Diminished crop growth or yield productions and carrying capacity for livestock
- Dust bowls, themselves a sign of erosion, which further erode the landscape
- Dust storms, when drought hits an area suffering from desertification and erosion
- Famine due to lack of water for irrigation
- Habitat damage, affecting both terrestrial and aquatic wildlife
- Hunger, drought provides too little water to support food crops.
- Malnutrition, dehydration and related diseases
- Mass migration, resulting in internal displacement and international refugees
- Reduced electricity production due to reduced water flow through hydroelectric dams
- Shortages of water for industrial users
- Snake migration, which results in snakebites
- Social unrest
- War over natural resources, including water and food
- Wildfires, such as Australian bushfires, are more common during times of drought and even death of people.
- Exposure and oxidation of acid sulfate soils due to falling surface and groundwater levels.
- Cyanotoxin accumulation within food chains and water supply, some of which are among

the most potent toxins known to science, can cause cancer with low exposure over long term. High levels of microcystin has been found in San Francisco Bay Area salt water shellfish and fresh water supplies throughout the state of California in 2016.

Globally

Drought is a normal, recurring feature of the climate in most parts of the world. It is among the earliest documented climatic events, present in the Epic of Gilgamesh and tied to the biblical story of Joseph's arrival in and the later Exodus from Ancient Egypt. Hunter-gatherer migrations in 9,500 BC Chile have been linked to the phenomenon, as has the exodus of early humans out of Africa and into the rest of the world around 135,000 years ago.

A South Dakota farm during the Dust Bowl, 1936

Examples

Well-known historical droughts include:

- 1900 India killing between 250,000 and 3.25 million.
- 1921–22 Soviet Union in which over 5 million perished from starvation due to drought
- 1928–30 Northwest China resulting in over 3 million deaths by famine.
- 1936 and 1941 Sichuan Province China resulting in 5 million and 2.5 million deaths respectively.
- The 1997–2009 Millennium Drought in Australia led to a water supply crisis across much of the country. As a result, many desalination plants were built for the first time.
- In 2006, Sichuan Province China experienced its worst drought in modern times with nearly 8 million people and over 7 million cattle facing water shortages.
- 12-year drought that was devastating southwest Western Australia, southeast South Australia, Victoria and northern Tasmania was "very severe and without historical precedent".

The Darfur conflict in Sudan, also affecting Chad, was fueled by decades of drought; combination of drought, desertification and overpopulation are among the causes of the Darfur conflict, because the Arab Baggara nomads searching for water have to take their livestock further south, to land mainly occupied by non-Arab farming people.

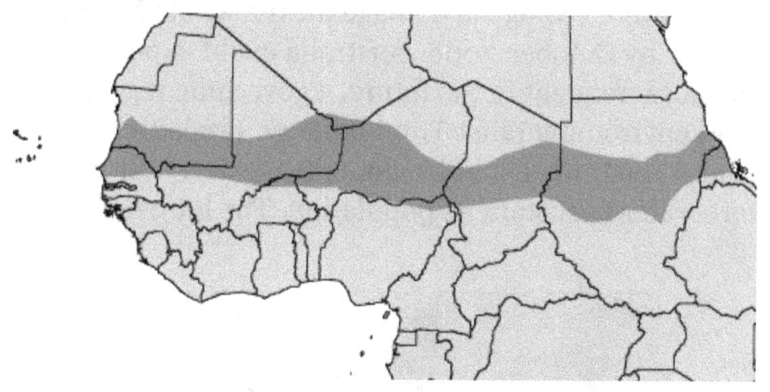

Affected areas in the western Sahel belt during the 2012 drought.

Approximately 2.4 billion people live in the drainage basin of the Himalayan rivers. India, China, Pakistan, Bangladesh, Nepal and Myanmar could experience floods followed by droughts in coming decades. Drought in India affecting the Ganges is of particular concern, as it provides drinking water and agricultural irrigation for more than 500 million people. The west coast of North America, which gets much of its water from glaciers in mountain ranges such as the Rocky Mountains and Sierra Nevada, also would be affected.

Drought affected area in Karnataka, India in 2012.

In 2005, parts of the Amazon basin experienced the worst drought in 100 years. A 23 July 2006 article reported Woods Hole Research Center results showing that the forest in its present form could survive only three years of drought. Scientists at the Brazilian National Institute of Amazonian Research argue in the article that this drought response, coupled with the effects of deforestation on regional climate, are pushing the rainforest towards a "tipping point" where it would irreversibly start to die. It concludes that the rainforest is on the brink of being turned into savanna or desert, with catastrophic consequences for the world's climate. According to the WWF, the combination of climate change and deforestation increases the drying effect of dead trees that fuels forest fires.

By far the largest part of Australia is desert or semi-arid lands commonly known as the outback. A 2005 study by Australian and American researchers investigated the desertification of the interior, and suggested that one explanation was related to human settlers who arrived about

50,000 years ago. Regular burning by these settlers could have prevented monsoons from reaching interior Australia. In June 2008 it became known that an expert panel had warned of long term, maybe irreversible, severe ecological damage for the whole Murray-Darling basin if it did not receive sufficient water by October 2008. Australia could experience more severe droughts and they could become more frequent in the future, a government-commissioned report said on July 6, 2008. Australian environmentalist Tim Flannery, predicted that unless it made drastic changes, Perth in Western Australia could become the world's first ghost metropolis, an abandoned city with no more water to sustain its population. The long Australian Millennial drought broke in 2010.

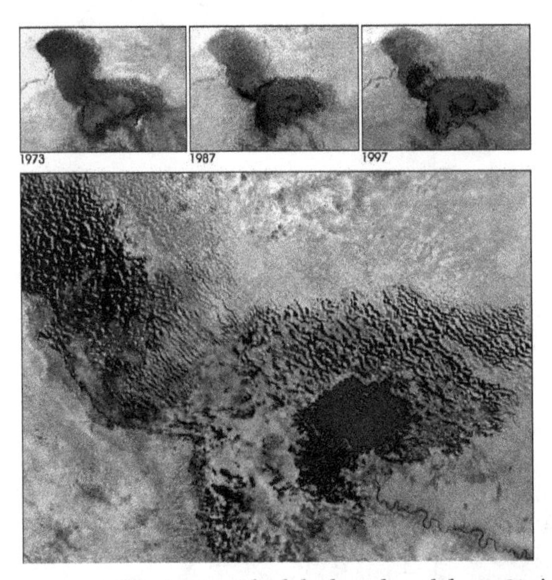

Lake Chad in a 2001 satellite image. The lake has shrunk by 95% since the 1960s.

Recurring droughts leading to desertification in East Africa have created grave ecological catastrophes, prompting food shortages in 1984–85, 2006 and 2011. During the 2011 drought, an estimated 50,000 to 150,000 people were reported to have died, though these figures and the extent of the crisis are disputed. In February 2012, the UN announced that the crisis was over due to a scaling up of relief efforts and a bumper harvest. Aid agencies subsequently shifted their emphasis to recovery efforts, including digging irrigation canals and distributing plant seeds.

In 2012, a severe drought struck the western Sahel. The Methodist Relief & Development Fund (MRDF) reported that more than 10 million people in the region were at risk of famine due to a month-long heat wave that was hovering over Niger, Mali, Mauritania and Burkina Faso. A fund of about £20,000 was distributed to the drought-hit countries.

Protection, Mitigation and Relief

Agriculturally, people can effectively mitigate much of the impact of drought through irrigation and crop rotation. Failure to develop adequate drought mitigation strategies carries a grave human cost in the modern era, exacerbated by ever-increasing population densities. President Roosevelt on April 27, 1935, signed documents creating the Soil Conservation Service (SCS)—now the Natural Resources Conservation Service (NRCS). Models of the law were sent to each state where they were enacted. These were the first enduring practical programs to curtail future susceptibility to drought, creating agencies that first began to stress soil conservation measures to protect farm lands today. It was not until the 1950s that there was an importance placed on water conservation was put into the existing laws (NRCS 2014).

Succulent plants are well-adapted to survive long periods of drought.

Water distribution on Marshall Islands during El Niño.

Strategies for drought protection, mitigation or relief include:

- Dams - many dams and their associated reservoirs supply additional water in times of drought.
- Cloud seeding - a form of intentional weather modification to induce rainfall. This remains a hotly debated topic, as the United States National Research Council released a report in 2004 stating that to date, there is still no convincing scientific proof of the efficacy of intentional weather modification.
- Desalination - of sea water for irrigation or consumption.
- Drought monitoring - Continuous observation of rainfall levels and comparisons with current usage levels can help prevent man-made drought. For instance, analysis of water usage in Yemen has revealed that their water table (underground water level) is put at grave risk by over-use to fertilize their Khat crop. Careful monitoring of moisture levels can also

help predict increased risk for wildfires, using such metrics as the Keetch-Byram Drought Index or Palmer Drought Index.

- Land use - Carefully planned crop rotation can help to minimize erosion and allow farmers to plant less water-dependent crops in drier years.
- Outdoor water-use restriction - Regulating the use of sprinklers, hoses or buckets on outdoor plants, filling pools, and other water-intensive home maintenance tasks. Xeriscaping yards can significantly reduce unnecessary water use by residents of towns and cities.
- Rainwater harvesting - Collection and storage of rainwater from roofs or other suitable catchments.
- Recycled water - Former wastewater (sewage) that has been treated and purified for reuse.
- Transvasement - Building canals or redirecting rivers as massive attempts at irrigation in drought-prone areas.

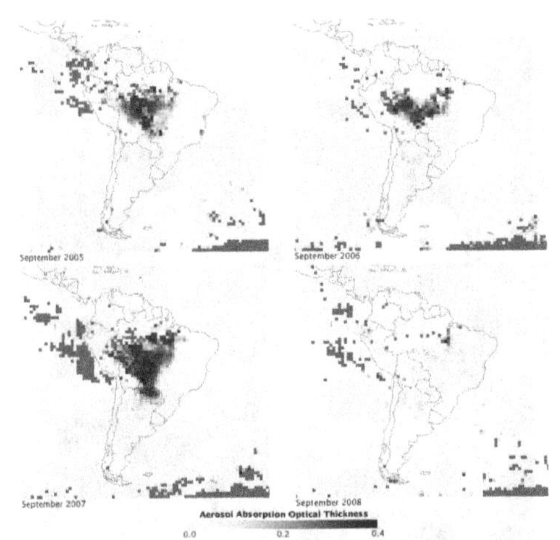

Aerosols over the Amazon each September for four burning seasons (2005 through 2008) during the Amazon basin drought. The aerosol scale (yellow to dark reddish-brown) indicates the relative amount of particles that absorb sunlight.

References

- Robert Penrose Pearce (2002). Meteorology at the Millennium. Academic Press. p. 66. ISBN 978-0-12-548035-2. Retrieved 2009-01-02.
- Robert A. Houze, Jr. (1994-06-28). Cloud Dynamics. Academic Press. p. 348. ISBN 0080502105. Retrieved 2015-02-18.
- Roland Paepe; Rhodes Whitmore Fairbridge; Saskia Jelgersma (1990). Greenhouse Effect, Sea Level and Drought. Springer Science & Business Media. p. 22. ISBN 0792310179. Retrieved 2015-02-18.
- Joseph S. D'Aleo; Pamela G. Grube (2002). The Oryx Resource Guide to El Niño and La Niña. Greenwood Publishing Group. pp. 48–49. ISBN 1573563781. Retrieved 2015-02-18.
- Bin Wang (2006-01-13). The Asian Monsoon. Springer Science & Business Media. p. 206. ISBN 3540406107. Retrieved 2008-05-03.
- Vijendra K. Boken; Arthur P. Cracknell; Ronald L. Heathcote (2005-03-24). Monitoring and Predicting Agricultural Drought : A Global Study: A Global Study. Oxford University Press. p. 349. ISBN 0198036787. Re-

trieved 2015-02-18.

- K.E.K. Neuendorf; J.P. Mehl, Jr.; J.A. Jackson (2005). Glossary of Geology. Springer-Verlag, New York. p. 779. ISBN 3-540-27951-2.
- Arthur Getis; Judith Getis and Jerome D. Fellmann (2000). Introduction to Geography, Seventh Edition. McGraw-Hill. p. 99. ISBN 0-697-38506-X.
- Prokurat, Sergiusz (2015). "Drought and water shortages in Asia as a threat and economic problem" (PDF). Journal of Modern Science. 26 (3). Retrieved 4 August 2016.
- Bill Gabbert (November 9, 2015). "Was the 2014 wildfire season in California affected by climate change?". Wildfire Today. Retrieved May 17, 2016.
- Yoon; et al. (2015). "Extreme Fire Season in California: A Glimpse Into the Future?". 96 (11). Bibcode:2015BAMS...96S...5Y. doi:10.1175/BAMS-D-15-00114.1.
- Alistair B. Fraser (1994-11-27). "Bad Meteorology: The reason clouds form when air cools is because cold air cannot hold as much water vapor as warm air". Retrieved 2015-02-17.
- Cooperative Extension Service (January 2014). Home Vegetable Gardening in Kentucky (PDF). University of Kentucky. p. 19. Retrieved 2015-02-18.
- National Oceanic and Atmospheric Administration (2002-05-16). "Warm Temperatures and Severe Drought Continued in April Throughout Parts of the United States; Global Temperature For April Second Warmest on Record". Retrieved 2015-02-18.

8

Plant Pathology: Effects of Biotic and Abiotic Stress

The study of the diseases found in plants is known as plant pathology. The diseases discussed in this section are blackleg (potatoes), soybean rust, clubroot, citrus canker and cherry X disease. The topics elaborated in this chapter will help in gaining a better perspective about the effects of biotic and abiotic stress.

Blackleg (Potatoes)

Blackleg of Potato complete plant wilt in field. These plants can sometimes be lost in the canopy.

Blackleg is a plant disease of potato caused by pectolytic bacteria that can result in stunting, wilting, chlorosis of leaves, necrosis of several tissues, a decline in yield, and at times the death of the potato plant. The term "blackleg" originates from the typical blackening and decay of the lower stem portion, or "leg", of the plant.

Blackleg in potatoes is most commonly caused by *Pectobacterium atrosepticum* (older synonym: *Erwinia carotovora subsp. astroseptica*), a gram-negative, nonsporulating, facultative anaerobe that is also associated with soft rot of potatoes. While other bacterial species such as *Pectobacterium carotovorum* and *Dickeya dadantii* can exhibit symptoms similar to blackleg of potato, these pathogens exhibit broader host ranges, are present in different climates, and typically are more associated with soft rot diseases.

Symptoms and Signs

Stem discoloration and darkening

Early blackleg symptoms develop in the growing season soon after the plants emerge. They are characterized by stunted, yellowish foliage that has a stiff, upright habit. The lower part of the below ground stem of such plants is dark brown to black in color and extensively decayed. When infected, the pith region of the stem is particularly susceptible to decay and may extend upward in the stem far beyond the tissue with externally visible symptoms. Young plants affected by blackleg are particularly susceptible, typically dying after a halt in development.

Wilting caused by blackleg

Blackleg symptoms may develop in more mature plants during the later part of the growing season, and are distinguished from those that develop earlier in the season. Blackleg appears as a black discoloration of previously healthy stems, accompanied by a rapid wilting, and sometimes yellowing, of the leaves. Starting below ground, black discoloration moves up the stem, often until the entire stem is black and wilted. However, in some cases of early disease development, mature stems may turn yellow and wilt even before black decay is evident. However, after the entire stem exhibits disease symptoms, the wilted plant can be lost from view in the healthy potato plant canopy.

Disease Cycle

1. A contaminated tuber can infect growing stems, or move into the vascular bundles of mature stems.

2. Infected stems can be symptomatic or asymptomatic, depending on environmental conditions, although the disease will remain and spread to other tubers on the same plant through the stolons.

3. In the field or during storage, they can contaminate and infect healthy tubers through wounds introduced during harvesting or through lenticels, and may also be spread through insects, wind, and rain. An important insect vector is the seed corn maggot (*Hylemya platura* and *H. florilega*), which spread the bacteria from diseased to healthy tissues. The bacteria are carried in the intestinal tracts of these insects, which spread the pathogen to healthy tissue by feeding on cut surfaces of healthy seed tissue. Another insect vector is the fruit fly (*Drosophila melanogaster*).

4. The pathogen will often survive in the infected tubers until the following planting season.

Environment and Biology

The pathogen *P. atrosepticum* thrives in moist, cool conditions, typically causing symptoms at temperatures below 25 °C. It is vulnerable to temperatures above 36 °C and dry conditions, and thus survives best in potato tuber tissues, although it is known to survive in other plant tissues. Unlike other pectolytic bacteria, evidence shows that *P. atrosepticum* does not survive well in soil outside its host tissue.

Disease symptoms are not necessarily uniformly exhibited from both shoots originating from a single tuber or in a field infested with *P. atrosepticum*. Additionally, presence of *P. atrosepticum* in the soil is not necessarily associated with disease symptoms. This is partly explained by the narrow environmental conditions needed for pathogenicity, although new findings in research are showing strong evidence of density dependent quorum sensing signals used by *P. atrosepticum* in exhibiting virulence.

Management

Cultural

Blackleg of potato has been successfully managed primarily using cultural techniques. These techniques generally rely on sterile propagation techniques, using knowledge *P. atrosepticum's* narrow environmental range to control planting timing, removing infected tissues and plants during the growing season, reducing tuber harvest damage, and proper storage.

Sterile Propagation

Given that tubers are the primary mechanism by which *P. atrosepticum* survives and spreads, clean seed potato stocks established using tissue cultures have been very successful in breaking the cycle of carrying disease forward from year to year. Buildup of tuber contamination is limited by reducing the number of field generations of these seed potatoes to 5 to 7 years. Some methods of sterile propagation include planting only healthy, whole seed potatoes. If healthy seed potatoes are to be cut, they should be first warmed to 12°-15 °C, cut, stored for 2 days at 12°-15 °C in a humid environment with good air flow. This warming and storing period ensures proper suberization of the tissue, which forms a barrier from *P. atrosepticum* infestation.

Planting Conditions

Given that *P. atrosepticum* thrives in cool, moist conditions, planting seed potatoes in well-drained soil after soil temperatures have increased well above 10 °C is very important to halting the onset of the disease early in the plant life cycle, when the plant is more susceptible to the worst effects of the disease.

Nutrition

Increasing application of nitrogen or complete fertilizers have shown reduced incidence of blackleg stem infection.

During the Growing Season

Although there is a risk of spreading the disease pathogen through injury of healthy plants, if proper techniques are followed, rogueing out all parts of the blackleg-diseased plants can be a useful way to reduce soil inoculum.

At Harvest and During Storage

Given that *P. atrosepticum* survives best in the tubers and additionally contributes to soft rot, it is critically important to reduce spread of the pathogen by removing tubers exhibiting soft rot decay before they are spread over grading lines and bin pilers for storage. Reducing post-harvest wounding is also important, especially for seed potatoes. Additionally, it is critically important to keep the potatoes at a low temperature with adequate aeration and humidity control in order to minimize development of the pathogen in infested stocks.

Biocontrol and Plant Resistance

New research on *P. atrosepticum* virulence pathways has elucidated the use of quorum sensing molecules to exhibit pathogenicity. These pathways include the control of the production of plant cell wall degrading enzymes in addition to other virulence factors. Research indicating the role of other soil microbes in degrading *P. atrosepticum* quorum sensing communication molecules provides the possibility for safe and effective control of the disease.

Plant defense mechanism studies on *P. atrosepticum*, used to better understand disease resistance, have focused more on the soft-rot symptoms that can sometimes be associated with *P. atrosepticum*. However, research is successfully identifying the quantity and type of plant resistance molecules that are produced in response to pathogen associated molecular patterns (PAMPs), and their effects on the activity and virulence of pathogens such as *P. atrosepticum*.

Importance

History

The symptoms of Blackleg of Potato were first described in Germany between 1878 and 1900, but the descriptions were incomplete and cannot definitively be linked to the particular disease. The first complete descriptions of Blackleg in potatoes were formed between 1901 and 1917 by several different scientists. These descriptions consisted of many different names, such as *Bacillus phytophthorus*, *Bacillus omnivorus*, *Bacillus oleraceae*, *Bacillus atrosepticus*, *Bacillus aroideae*, *Bacillus solanisaprus*, and *Bacillus melanogenes*. Investigations between 1918 and 1958 confirmed that these bacteria were of a single species, and were officially appointed the name *Pectobacterium carotovorum*. A variety of *Pectobacterium* (*P. carotovorum var. atrosepticum*, which includes *B. melanogenes* and *B. phytophthorus*) can be differentiated from the rest, although it is considered the same species of bacteria.

Although it was an important disease historically, Blackleg of potato is less of an issue today due to very successful results from changes in cultural practices regarding seed potato certification programs. As a major problem in wet, cool seasons and irrigated fields, historically it has more heavily impacted northern U.S. states with climates amenable to disease development, with disease incidence levels as high as 10%. In places like Scotland, it historically has had disease incidence levels of up to 30%. Victoria, Australia also had issues with this disease in the past. In terms of the impact of the disease on yields, one past study indicated that for every 1% increase in disease incidence, yields generally trended down at 0.8%.

Resistance

Given the success with cultural control practices in managing the disease, cultivars resistance is better characterized in the U.S. by susceptible varieties. Washington State University, which has posted a large comprehensive list of potato cultivars available in North America, only calls out two blackleg susceptible varieties: Monona and Superior.

In the U.K., and more specifically in Scotland, where the disease has been an issue, they better characterize blackleg-resistant varieties. Varieties with resistance values of 6-9 on a scale of 1-9 include Avondale, Axona, Bonnie, Cara, Emma, Isle Of Jura, Orla, Osprey, Sarpo Mira, Saxon, Sebastian, Vales Sovereign.

Soybean Rust

Soybean rust is a disease that affects soybeans and other legumes. It is caused by two types of fungi, Phakopsora pachyrhizi, commonly known as Asian soybean rust and Phakopsora meibomiae, commonly known as New World soybean rust. P. meibomiae is the weaker pathogen of the two and generally does not cause widespread problems. The disease has been reported across Asia, Australia, Africa, South America and the United States.

Importance

Soybean is one of the most important commercial crops around the world and in the United States.

Asian soybean rust is the major disease that affects soybeans. It causes lesions on the leaves of soybean plants and eventually kills the plants. The disease has caused serious yield loss of soybeans. In the areas where this disease is common, the yield losses can be up to 80%. In 2002, USDA reported 10-60% of yield losses in South America and Africa.

Host and Symptoms

Soybean rust is caused by two types of fungi, *Phakopsora pachyrhizi* and *Phakopsora meibomiae*. It affects several important commercial plants, however, most notable for soybeans. Asian Soybean Rust can infect and reproduce on 90 known plant species, 20 of which are found in the United States, such as, soybeans, dry beans, kidney beans, peas, leguminous forage crops such as trefoil and sweet clover and weeds such as kudzu.

Soybean leaves infected with ASR (photo from USDA)

At the early stage of Asian Soybean Rust, it causes yellow mosaic discoloration on the upper surfaces of older foliage. At this stage, it is usually hard to identify since the symptoms are relatively small and poorly defined.

Later as the disease continues to progress, the leaves will turn yellow and there will be lesions mostly on the undersides of the leaves and sometimes on petioles, stems or pods and premature defoliation can also be observed.

Asian Soybean Rust produces two types of lesions. Lesions at the later stage will turn from gray to tan or reddish brown. Mature tan lesion consists of small pustules which surrounded by discolored necrotic areas. Tan spores can be found at the necrotic areas on the underside of the leaf. For Reddish brown lesion, it has larger reddish brown necrotic areas with few pustules and visible spores on the underside of the leaf. A good way to distinguish Asian Soybean Rust from other diseases is to look at the pustules it procudes. ASR pustules usually do not have the yellow halo which is related to bacterial pustule. Besides, ASR pustules are raised and can be commonly found on the underside of the leaf which makes it different from the lesions caused by spot diseases.

As one of ASR's most known hosts, soybean plants are susceptible at any stage in the life cycle. However, symptoms are most commonly found during or after flowering. Soybean plants infected by Asian Soybean Rust will result in declining of pod production and fill.

Environment

Asian Soybean Rust (ASR) was first detected in Asia. It has been found in many countries around the world since then. For example, Australia, China, Korea, India, Japan, Nepal, Taiwan, Thailand, the Philippines, Mozambique, Nigeria, Rwanda, Uganda, Zimbabwe, South Africa, Brazil, Argentina, and Paraguay. This disease was first detected in the United States in Puerto Rico in 1976 and firstly reported in the continental United States in 2004.

Asian Soybean Rust favors the environments that are humid and warm. Continuous period of wetness on leaves will help the growth of this disease since this situation is required for spores to germinate. Therefore, is most likely to appear under the condition which the temperature is between 60 and 85 degree Fahrenheit and relative humidity of 75% to 80%. Therefore, ASR is a more serious problem in tropical and subtropical areas in Asia, Africa, Australia and South America. It is unable to survive the cold winters of northern habitats.

Disease Cycle

Soybean rust is spread by windblown spores and has caused significant crop losses in many soybean-growing regions of the world. Windblown spores can travel for great distances and are released in cycles of seven days to two weeks. It is likely that ASR will survive on vast acreages of naturalized kudzu in the southern U.S. and thereby establish a permanent presence in the continental U.S. It is commonly believed that the disease was carried from Venezuela to the United States by Hurricane Ivan.

Phakopsora pachyrhizi is an obligate parasite, meaning that it must have live, green tissue to survive. For this reason ASR is something that will blow in every year, as cold winters will push it back. It can overwinter in southern states, so long as it has a living host.

ASR overwinters on live host legumes and sporulates the following spring. It cannot survive on dead tissue or crop residues.

Additional hosts can serve as overwintering reservoirs for the pathogen and allow for build-up of inoculum, in those environs free from freezing temperatures. The pathogen is well adapted for long-distance dispersal, because spores can be readily carried long distances by the wind to new, rust-free regions.

Overwintering sites of soybean rust are restricted to areas with very mild winters, such as the gulf coasts of Florida, the very southernmost areas of Texas, or in Mexico. Soybean rust will not survive over the winter in the North Central region because it can't live and reproduce without green living tissue.

Spores of the soybean rust pathogen are transported readily by air currents and can be carried hundreds of miles in a few days. Weather conditions will determine when and where the spores travel from south to north.

Rust spores, called Urediniospores, are able to penetrate the plant cells directly, rather than through natural openings or through wounds in the leaf tissue. Thus infection is relatively quick: about 9 to 10 days from initial infection to the next cycle of spore production.

Rust is a multi-cyclic disease. After the initial infection is established, the infection site can produce spores for 10 to 14 days. Abundant spore production occurs during wet leaf periods (in the form of rain or dew) of at least 8 hours and moderate temperatures of 60 to 80°F (15.6 to 26.7°C).

The Process:

The infection process starts when urediniospores germinate to produce a single germ tube that grows across the leaf surface, until an appressorium forms. Appressoria form over anticlinal walls or over the center of epidermal cells, but rarely over stomata. Penetration of epidermal cells is by direct penetration through the cuticle by an appressorial peg. When appressoria form over stomata, the hyphae penetrate one of the guard cells rather than entering the leaf through the stomatal opening. This rust and related species are unique in their ability to directly penetrate the epidermis; most rust pathogens enter the leaf through stomatal openings and penetrate cells once inside the leaf. The direct penetration of the epidermal cells and the non-specific induction of appressoria in the infection process of *P. pachyrhizi* may aid in understanding the broad host range of the pathogen and may have consequences in the development of resistant cultivars.

Uredinia can develop 5 to 8 days after infection by urediniospores. The first urediniospores can be produced as early as 9 days after infection, and spore production can continue for up to 3 weeks. Uredinia may develop for up to 4 weeks after a single inoculation, and secondary uredinia will arise on the margins of the initial infections for an additional 8 weeks. Thus, from an initial infection, there could be first generation pustules that maintain sporulation for up to 15 weeks. Even under dry conditions this extended sporulation capacity allows the pathogen to persist and remain a threat. If conditions for re-infection are sporadic throughout the season, significant inoculum potential still remains from the initial infection to reestablish an epidemic. Successful infection is dependent on the availability of moisture on plant surfaces. At least 6 hours of free moisture is needed for infection with maximum infections occurring with 10 to 12 hours of free moisture. Temperatures between 15 and 28°C are ideal for infection.

Management and Control

Disease control options for ASR are limited. Rust descends in clouds of spores across the countryside. Cultural practices such as row spacing and crop rotations have little effect. Resistant cultivars do not exist. When weather and disease infection conditions are favorable, the occurrence of ASR can be widespread. Thus, remedial control measures—using fungicides as protective sprays—are the only effective disease control method.

Synthetic fungicides are the primary disease control option for protection against Asian soybean rust. The cost of spraying is estimated to be about $15 to $20 per acre; however, two or three sprays may be needed over the course of the growing season. These are significant additional production costs for soybean growers.

Fungicide screening trials to determine disease control efficacy have been field conducted in South America and South Africa. These reports are available on the Web through USDA's Integrated Pest Management Information Centers. These research trials form the basis for fungicidal recommendations in the U.S.

Recent research from Washington State University indicates that the herbicide Glyphosate may be effective in dealing with the fungus.

Rust-resistant varieties of soybeans are currently in development by both public universities and private industry.

In some regions, the selection of winter cover crops and forage legumes may be effected, since they can serve as host plants. Resistant soybean varieties are not yet available. However, resistance genes have been identified and host resistance is expected to be an effective, long-term solution for soybean rust. Until resistant commercial varieties are in place, the management of rust depends on judicious use of fungicides.

When untreated, soybean rust, causes yield losses due to premature defoliation, fewer seeds per pod and decreased number of filled pods per plant.

Clubroot

Clubroot on cauliflower

Clubroot is a common disease of cabbages, broccoli, cauliflower, Brussels sprouts, radishes, turnips, stocks, wallflowers and other plants belonging to the family Brassicaceae (Cruciferae). It is caused by *Plasmodiophora brassicae*, which was once considered a slime mold but is now put in the group Phytomyxea. It is the first Phytomyxea for which the genome has been sequenced. It has as many as thirteen races. Gall formation or distortion takes place on latent roots and gives the shape of a club or spindle. In the cabbage such attacks on the roots cause undeveloped heads or a failure to head at all, followed often by decline in vigor or by death. It is an important disease, affecting an estimated 10% of the total cultured area worldwide.

Historical reports of clubroot date back to the 13th century in Europe. In the late 19th century, a severe epidemic of clubroot destroyed large proportions of the cabbage crop in St. Petersburg. The Russian scientist Mikhail Woronin eventually identified the cause of clubroot as a "plasmodiophorous organism" in 1875, and gave it the name *Plasmodiophora brassicae*.

In 18th, 19th and early 20th century Britain clubroot was sometimes called *finger and toe, fingers and toes, anbury*, or *ambury*, these last two also meaning a soft tumor on a horse.

The potential of cultural practices to reduce crop losses due to clubroot is limited, and chemical treatments to control the fungus are either banned due to environmental regulations or are not cost effective. Breeding of resistant cultivars therefore is a promising alternative.

In Cabbages

Cabbage Clubroot is a disease of Brassicaceae (mustard family or cabbage family) caused by the soil-borne *Plasmodiophora brassicae*. The disease first appears scattered in fields, but in successive seasons it will infect the entire field, reducing the yield significantly and sometimes result-ing in no yield at all. Symptoms appear as yellowing, wilting, stunting, and galls on the roots. It is transmitted by contaminated transplants, animals, surface water runoff, contaminated equip-ment, and irrigation water. The pathogen can survive in a field for years as resting spores without a host present and will infect the next crop planted if it is a susceptible host. This pathogen prefers a wet climate and a pH around 5.7, so proper irrigation and the addition of compounds that raise the pH can be used to control this disease. Other control methods include sanitation to prevent transmission, chemical control, and resistant varieties.

Hosts and Symptoms

Cabbage Clubroot affects cabbage, Chinese cabbage, and Brussels sprouts most severely, but it has a range of hosts that it affects less severely like kohlrabi, kale, cauliflower, collards, broccoli, rutabaga, sea kale, turnips, and radishes.

Wilting and yellowing of plants in cabbage field.

Developing plants may not show any symptoms but as the plants get older they will start to show symptoms of chlorosis or yellowing, wilting during hot days, and exhibit stunted growth. Below ground, the roots experience cell proliferation due to increased auxin or growth hormone production from the plant as well as the pathogen. This causes the formation of galls that can grow big enough to restrict the xylem tissue inhibiting efficient water uptake by the plant. Galls appear like clubs or spindles on the roots. Eventually the roots will rot and the plant will die.

Galls on plant roots.

Disease Cycle

In the spring, resting spores in the soil germinate and produce zoospores. These zoospores swim through the moist soil and enter host plants through wounds or root hairs. A plasmodium is formed from the division of many amoeba-like cells. This plasmodium eventually divides and forms secondary zoospores that are once again released into the soil. The secondary infection by the zoospores can infect the first host or surrounding hosts. These secondary zoospores can be transmitted to other fields through farm machinery or water erosion. They form a secondary plasmodium that affects plant hormones to cause swelling in root cells. These cells eventually turn into galls or "clubs". The secondary plasmodium forms the overwintering resting spores which get released into the soil as the "clubs" rot and disintegrate. These resting spores can live in the soil for up to 20 years while they wait for a root tip to come in close proximity for them to infect.

Environment

Clubroot is a disease that prefers warmer temperatures and moist conditions. Ideal conditions for the proliferation of this disease would be a soil temperature between 20–24 °C and a pH less than 6.5; therefore, this disease tends to be prominent in lower fields where water tends to collect.

Management

Clubroot is very hard to control. The primary step for management and long-term control is exclusion of the disease. Good sanitation practice is important with regard to the use of tools and machinery in order to prevent the introduction of the pathogen to a disease-free field. It is not uncommon for an inattentive farmer or gardener to unknowingly carry in the pathogen after being previously exposed to it at a different time. One should avoid purchasing infected transplants of cabbage so as to prohibit the infestation of *P. brassicae*. Soil type is also an important factor in the development and spread of cabbage clubroot; the use of sand will allow for the plants to grow in well-drained soil, thereby eliminating the possibility of the pathogen to proliferate in a hospitable environment.

Although it is difficult to eradicate the pathogen once it is introduced to a field, there are several methods for its control. Keeping the soil at a slightly basic pH of 7.1–7.2 by the addition of agricultural lime as well as the integration of crop rotation will reduce the occurrence of cabbage clubroot in already infected fields. Fumigation using metam sodium in a field containing diseased cabbages is yet another way to decrease the buildup of the pathogen. Control and management practices on already infected fields help to reduce the overall impact that *P. brassicae* has on a field of cabbage and other cruciferous plants, but it is extremely difficult to rid an individual plant of the disease once it is already infected.

Importance

Clubroot can be a reoccurring problem for years because it is easily spread from plant to plant. *P. brassicae* is able to infect 300 species of cruciferous plants, making this disease a recurring problem even with crop rotation. This wide host range allows the pathogen to continue its infection cycle in the absence of cabbages. Additionally, cabbage clubroot may be a stubborn disease due to its ability to form a microbial cyst as an overwintering structure. These cysts may last many years in the soil until it comes into contact with a suitable host, making it difficult to entirely avoid the introduction of the disease. Those growing cabbage need to be aware of the possibility of *Plasmodiophora* infestation by simply growing in particular fields that may have had cabbage clubroot previously.

Canola Infestation in Alberta

In 2003 clubroot was identified in Alberta, Canada, as an outbreak in canola crops in the central area of the province mainly isolated to the Edmonton area. Clubroot is a soilborne disease caused by the biotrophic protist *Plasmodiophora brassica*. The infection causes the formation of large galls on the roots which look like clubs. These formations impede nutrient and water uptake and can cause plant death, wiping out important money generating canola crops. Initially 12 commercial fields of canola were identified, but that number grew to over 400 by 2008.

In 2007, Alberta declared *P.brassicae* a pest via the foundation legislation in hopes to help contain spread of the disease.

The Pathotype 3, is the predominant source for Alberta outbreaks. Studies showed that out of the 13 strains of *P. brassicae*, the most virulent form is dominant in Alberta.

Studies have shown that infestation numbers are highest at common field entrances and decline as you move further into the field, away from the entrance. From these results, it was concluded that infested soil on farm machinery was increasing spread of the pathogen. Some natural field to field spread is starting to be seen

Liming has been an effective control measure to curb clubroot since the 19th century. This method does not eradicate clubroot but it will slow its development by creating unfavorable conditions. In addition, calcium and magnesium can be added to the nutrition profile of the soil to help control clubroot. To get efficient results the field soil, [pH] must be kept above 7.5. This takes massive applications to field soil in order to treat all of the soil where spores of clubroot are found. Combining lime with one other treatment has shown most effective.

Several strains of canola have been tried, including European winter canola cv. Mendel (Brassica napus L.), as a clubroot resistant crop. It has been found that few cultivators exist. Specific genotypes do exist, of the Mendel strain, which could be a solution for canola crops in the Canadian prairies.

Crop rotation with non-host crops is another method to help prevent clubroot. The half life of P. Brassicae is 3.6 years. Unfortunately, long rotations of approximately 20 years are required in order to be effective. This is very difficult with typical canola rotations not being more than three years. Canola crop brings in high revenue to farmers. This would also require complete removal of Cruciferae crops, such as wild radish and mustard.

Some fungicide has been found to help with clubroot but it is very pricey and would take huge amounts to saturate the soil. The best way to prevent contamination between fields is to clean agricultural equipment and vehicles which have come in contact with club root before moving to a new field. All contaminated soil, equipment and tools must not be moved to clean, disease-free fields. The best preventative method is field monitoring. Throughout the season, plants should be monitored for early symptoms of club root. More research is being conducted for early detection of club root in fall soils.

Citrus Canker

Citrus canker is a disease affecting *Citrus* species caused by the bacterium *Xanthomonas axonopodis*. Infection causes lesions on the leaves, stems, and fruit of citrus trees, including lime, oranges, and grapefruit. While not harmful to humans, canker significantly affects the vitality of citrus trees, causing leaves and fruit to drop prematurely; a fruit infected with canker is safe to eat, but too unsightly to be sold.

The disease, which is believed to have originated in Southeast Asia, is extremely persistent when it becomes established in an area. Citrus groves have been destroyed in attempts to eradicate the disease. Brazil and the United States are currently suffering from canker outbreaks.

Biology

Xanthomonas axonopodis is a rod-shaped Gram-negative bacterium with polar flagella. The bacterium has a genome length around 5 megabase pairs. A number of types of citrus canker diseases are caused by different pathovars and variants of the bacterium:

- The Asiatic type of canker (canker A), *X. axonopodis* pv. *citri*, caused by a group of strains originally found in Asia, is the most widespread and severe form of the disease.
- Cancrosis B, caused by a group of *X. axonopodis* pv. *aurantifolii* strains originally found in South America is a disease of lemons, key lime, bitter orange, and pomelo.
- Cancrosis C, also caused by strains within *X. axonopodis* pv. *aurantifolii*, only infects key lime and bitter orange.
- A* strains, discovered in Oman, Saudi Arabia, Iran, and India, only infect key lime.

Pathology

Plants infected with citrus canker have characteristic lesions on leaves, stems, and fruit with raised, brown, water-soaked margins, usually with a yellow halo or ring effect around the lesion. Older lesions have a corky appearance, still in many cases retaining the halo effect. The bacterium propagates in lesions in leaves, stems, and fruit. The lesions ooze bacterial cells that, when dispersed by windblown rain, can spread to other plants in the area. Infection may spread further by hurricanes. The disease can also be spread by contaminated equipment, and by transport of infected or apparently healthy plants. Due to latency of the disease, a plant may appear to be healthy, but actually be infected.

Citrus canker bacteria can enter through a plant's stomata or through wounds on leaves or other green parts. In most cases, younger leaves are considered to be the most susceptible. Also, damage caused by citrus leaf miner larvae (*Phyllocnistis citrella*) can be sites for infection to occur. Within a controlled laboratory setting, symptoms can appear in 14 days following inoculation into a susceptible host. In the field environment, the time for symptoms to appear and be clearly discernible from other foliar diseases varies; it may be on the order of several months after infection. Lower temperatures increase the latency of the disease. Citrus canker bacteria can stay viable in old lesions and other plant surfaces for several months.

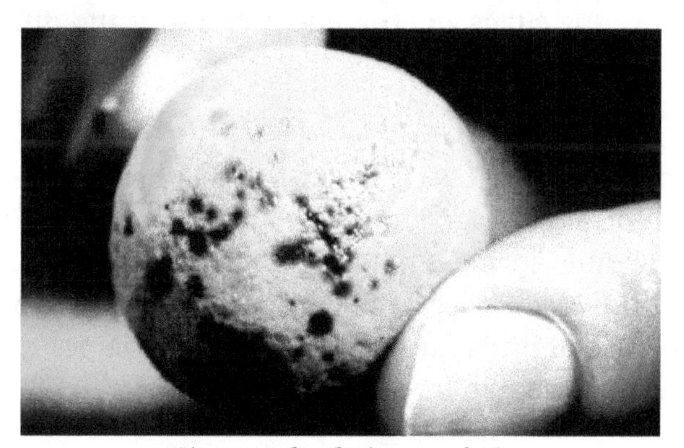

Citrus canker lesions on fruit

Detection

The disease can be detected in groves and on fruit by the appearance of lesions. Early detection is critical in quarantine situations. Bacteria can be tested for pathogenicity by inoculating multiple citrus species with them. Additional diagnostic tests (antibody detection), fatty-acid profiling, and genetic procedures using polymerase chain reaction can be conducted to confirm diagnosis and may help to identify the particular canker strain. Clara H. Hasse detected that citrus canker was not of fungoid origin but caused by bacteria. Her research published in the 1915 *Journal of Agricultural Research* played a major part in saving citrus crops in multiple states.

Susceptibility

Not all species and varieties of citrus have been tested for citrus canker. Most of the common species and varieties of citrus are susceptible to it. Some species are more susceptible than others, while a few species are resistant to infection.

Susceptibility	Variety
Highly susceptible	Grapefruit (*Citrus* x *paradisi*), Key lime (*C. aurantiifolia*), Pointed leaf hystrix (*C. hystrix*), lemon (*C. limon*)

Susceptible	Limes (*C. latifolia*) including Tahiti lime, Palestine sweet lime; trifoliate orange (*Poncirus trifoliata*); citranges/citrumelos (*P. trifoliata* hybrids); tangerines, tangors, tangelos (*C. reticulata* hybrids); sweet oranges (*C. sinensis*); bitter oranges (*C. aurantium*)
Resistant	Citron (*C. medica*), Mandarins (*C. reticulata*)
Highly resistant	Calamondin (*X Citrofortunella*), kumquat (*Fortunella* spp.)
Modified from: Gottwald, T.R. et al. (2002). Citrus canker: The pathogen and its impact. Online. *Plant Health Progress*	

Management

Scientists have not been able to come up with a proper system to help treat outbreaks. If this disease continues to spread, farming citrus will become very costly and difficult.

Distribution and Economic Impact

Citrus canker is thought to have originated in the area of Southeast Asia-India. It is now also present in Japan, South and Central Africa, the Middle East, Bangladesh, the Pacific Islands, some countries in South America, and Florida. Some areas of the world have eradicated citrus canker and others have ongoing eradication programs, but the disease remains endemic in most areas where it has appeared. Because of its rapid spread, high potential for damage, and impact on export sales and domestic trade, citrus canker is a significant threat to all citrus-growing regions.

Australia

The citrus industry is the largest fresh-fruit exporting industry in Australia. Australia has had three outbreaks of citrus canker, all of which have been successfully eradicated. The disease was found twice during the 1900s in the Northern Territory and was eradicated each time. In 2004, an unexplained outbreak occurred in central Queensland. The state and federal governments ordered all commercial groves, all noncommercial citrus trees, and all native lime trees (*C. glauca*) in the vicinity of Emerald to be destroyed rather than trying to isolate infected trees. Eradication was successful, with permission to replant being granted to farmers by the biosecurity unit of the Queensland Department of Primary Industries in early 2009.

Brazil

Citrus is an important domestic and export crop for Brazil. Citrus agriculture is the second-most important agricultural activity in the state of São Paulo, the largest sweet orange production area in the world. Over 100,000 groves are in São Paulo, and the area planted with citrus is increasing. Of the estimated 2 million trees, greater than 80% are a single variety of orange, and the remainder is made up of tangerine and lemon trees. Because of the uniformity in citrus variety the state has been adversely affected by canker, causing crop and monetary losses. In Brazil, rather than destroying entire groves to eradicate the disease, contaminated trees and trees within a 30-m radius are destroyed; by 1998, over half a million trees had been destroyed.

United States

Citrus canker was first found in the United States in 1910 not far from the Georgia – Florida border. Subsequently, canker was discovered in 1912 in Dade County, more than 400 mi (600 km) away. Beyond Florida, the disease was discovered in the Gulf states and reached as far north as South Carolina. It took more than 20 years to eradicate that outbreak of citrus canker, from 1913 through 1931, $2.5 million in state and private funds were spent to control it—a sum equivalent to $28 million in 2000 dollars. In 26 counties, some 257,745 grove trees and 3,093,110 nursery trees were destroyed by burning. Citrus canker was detected again on the Gulf Coast of Florida in 1986 and declared eradicated in 1994.

The most recent outbreak of citrus canker was discovered in Miami, Dade County, Florida, on Sept. 28, 1995, by Louis Willio Francillon, a Florida Department of Agriculture agronomist. Despite eradication attempts, by late 2005, the disease had been detected in many places distant from the original discovery, for example, in Orange Park, 315 miles (500 km) away. In January 2000, the Florida Department of Agriculture adopted a policy of removing all infected trees and all citrus trees within a 1900-ft radius of an infected tree in both residential areas and commercial groves. Previous to this eradication policy, the department eradicated all citrus trees within 125 ft of an infected one. The program ended in January 2006 following a statement from the USDA that eradication was not feasible.

Cherry X Disease

Cherry trees infected with X-disease yield smaller and paler fruit (upper left).

Cherry X disease also known as Cherry Buckskin disease is caused by a plant pathogenic phytoplasma. Phytoplasma's are obligate parasites of plants and insects. They are specialized bacteria, characterized by their lack of a cell wall, often transmitted through insects, and are responsible for large losses in crops, fruit trees, and ornamentals. The phytoplasma causing Cherry X disease has a fairly limited host range mostly of stone fruit trees. Hosts of the pathogen include sweet/sour cherries, choke cherry, peaches, nectarines, almonds, clover, and dandelion. Most commonly the pathogen is introduced into economical fruit orchards from wild choke cherry and herbaceous weed hosts. The pathogen is vectored by mountain and cherry leafhoppers. The mountain leafhopper vectors the pathogen from wild hosts to cherry orchards but does not feed on the other hosts. The cherry leafhopper which feeds on the infected cherry trees then becomes the next vector that transmits from cherry orchards to peach, nectarine, and other economic crops. Control of Cherry X disease is limited to controlling the spread, vectors, and weed hosts of the pathogen. Once the pathogen has infected a tree it is fatal and removal is necessary to stop it from becoming a reservoir for vectors.

Hosts

For Cherry X disease there are two types of hosts for the phytoplasma, reservoir and non-reservoir hosts. Reservoir hosts can survive for long periods while being infected with the disease. This allows them to be a constant food source for the leafhoppers which act to vector the phytoplasma from these hosts to other hosts in the area. Choke cherry is the most common reservoir host and a favorite food for the cherry leafhoppers. Other reservoir hosts include clovers and dandelions. Sweet/sour cherries, as well as almonds and Japanese plums are all fruit tree reservoir hosts for the Cherry X disease. All of these, once infected, can act as a source for the disease to be vectored from to other hosts. While non-cherry hosts can become infected they are not the preferred host of the phytoplasma. Because of the vectors preference for cherry trees, choke cherry which is a wild growing cherry species is the most common host of the disease. The range that Cherry X disease is distributed over is directly linked to the distribution of wild choke cherry populations.

Non-reservoir hosts are hosts that once infected do not allow for the disease to be spread. Peach and nectarine trees can be infected but they do not allow for the spread of the disease. This process which causes them to halt the spread of the pathogen is still not well understood. Peaches are commonly infected when near cherry orchards. Non-reservoir hosts are infected when cherry leafhoppers that are carrying the phytoplasma feed on non-reservoir hosts that are near a cherry orchard that has the pathogen.

Symptoms

The symptoms of Cherry X disease vary greatly depending on the host. On cherry hosts symptoms can usually first be seen on the fruits, causing them to be smaller in size with a leathery skin. Pale fruit is common at harvest time. It is common for symptoms to first be seen in a single branch. The branch may lose its older leaves, and the leaves tend to be smaller with a bronzed complexion.

The rootstock that the cherry is grafted onto can play a significant role in the disease symptoms seen. Rootstocks of Mahaleb cherry exhibit different symptoms from stocks of Colt, Mazzard, or Stockton Morello. When the scion is grafted onto Mahaleb, symptoms consistent with Phytophthora root rot can be seen. To distinguish between root rot and x-disease the wood under the bark at the graft union should be examined. If it is x-disease the wood at the union will have grooves and pits this causes a browning of the phloem and shows the cells in decline. This rapid decline is caused by the rootstock cells near the graft union dying in large quantities. Foliage begins to turn yellow and the curl upward and inward toward the leaf midrib. Trees infected with Mahaleb rootstock die by late summer or early the following year.

When Cherries are grafted onto Colt, Mazzard, or Stockton Morello rootstocks, there is a different range of symptoms. Affected leaves are smaller than normal and the foliage may be sparse. Dieback of shoot tips is common as the disease progresses. Fruit on branches are smaller, lighter, pointed, low sugar content, poor flavor, and a bitter taste.

Peaches are the next most common economic fruit host of the X-disease. Symptoms can be seen after about two months single branches will begin to show symptoms of their individual leaves. These leaves curl up and inward with irregular yellow to reddish-purple spots. These spots can drop out leaving "shotholes". Leaves that are affected by the disease will fall prematurely. After 2–3 years the entire tree will show symptoms.

Disease Cycle

Mountain leafhopper (Colladonus montanus) overwinters on winter annual weeds, particularly near streams and canals. Adults can be plentiful on sugarbeet during late winter/spring and migrate to favored weed hosts such as curly dock or burclovers in orchards. The Mountain leafhopper is most abundant vector found on cherry but does not reproduce on cherry. The mountain leafhopper (Colladonus montanus) spreads the disease from wild herbaceous hosts to woody hosts. It is believed that it is more responsible for the introduction of the disease into cherry trees, then in transferring them from cherry tree to cherry tree in an orchard. The cherry leafhopper (Fieberiella florii) reproduces on a broad range of woody hosts. The cherry leafhopper is more important in vectoring the disease from tree to tree within an orchard, since cherry is a favored host. After a leafhopper feeds on an infected host the pathogen has to undergo a latent period. During the latent period the pathogen spreads and multiplies inside the vector. Depending on temperature and the vector, the average latent period for the cherry x disease is about a month or longer. The phytoplasma is then transmitted from the leafhopper to the tree when the leafhopper is feeding on the trees phloem. It's then spread throughout the tree becoming systemic. July through October is when the highest concentrations of pathogen are present in leaves of infected trees.

Importance of the Disease

The disease is fatal and will always yield damaged fruit (choke cherries) as well as a dying/dead tree. If left unattended to, the leafhoppers can become life-time transmitters/vectors for the disease following about a 1-month latent period. The disease can take as quickly as 2–3 months to develop symptoms but more commonly 6–9 months, but the symptoms are usually first seen in the next growing season after the infection, with the rare exception that the infection and first symptoms occur both in the same spring season. In high cherry producing areas, such as California, Washington, and Oregon, this disease could be devastating if left unchecked. For instance in 2002, 57,000 tons of cherries were harvested from 24,000 acres in California. The year grossed a total of over $152 million. If, in 2002, this disease was allowed to incubate, the results would show a drastic decline of production and huge loss of revenue as early as 2003. This disease does not take long to develop and since fatality is always the endgame, high producing areas such as these would see results of epidemic proportions.

Environment

Leafhoppers are the only known vectors that can carry the X-disease from a wild host into peach and cherry orchards. Orchard trees are most often infected by insect vectors. In California where it was first noted, the two most important vectors were the mountain leafhopper, Collandonus montanus, and the cherry leafhopper, Fieberiella florii.

Mountain Leafhopper (Collandonus Montanus)

The mountain leafhopper survives on winter annual weeds during winter, usually near stream banks or canals. In late winter or spring, adults can be found in sugar beet fields and can then migrate to favored weed hosts (curly dock, burclovers) in orchards. The mountain leafhopper is most often the abundant vector found on cherry, however, cherry is not the preferred host and the leafhopper does not reproduce on cherry. Preferred hosts for the mountain leafhopper are; alfalfa, California burclover, clovers, curly dock, and sweet clovers. Of the preferred hosts alfalfa and curly dock cannot become infected with the disease itself but are just a host for the leafhopper. Occasional hosts are; vetches (in legume family) and sweet cherry. It's believed that the role of this leafhopper is introducing the disease into cherry orchards rather than spreading the disease between cherry trees within an orchard.(http://ucanr.org/sites/cccoopext/files/80935.pdf)

Cherry Leafhopper (Fieberiella Florii)

The cherry leafhopper has a more significant role in spreading the disease between cherry trees because cherry is a favored host. The leafhopper feeds and reproduces on a wide range of woody hosts. Preferred hosts for the cherry leafhopper are; box wood, lilac, myrtle, privet, pyracantha, sweet cherry, and viburnum. Of these preferred hosts only sweet cherry can become infected with the disease itself. Occasional hosts are; almond, apple and crabapple, apricot, bitter cherry, ceanothus, chokecherry, hawthorn, peach, pear, Japanese plum, and prune. Of these occasional hosts only chokecherry and bittercherry and occasionally almond, peach and Japanese plum can become infected with the disease itself. (http://ucanr.org/sites/cccoopext/files/80935.pdf)

There are seven known vectors that transmit the disease in western United States. These leafhoppers are *Colladonus geminatus, Fieberiella florii, Keonolla confluens, Scaphytopius delongi, Osbornellus borealis, Colladonus montanus,* and *Euscelidius variegatus*. Other possible leafhopper vectors are *Scaphytopius aculus, Paraphlepsius irroratus, Colladonus clitellarius,* and *Norvellina seminude*. Not a lot of information is available for ideal environmental conditions for the disease. However, conditions conducive to leafhoppers is most likely the key for the greatest spread of disease.

Management

There are numerous steps one has to take to try to manage the disease as best as possible. The aim is at prevention because once the pathogen reaches the cherry trees, disease will surely ensue and there is no cure or remedy to prevent the loss of fruit production as well as the ultimate death of the tree.

Pest Management

The first approach, which is the best approach at an effective management practice would be to eradicate or severely damage the Mountain and Cherry Leafhopper population because the leafhoppers are the number one vectors for this pathogen. To do this, pesticides (i.e. acephate, bifenthrin, cyfluthrin) could be applied or biological control (predators of the leafhopper) could be used. There should be a pre-season application of control measures as well as a post-season application. This is to maximize the effort at controlling both types of leafhoppers (Cherry and Mountain), thus cutting down the starting inoculum at both stages in the life cycle.

Weed Host Management

Some herbaceous hosts naturally have the Cherry X Disease. Once the spreads to the cherry hosts, with the help of the mountain leafhoppers, the cherry leafhoppers can spread the disease around to other woody hosts. Here are some approaches at management with each host type:

Herbaceous Hosts

The herbaceous hosts are common weeds (i.e. clovers, dandelions, alfalfa) that serve as a feeding ground for the mountain leafhoppers. The herbaceous hosts are the source of the X Disease, which is picked up and transmitted to the cherry hosts by the mountain leafhopper. For a control, conventional herbicides are effective. There exists a common herbaceous host, curly dock, which serves as the mountain leafhopper's main breeding ground. Getting rid of curly dock with an herbicide would be key to limit the population, thus limiting the spread of the X Disease to the cherry hosts.

Woody Hosts

After the disease moves on from the herbaceous host with the help of the mountain leafhoppers, it moves to the cherry hosts (i.e. bitter cherry and chokecherry). Once there, the infected trees should be destroyed and removed, along with all infected fruits. This is to prevent further spreading into other woody hosts such as peach, plum, apple etc., because once a tree is infected, it cannot be saved and it will become a source of the X Disease which the cherry leafhoppers can pick up and spread to the other woody hosts. In conclusion, all infected woody hosts should be removed and destroyed along with all infected fruits.

References

- Hybrid Vegetable Development, by P. K. Singh (Ed.), S. K. Dasgupta (Ed.), S. K. Tripathi (Ed.), Haworth Press. ISBN 1-56022-118-6
- "Cherry Buckskin Disease (X-Disease)" (PDF). University of California Cooperative Extension Contra Costa. Retrieved 21 October 2013.
- Ellis, Michael A. "X-Disease (Mycoplasma disease of peaches and nectarines)" (PDF). Department of Plant pathology, The Ohio State University extension. Retrieved 13 November 2013.
- "Cherry Buckskin Symptoms & Vectors" (PDF). University of California Cooperative Extension Contra Costa. Retrieved 21 October 2013.
- "Identifying Choke Cherry – Source of X Disease" (PDF). University of New Hampshire cooperative extension. Retrieved 22 October 2013.
- Taboada, O.; Rosenberger, D. A.; Jones, A. L. (1975). "Leafhopper Fauna of X-Diseased Peach and Cherry Orchards in Southwest Michigan". Journal of Economic Entomology: 255–257. Retrieved October 20, 2013.
- Christianson, J. (2008). Club Root of Crucifers. University of Nebraska-Lincoln. Retrieved from http://nu-distance.unl.edu/homer/disease/hort/crucifer/CrClbRoot.html Mar 2012
- Kowata-Dresch, L.S., May-De Milo, L.L. (2012). Clubroot Management of Highly Infested Soils. "Crop Protection, 35, p47-52." doi:10.1016/j.cropro.2011.12.012
- Trembly, N., et al. (1999). Clubroot of Crucifiers: Control Strategies. "Agriculture and Agrifood Canada." Retrieved from http://publications.gc.ca/collections/Collection/A42-85-1999E.pdf March 2012.

9

Conservation Methods of Biotic and Abiotic Stress

Precautions need to be taken to prevent plants from biotic as well as abiotic stress. Some of these conservation methods are pesticide, insecticide, herbicide, fungicide and bactericide. Insecticides are used to kill insects, whereas pesticides are meant for attracting and then destroying any pests. This chapter discusses in detail the conservation methods of biotic and abiotic stress.

Pesticide

A crop-duster spraying pesticide on a field

Pesticides are substances meant for attracting, seducing, and then destroying any pest. They are a class of biocide. The most common use of pesticides is as plant protection products (also known as crop protection products), which in general protect plants from damaging influences such as weeds, fungi, or insects. This use of pesticides is so common that the term *pesticide* is often treated as synonymous with *plant protection product*, although it is in fact a broader term, as pesticides are also used for non-agricultural purposes. The term pesticide includes all of the following: herbicide, insecticide, insect growth regulator, nematicide, termiticide, molluscicide, piscicide, avicide, rodenticide, predacide, bactericide, insect repellent, animal repellent, antimicrobial, fungicide, disinfectant (antimicrobial), and sanitizer.

In general, a pesticide is a chemical or biological agent (such as a virus, bacterium, antimicrobial, or disinfectant) that deters, incapacitates, kills, or otherwise discourages pests. Target pests can include insects, plant pathogens, weeds, mollusks, birds, mammals, fish, nematodes (roundworms), and microbes that destroy property, cause nuisance, or spread disease, or are disease vectors. Although pesticides have benefits, some also have drawbacks, such as potential toxicity to humans and other species. According to the Stockholm Convention on Persistent Organic Pollutants, 9 of the 12 most dangerous and persistent organic chemicals are organochlorine pesticides.

A Lite-Trac four-wheeled self-propelled crop sprayer spraying pesticide on a field

Definition

Type of pesticide	Target pest group
Herbicides	Plant
Algicides or Algaecides	Algae
Avicides	Birds
Bactericides	Bacteria
Fungicides Fungi and Oomycetes	
Insecticides	Insects
Miticides or Acaricides	Mites
Molluscicides	Snails
Nematicides	Nematodes
Rodenticides	Rodents
Virucides	Viruses

The Food and Agriculture Organization (FAO) has defined *pesticide* as:

> any substance or mixture of substances intended for preventing, destroying, or controlling any pest, including vectors of human or animal disease, unwanted species of plants or animals, causing harm during or otherwise interfering with the production, processing, storage, transport, or marketing of food, agricultural commodities, wood and wood products or animal feedstuffs, or substances that may be administered to animals for the control of insects, arachnids, or other pests in or on their bodies. The term includes substances intended for use as a plant growth regulator, defoliant, desiccant, or agent for thinning fruit or preventing the premature fall of fruit. Also used as substances applied to crops either before or after harvest to protect the commodity from deterioration during storage and transport.

Pesticides can be classified by target organism (e.g., herbicides, insecticides, fungicides, rodenticides, and pediculicides - see table), chemical structure (e.g., organic, inorganic, synthetic, or biological (biopesticide), although the distinction can sometimes blur), and physical state (e.g. gaseous (fumigant)). Biopesticides include microbial pesticides and biochemical pesticides. Plant-derived pesticides, or "botanicals", have been developing quickly. These include the pyrethroids, rotenoids, nicotinoids, and a fourth group that includes strychnine and scilliroside.

Many pesticides can be grouped into chemical families. Prominent insecticide families include organochlorines, organophosphates, and carbamates. Organochlorine hydrocarbons (e.g., DDT) could be separated into dichlorodiphenylethanes, cyclodiene compounds, and other related compounds. They operate by disrupting the sodium/potassium balance of the nerve fiber, forcing the nerve to transmit continuously. Their toxicities vary greatly, but they have been phased out because of their persistence and potential to bioaccumulate. Organophosphate and carbamates largely replaced organochlorines. Both operate through inhibiting the enzyme acetylcholinesterase, allowing acetylcholine to transfer nerve impulses indefinitely and causing a variety of symptoms such as weakness or paralysis. Organophosphates are quite toxic to vertebrates, and have in some cases been replaced by less toxic carbamates. Thiocarbamate and dithiocarbamates are subclasses of carbamates. Prominent families of herbicides include phenoxy and benzoic acid herbicides (e.g. 2,4-D), triazines (e.g., atrazine), ureas (e.g., diuron), and Chloroacetanilides (e.g., alachlor). Phenoxy compounds tend to selectively kill broad-leaf weeds rather than grasses. The phenoxy and benzoic acid herbicides function similar to plant growth hormones, and grow cells without normal cell division, crushing the plant's nutrient transport system. Triazines interfere with photosynthesis. Many commonly used pesticides are not included in these families, including glyphosate.

Pesticides can be classified based upon their biological mechanism function or application method. Most pesticides work by poisoning pests. A systemic pesticide moves inside a plant following absorption by the plant. With insecticides and most fungicides, this movement is usually upward (through the xylem) and outward. Increased efficiency may be a result. Systemic insecticides, which poison pollen and nectar in the flowers may kill bees and other needed pollinators

In 2009, the development of a new class of fungicides called paldoxins was announced. These work by taking advantage of natural defense chemicals released by plants called phytoalexins, which fungi then detoxify using enzymes. The paldoxins inhibit the fungi's detoxification enzymes. They are believed to be safer and greener.

Uses

Pesticides are used to control organisms that are considered to be harmful. For example, they are used to kill mosquitoes that can transmit potentially deadly diseases like West Nile virus, yellow fever, and malaria. They can also kill bees, wasps or ants that can cause allergic reactions. Insecticides can protect animals from illnesses that can be caused by parasites such as fleas. Pesticides can prevent sickness in humans that could be caused by moldy food or diseased produce. Herbicides can be used to clear roadside weeds, trees and brush. They can also kill invasive weeds that may cause environmental damage. Herbicides are commonly applied in ponds and lakes to control algae and plants such as water grasses that can interfere with activities like swimming and fishing and cause the water to look or smell unpleasant. Uncontrolled pests such as termites and mold can damage structures such as houses. Pesticides are used in grocery stores and food storage facilities to manage rodents and insects that infest food such as grain. Each use of a pesticide carries some associated risk. Proper pesticide use decreases these associated risks to a level deemed acceptable by pesticide regulatory agencies such as the United States Environmental Protection Agency (EPA) and the Pest Management Regulatory Agency (PMRA) of Canada.

DDT, sprayed on the walls of houses, is an organochlorine that has been used to fight malaria since the 1950s. Recent policy statements by the World Health Organization have given stronger support to this approach. However, DDT and other organochlorine pesticides have been banned in most countries worldwide because of their persistence in the environment and human toxicity. DDT use is not always effective, as resistance to DDT was identified in Africa as early as 1955, and by 1972 nineteen species of mosquito worldwide were resistant to DDT.

Amount Used

In 2006 and 2007, the world used approximately 2.4 megatonnes (5.3×10^9 lb) of pesticides, with herbicides constituting the biggest part of the world pesticide use at 40%, followed by insecticides (17%) and fungicides (10%). In 2006 and 2007 the U.S. used approximately 0.5 megatonnes (1.1×10^9 lb) of pesticides, accounting for 22% of the world total, including 857 million pounds (389 kt) of conventional pesticides, which are used in the agricultural sector (80% of conventional pesticide use) as well as the industrial, commercial, governmental and home & garden sectors. Pesticides are also found in majority of U.S. households with 78 million out of the 105.5 million households indicating that they use some form of pesticide. As of 2007, there were more than 1,055 active ingredients registered as pesticides, which yield over 20,000 pesticide products that are marketed in the United States.

The US used some 1 kg (2.2 pounds) per hectare of arable land compared with: 4.7 kg in China, 1.3 kg in the UK, 0.1 kg in Cameroon, 5.9 kg in Japan and 2.5 kg in Italy. Insecticide use in the US has declined by more than half since 1980, (.6%/yr) mostly due to the near phase-out of organophosphates. In corn fields, the decline was even steeper, due to the switchover to transgenic Bt corn.

For the global market of crop protection products, market analysts forecast revenues of over 52 billion US$ in 2019.

Benefits

Pesticides can save farmers' money by preventing crop losses to insects and other pests; in the U.S., farmers get an estimated fourfold return on money they spend on pesticides. One study found that not using pesticides reduced crop yields by about 10%. Another study, conducted in 1999, found that a ban on pesticides in the United States may result in a rise of food prices, loss of jobs, and an increase in world hunger.

There are two levels of benefits for pesticide use, primary and secondary. Primary benefits are direct gains from the use of pesticides and secondary benefits are effects that are more long-term.

Primary Benefits

1. Controlling pests and plant disease vectors

- Improved crop/livestock yields
- Improved crop/livestock quality
- Invasive species controlled

2. Controlling human/livestock disease vectors and nuisance organisms
    - Human lives saved and suffering reduced
    - Animal lives saved and suffering reduced
    - Diseases contained geographically
3. Controlling organisms that harm other human activities and structures
    - Drivers view unobstructed
    - Tree/brush/leaf hazards prevented
    - Wooden structures protected

Monetary

Every dollar ($1) that is spent on pesticides for crops yields four dollars ($4) in crops saved. This means based that, on the amount of money spent per year on pesticides, $10 billion, there is an additional $40 billion savings in crop that would be lost due to damage by insects and weeds. In general, farmers benefit from having an increase in crop yield and from being able to grow a variety of crops throughout the year. Consumers of agricultural products also benefit from being able to afford the vast quantities of produce available year-round. The general public also benefits from the use of pesticides for the control of insect-borne diseases and illnesses, such as malaria. The use of pesticides creates a large job market within the agrichemical sector.

Costs

On the cost side of pesticide use there can be costs to the environment, costs to human health, as well as costs of the development and research of new pesticides.

Health Effects

Pesticides may cause acute and delayed health effects in people who are exposed. Pesticide exposure can cause a variety of adverse health effects, ranging from simple irritation of the skin and eyes to more severe effects such as affecting the nervous system, mimicking hormones causing reproductive problems, and also causing cancer. A 2007 systematic review found that "most studies on non-Hodgkin lymphoma and leukemia showed positive associations with pesticide exposure" and thus concluded that cosmetic use of pesticides should be decreased. There is substantial evidence of associations between organophosphate insecticide exposures and neurobehavioral alterations. Limited evidence also exists for other negative outcomes from pesticide exposure including neurological, birth defects, and fetal death.

The American Academy of Pediatrics recommends limiting exposure of children to pesticides and using safer alternatives:

The World Health Organization and the UN Environment Programme estimate that each year, 3 million workers in agriculture in the developing world experience severe poisoning from pesticides, about 18,000 of whom die. Owing to inadequate regulation and safety precautions, 99% of pesticide related deaths occur in developing countries that account for only 25% of pesticide usage. According to one study, as many as 25 million workers in developing countries may suffer mild pesticide poisoning yearly. There are several careers aside from agriculture that may also put individuals at risk of health effects from pesticide exposure including pet groomers, groundskeepers, and fumigators.

A sign warning about potential pesticide exposure.

One study found pesticide self-poisoning the method of choice in one third of suicides worldwide, and recommended, among other things, more restrictions on the types of pesticides that are most harmful to humans.

A 2014 epidemiological review found associations between autism and exposure to certain pesticides, but noted that the available evidence was insufficient to conclude that the relationship was causal.

Environmental Effect

Pesticide use raises a number of environmental concerns. Over 98% of sprayed insecticides and 95% of herbicides reach a destination other than their target species, including non-target species, air, water and soil. Pesticide drift occurs when pesticides suspended in the air as particles are carried by wind to other areas, potentially contaminating them. Pesticides are one of the causes of water pollution, and some pesticides are persistent organic pollutants and contribute to soil contamination.

In addition, pesticide use reduces biodiversity, contributes to pollinator decline, destroys habitat (especially for birds), and threatens endangered species.

Pests can develop a resistance to the pesticide (pesticide resistance), necessitating a new pesticide. Alternatively a greater dose of the pesticide can be used to counteract the resistance, although this will cause a worsening of the ambient pollution problem.

Since chlorinated hydrocarbon pesticides dissolve in fats and are not excreted, organisms tend to retain them almost indefinitely. Biological magnification is the process whereby these chlorinated hydrocarbons (pesticides) are more concentrated at each level of the food chain. Among marine animals, pesticide concentrations are higher in carnivorous fishes, and even more so in the fish-eating birds and mammals at the top of the ecological pyramid. Global distillation is the process whereby pesticides are transported from warmer to colder regions of the Earth, in particular the Poles and mountain tops. Pesticides that evaporate into the atmosphere at relatively high temperature can be carried considerable distances (thousands of kilometers) by the wind to an area of lower temperature, where they condense and are carried back to the ground in rain or snow.

In order to reduce negative impacts, it is desirable that pesticides be degradable or at least quickly deactivated in the environment. Such loss of activity or toxicity of pesticides is due to both innate chemical properties of the compounds and environmental processes or conditions. For example, the presence of halogens within a chemical structure often slows down degradation in an aerobic environment. Adsorption to soil may retard pesticide movement, but also may reduce bioavailability to microbial degraders.

Economics

Harm	Annual US cost
Public health	$1.1 billion
Pesticide resistance in pest	$1.5 billion
Crop losses caused by pesticides	$1.4 billion
Bird losses due to pesticides	$2.2 billion
Groundwater contamination	$2.0 billion
Other costs	$1.4 billion
Total costs	**$9.6 billion**

Human health and environmental cost from pesticides in the United States is estimated at $9.6 billion offset by about $40 billion in increased agricultural production:

Additional costs include the registration process and the cost of purchasing pesticides. The registration process can take several years to complete (there are 70 different types of field test) and can cost $50–70 million for a single pesticide. Annually the United States spends $10 billion on pesticides.

Alternatives

Alternatives to pesticides are available and include methods of cultivation, use of biological pest controls (such as pheromones and microbial pesticides), genetic engineering, and methods of interfering with insect breeding. Application of composted yard waste has also been used as a way of controlling pests. These methods are becoming increasingly popular and often are safer than traditional chemical pesticides. In addition, EPA is registering reduced-risk conventional pesticides in increasing numbers.

Cultivation practices include polyculture (growing multiple types of plants), crop rotation, planting crops in areas where the pests that damage them do not live, timing planting according to when pests will be least problematic, and use of trap crops that attract pests away from the real crop. In the U.S., farmers have had success controlling insects by spraying with hot water at a cost that is about the same as pesticide spraying.

Release of other organisms that fight the pest is another example of an alternative to pesticide use. These organisms can include natural predators or parasites of the pests. Biological pesticides based on entomopathogenic fungi, bacteria and viruses cause disease in the pest species can also be used.

Interfering with insects' reproduction can be accomplished by sterilizing males of the target species and releasing them, so that they mate with females but do not produce offspring. This technique was first used on the screwworm fly in 1958 and has since been used with the medfly, the tsetse fly, and the gypsy moth. However, this can be a costly, time consuming approach that only works on some types of insects.

Agroecology emphasize nutrient recycling, use of locally available and renewable resources, adaptation to local conditions, utilization of microenvironments, reliance on indigenous knowledge and yield maximization while maintaining soil productivity. Agroecology also emphasizes empowering people and local communities to contribute to development, and encouraging "multi-directional" communications rather than the conventional "top-down" method.

Push Pull Strategy

The term "push-pull" was established in 1987 as an approach for integrated pest management (IPM). This strategy uses a mixture of behavior-modifying stimuli to manipulate the distribution and abundance of insects. "Push" means the insects are repelled or deterred away from whatever resource that is being protected. "Pull" means that certain stimuli (semiochemical stimuli, pheromones, food additives, visual stimuli, genetically altered plants, etc.) are used to attract pests to trap crops where they will be killed. There are numerous different components involved in order to implement a Push-Pull Strategy in IPM.

Many case studies testing the effectiveness of the push-pull approach have been done across the world. The most successful push-pull strategy was developed in Africa for subsistence farming. Another successful case study was performed on the control of *Helicoverpa* in cotton crops in Australia. In Europe, the Middle East, and the United States, push-pull strategies were successfully used in the controlling of *Sitona lineatus* in bean fields.

Some advantages of using the push-pull method are less use of chemical or biological materials and better protection against insect habituation to this control method. Some disadvantages of the push-pull strategy is that if there is a lack of appropriate knowledge of behavioral and chemical ecology of the host-pest interactions then this method becomes unreliable. Furthermore, because the push-pull method is not a very popular method of IPM operational and registration costs are higher.

Effectiveness

Some evidence shows that alternatives to pesticides can be equally effective as the use of chemicals. For example, Sweden has halved its use of pesticides with hardly any reduction in crops. In Indonesia, farmers have reduced pesticide use on rice fields by 65% and experienced a 15% crop increase. A study of Maize fields in northern Florida found that the application of composted yard waste with high carbon to nitrogen ratio to agricultural fields was highly effective at reducing the population of plant-parasitic nematodes and increasing crop yield, with yield increases ranging from 10% to 212%; the observed effects were long-term, often not appearing until the third season of the study.

However, pesticide resistance is increasing. In the 1940s, U.S. farmers lost only 7% of their crops to pests. Since the 1980s, loss has increased to 13%, even though more pesticides are being used.Between 500 and 1,000 insect and weed species have developed pesticide resistance since 1945.

Types

Pesticides are often referred to according to the type of pest they control. Pesticides can also be considered as either biodegradable pesticides, which will be broken down by microbes and other living beings into harmless compounds, or persistent pesticides, which may take months or years before they are broken down: it was the persistence of DDT, for example, which led to its accumulation in the food chain and its killing of birds of prey at the top of the food chain. Another way to think about pesticides is to consider those that are chemical pesticides or are derived from a common source or production method.

Some examples of chemically-related pesticides are:

Organophosphate Pesticides

Organophosphates affect the nervous system by disrupting, acetylcholinesterase activity, the enzyme that regulates acetylcholine, a neurotransmitter. Most organophosphates are insecticides. They were developed during the early 19th century, but their effects on insects, which are similar to their effects on humans, were discovered in 1932. Some are very poisonous. However, they usually are not persistent in the environment.

Carbamate Pesticides

Carbamate pesticides affect the nervous system by disrupting an enzyme that regulates acetylcholine, a neurotransmitter. The enzyme effects are usually reversible. There are several subgroups within the carbamates.

Organochlorine Insecticides

They were commonly used in the past, but many have been removed from the market due to their health and environmental effects and their persistence (e.g., DDT, chlordane, and toxaphene).

Pyrethroid Pesticides

They were developed as a synthetic version of the naturally occurring pesticide pyrethrin, which is found in chrysanthemums. They have been modified to increase their stability in the environment. Some synthetic pyrethroids are toxic to the nervous system.

Sulfonylurea Herbicides

The following sulfonylureas have been commercialized for weed control: amidosulfuron, azimsulfuron, bensulfuron-methyl, chlorimuron-ethyl, ethoxysulfuron, flazasulfuron, flupyrsulfuron-methyl-sodium, halosulfuron-methyl, imazosulfuron, nicosulfuron, oxasulfuron, primisulfuron-methyl, pyrazosulfuron-ethyl, rimsulfuron, sulfometuron-methyl Sulfosulfuron, terbacil, bispyribac-sodium, cyclosulfamuron, and pyrithiobac-sodium. Nicosulfuron, triflusulfuron methyl, and chlorsulfuron are broad-spectrum herbicides that kill plants by inhibiting the enzyme acetolactate synthase. In the 1960s, more than 1 kg/ha (0.89 lb/acre) crop protection chemical was typically applied, while sulfonylureates allow as little as 1% as much material to achieve the same effect.

Biopesticides

Biopesticides are certain types of pesticides derived from such natural materials as animals, plants, bacteria, and certain minerals. For example, canola oil and baking soda have pesticidal applications and are considered biopesticides. Biopesticides fall into three major classes:

- Microbial pesticides which consist of bacteria, entomopathogenic fungi or viruses (and sometimes includes the metabolites that bacteria or fungi produce). Entomopathogenic nematodes are also often classed as microbial pesticides, even though they are multi-cellular.
- Biochemical pesticides or herbal pesticides are naturally occurring substances that control (or monitor in the case of pheromones) pests and microbial diseases.
- Plant-incorporated protectants (PIPs) have genetic material from other species incorporated into their genetic material (*i.e.* GM crops). Their use is controversial, especially in many European countries.

Classified by Type of Pest

Pesticides that are related to the type of pests are:

Type	Action
Algicides	Control algae in lakes, canals, swimming pools, water tanks, and other sites
Antifouling agents	Kill or repel organisms that attach to underwater surfaces, such as boat bottoms
Antimicrobials	Kill microorganisms (such as bacteria and viruses)
Attractants	Attract pests (for example, to lure an insect or rodent to a trap). (However, food is not considered a pesticide when used as an attractant.)
Biopesticides	Biopesticides are certain types of pesticides derived from such natural materials as animals, plants, bacteria, and certain minerals
Biocides	Kill microorganisms
Disinfectants and sanitizers	Kill or inactivate disease-producing microorganisms on inanimate objects
Fungicides	Kill fungi (including blights, mildews, molds, and rusts)

Fumigants	Produce gas or vapor intended to destroy pests in buildings or soil
Herbicides	Kill weeds and other plants that grow where they are not wanted
Insecticides	Kill insects and other arthropods
Miticides	Kill mites that feed on plants and animals
Microbial pesticides	Microorganisms that kill, inhibit, or out compete pests, including insects or other micro-organisms
Molluscicides	Kill snails and slugs
Nematicides	Kill nematodes (microscopic, worm-like organisms that feed on plant roots)
Ovicides	Kill eggs of insects and mites
Pheromones	Biochemicals used to disrupt the mating behavior of insects
Repellents	Repel pests, including insects (such as mosquitoes) and birds
Rodenticides	Control mice and other rodents

Further Types of Pesticides

The term pesticide also include these substances:

Defoliants : Cause leaves or other foliage to drop from a plant, usually to facilitate harvest.

Desiccants : Promote drying of living tissues, such as unwanted plant tops.

Insect growth regulators : Disrupt the molting, maturity from pupal stage to adult, or other life processes of insects.

Plant growth regulators : Substances (excluding fertilizers or other plant nutrients) that alter the expected growth, flowering, or reproduction rate of plants.

Regulation

International

In most countries, pesticides must be approved for sale and use by a government agency.

In Europe, recent EU legislation has been approved banning the use of highly toxic pesticides including those that are carcinogenic, mutagenic or toxic to reproduction, those that are endocrine-disrupting, and those that are persistent, bioaccumulative and toxic (PBT) or very persistent and very bioaccumulative (vPvB). Measures were approved to improve the general safety of pesticides across all EU member states.

Though pesticide regulations differ from country to country, pesticides, and products on which they were used are traded across international borders. To deal with inconsistencies in regulations among countries, delegates to a conference of the United Nations Food and Agriculture Organization adopted an International Code of Conduct on the Distribution and Use of Pesticides in 1985 to create voluntary standards of pesticide regulation for different countries. The Code was updated in 1998 and 2002. The FAO claims that the code has raised awareness about pesticide hazards and decreased the number of countries without restrictions on pesticide use.

Three other efforts to improve regulation of international pesticide trade are the United Nations London Guidelines for the Exchange of Information on Chemicals in International Trade and the

United Nations Codex Alimentarius Commission. The former seeks to implement procedures for ensuring that prior informed consent exists between countries buying and selling pesticides, while the latter seeks to create uniform standards for maximum levels of pesticide residues among participating countries. Both initiatives operate on a voluntary basis.

Pesticides safety education and pesticide applicator regulation are designed to protect the public from pesticide misuse, but do not eliminate all misuse. Reducing the use of pesticides and choosing less toxic pesticides may reduce risks placed on society and the environment from pesticide use. Integrated pest management, the use of multiple approaches to control pests, is becoming widespread and has been used with success in countries such as Indonesia, China, Bangladesh, the U.S., Australia, and Mexico. IPM attempts to recognize the more widespread impacts of an action on an ecosystem, so that natural balances are not upset. New pesticides are being developed, including biological and botanical derivatives and alternatives that are thought to reduce health and environmental risks. In addition, applicators are being encouraged to consider alternative controls and adopt methods that reduce the use of chemical pesticides.

Pesticides can be created that are targeted to a specific pest's lifecycle, which can be environmentally more friendly. For example, potato cyst nematodes emerge from their protective cysts in response to a chemical excreted by potatoes; they feed on the potatoes and damage the crop. A similar chemical can be applied to fields early, before the potatoes are planted, causing the nematodes to emerge early and starve in the absence of potatoes.

United States

Preparation for an application of hazardous herbicide in USA.

In the United States, the Environmental Protection Agency (EPA) is responsible for regulating pesticides under the Federal Insecticide, Fungicide, and Rodenticide Act (FIFRA) and the Food Quality Protection Act (FQPA). Studies must be conducted to establish the conditions in which the material is safe to use and the effectiveness against the intended pest(s). The EPA regulates pesticides to ensure that these products do not pose adverse effects to humans or the environment. Pesticides produced before November 1984 continue to be reassessed in order to meet the current scientific and regulatory standards. All registered pesticides are reviewed every 15 years to ensure they meet the proper standards. During the registration process, a label is created. The label contains directions for proper use of the material in addition to safety restrictions. Based on acute toxicity, pesticides are assigned to a Toxicity Class.

Some pesticides are considered too hazardous for sale to the general public and are designated restricted use pesticides. Only certified applicators, who have passed an exam, may purchase or supervise the application of restricted use pesticides. Records of sales and use are required to be maintained and may be audited by government agencies charged with the enforcement of pesticide regulations. These records must be made available to employees and state or territorial environmental regulatory agencies.

The EPA regulates pesticides under two main acts, both of which amended by the Food Quality Protection Act of 1996. In addition to the EPA, the United States Department of Agriculture (USDA) and the United States Food and Drug Administration (FDA) set standards for the level of pesticide residue that is allowed on or in crops. The EPA looks at what the potential human health and environmental effects might be associated with the use of the pesticide.

In addition, the U.S. EPA uses the National Research Council's four-step process for human health risk assessment: (1) Hazard Identification, (2) Dose-Response Assessment, (3) Exposure Assessment, and (4) Risk Characterization.

Recently Kaua'i County (Hawai'i) passed Bill No. 2491 to add an article to Chapter 22 of the county's code relating to pesticides and GMOs. The bill strengthens protections of local communities in Kaua'i where many large pesticide companies test their products.

History

Since before 2000 BC, humans have utilized pesticides to protect their crops. The first known pesticide was elemental sulfur dusting used in ancient Sumer about 4,500 years ago in ancient Mesopotamia. The Rig Veda, which is about 4,000 years old, mentions the use of poisonous plants for pest control. By the 15th century, toxic chemicals such as arsenic, mercury, and lead were being applied to crops to kill pests. In the 17th century, nicotine sulfate was extracted from tobacco leaves for use as an insecticide. The 19th century saw the introduction of two more natural pesticides, pyrethrum, which is derived from chrysanthemums, and rotenone, which is derived from the roots of tropical vegetables. Until the 1950s, arsenic-based pesticides were dominant. Paul Müller discovered that DDT was a very effective insecticide. Organochlorines such as DDT were dominant, but they were replaced in the U.S. by organophosphates and carbamates by 1975. Since then, pyrethrin compounds have become the dominant insecticide. Herbicides became common in the 1960s, led by "triazine and other nitrogen-based compounds, carboxylic acids such as 2,4-dichlorophenoxyacetic acid, and glyphosate".

The first legislation providing federal authority for regulating pesticides was enacted in 1910; however, decades later during the 1940s manufacturers began to produce large amounts of synthetic pesticides and their use became widespread. Some sources consider the 1940s and 1950s to have been the start of the "pesticide era." Although the U.S. Environmental Protection Agency was established in 1970 and amendments to the pesticide law in 1972, pesticide use has increased 50-fold since 1950 and 2.3 million tonnes (2.5 million short tons) of industrial pesticides are now used each year. Seventy-five percent of all pesticides in the world are used in developed countries, but use in developing countries is increasing. A study of USA pesticide use trends through 1997 was published in 2003 by the National Science Foundation's Center for Integrated Pest Management.

In the 1960s, it was discovered that DDT was preventing many fish-eating birds from reproducing, which was a serious threat to biodiversity. Rachel Carson wrote the best-selling book *Silent Spring* about biological magnification. The agricultural use of DDT is now banned under the Stockholm Convention on Persistent Organic Pollutants, but it is still used in some developing nations to prevent malaria and other tropical diseases by spraying on interior walls to kill or repel mosquitoes.

Insecticide

FLIT manual spray pump for insecticides from 1928

An insecticide is a substance used to kill insects. They include ovicides and larvicides used against insect eggs and larvae, respectively. Insecticides are used in agriculture, medicine, industry and by consumers. Insecticides are claimed to be a major factor behind the increase in agricultural 20th century's productivity. Nearly all insecticides have the potential to significantly alter ecosystems; many are toxic to humans; some concentrate along the food chain.

Insecticides can be classified in two major groups: systemic insecticides, which have residual or long term activity; and contact insecticides, which have no residual activity.

Furthermore, one can distinguish three types of insecticide. 1. Natural insecticides, such as nicotine, pyrethrum and neem extracts, made by plants as defenses against insects. 2. Inorganic insecticides, which are metals. 3. Organic insecticides, which are organic chemical compounds, mostly working by contact.

The mode of action describes how the pesticide kills or inactivates a pest. It provides another way of classifying insecticides. Mode of action is important in understanding whether an insecticide will be toxic to unrelated species, such as fish, birds and mammals.

Insecticides are distinct from insect repellents, which do not kill.

Type of Activity

Systemic insecticides become incorporated and distributed systemically throughout the whole plant. When insects feed on the plant, they ingest the insecticide. Systemic insecticides produced by transgenic plants are called plant-incorporated protectants (PIPs). For instance, a gene that codes for a specific Bacillus thuringiensis biocidal protein was introduced into corn and other species. The plant manufactures the protein, which kills the insect when consumed.

Contact insecticides are toxic to insects upon direct contact. These can be inorganic insecticides, which are metals and include arsenates, copper and fluorine compounds, which are less commonly used, and the commonly used sulfur. Contact insecticides can be organic insecticides, i.e. organic chemical compounds, synthetically produced, and comprising the largest numbers of pesticides used today. Or they can be natural compounds like pyrethrum, neem oil etc. Contact insecticides usually have no residual activity.

Efficacy can be related to the quality of pesticide application, with small droplets, such as aerosols often improving performance.

Biological Pesticides

Many organic compounds are produced by plants for the purpose of defending the host plant from predation. A trivial case is tree rosin, which is a natural insecticide. Specific, the production of oleoresin by conifer species is a component of the defense response against insect attack and fungal pathogen infection. Many fragrances, e.g. oil of wintergreen, are in fact antifeedants.

Four extracts of plants are in commercial use: pyrethrum, rotenone, neem oil, and various essential oils

Other Biological Approaches

Plant-incorporated Protectants

Transgenic crops that act as insecticides began in 1996 with a genetically modified potato that produced the Cry protein, derived from the bacterium Bacillus thuringiensis, which is toxic to beetle larvae such as the Colorado potato beetle. The technique has been expanded to include the use of RNA interference RNAi that fatally silences crucial insect genes. RNAi likely evolved as a defense against viruses. Midgut cells in many larvae take up the molecules and help spread the signal. The technology can target only insects that have the silenced sequence, as was demonstrated when a particular RNAi affected only one of four fruit fly species. The technique is expected to replace many other insecticides, which are losing effectiveness due to the spread of pesticide resistance.

Enzymes

Many plants exude substances to repel insects. Premier examples are substances activated by the enzyme myrosinase. This enzyme converts glucosinolates to various compounds that are toxic to herbivorous insects. One product of this enzyme is allyl isothiocyanate, the pungent ingredient in horseradish sauces.

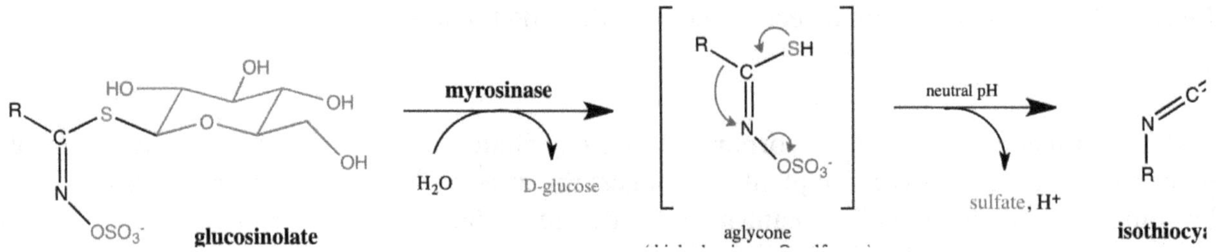

Biosynthesis of antifeedants by the action of myrosinase.

The myrosinase is released only upon crushing the flesh of horseradish. Since allyl isothiocyanate is harmful to the plant as well as the insect, it is stored in the harmless form of the glucosinolate, separate from the myrosinase enzyme.

Bacterial

Bacillus thuringiensis is a bacterial disease that affects Lepidopterans and some other insects. Toxins produced by strains of this bacterium are used as a larvicide against caterpillars, beetles, and mosquitoes. Toxins from *Saccharopolyspora spinosa* are isolated from fermentations and sold as Spinosad. Because these toxins have little effect on other organisms, they are considered more environmentally friendly than synthetic pesticides. The toxin from *B. thuringiensis* (Bt toxin) has been incorporated directly into plants through the use of genetic engineering. Other biological insecticides include products based on entomopathogenic fungi (e.g., *Beauveria bassiana, Metarhizium anisopliae*), nematodes (e.g., *Steinernema feltiae*) and viruses (e.g., *Cydia pomonella* granulovirus).

Synthetic Insecticide

A major emphasis of organic chemistry is the development of chemical tools to enhance agricultural productivity. Insecticides represent a major area of emphasis. Many of the major insecticides are inspired by biological analogues. Many others are completely alien to nature.

Organochlorides

The best known organochloride, DDT, was created by Swiss scientist Paul Müller. For this discovery, he was awarded the 1948 Nobel Prize for Physiology or Medicine. DDT was introduced in 1944. It functions by opening sodium channels in the insect's nerve cells. The contemporaneous rise of the chemical industry facilitated large-scale production of DDT and related chlorinated hydrocarbons.

Organophosphates and Carbamates

Organophosphates are another large class of contact insecticides. These also target the insect's nervous system. Organophosphates interfere with the enzymes acetylcholinesterase and other cholinesterases, disrupting nerve impulses and killing or disabling the insect. Organophosphate insecticides and chemical warfare nerve agents (such as sarin, tabun, soman, and VX) work in the same way. Organophosphates have a cumulative toxic effect to wildlife, so multiple exposures to the chemicals amplifies the toxicity. In the US, organophosphate use declined with the rise of substitutes.

Carbamate insecticides have similar mechanisms to organophosphates, but have a much shorter duration of action and are somewhat less toxic.

Pyrethroids

Pyrethroid pesticides mimic the insecticidal activity of the natural compound pyrethrum, the biopesticide found in pyrethrins. These compounds are nonpersistent sodium channel modulators and are less toxic than organophosphates and carbamates. Compounds in this group are often applied against household pests.

Neonicotinoids

Neonicotinoids are synthetic analogues of the natural insecticide nicotine (with much lower acute mammalian toxicity and greater field persistence). These chemicals are acetylcholine receptor agonists. They are broad-spectrum systemic insecticides, with rapid action (minutes-hours). They are applied as sprays, drenches, seed and soil treatments. Treated insects exhibit leg tremors, rapid wing motion, stylet withdrawal (aphids), disoriented movement, paralysis and death. Imidacloprid may be the most common. It has recently come under scrutiny for allegedly pernicious effects on honeybees and its potential to increase the susceptibility of rice to planthopper attacks.

Ryanoids

Ryanoids are synthetic analogues with the same mode of action as ryanodine, a naturally occurring insecticide extracted from *Ryania speciosa* (Flacourtiaceae). They bind to calcium channels in cardiac and skeletal muscle, blocking nerve transmission. Only one such insecticide is currently registered, Rynaxypyr, generic name chlorantraniliprole.

Insect Growth Regulators

Insect growth regulator (IGR) is a term coined to include insect hormone mimics and an earlier class of chemicals, the benzoylphenyl ureas, which inhibit chitin(exoskeleton) biosynthesis in insects. Diflubenzuron is a member of the latter class, used primarily to control caterpillars that are pests. The most successful insecticides in this class are the juvenoids (juvenile hormone analogues). Of these, methoprene is most widely used. It has no observable acute toxicity in rats and is approved by World Health Organization (WHO) for use in drinking water cisterns to combat malaria. Most of its uses are to combat insects where the adult is the pest, including mosquitoes, several fly species, and fleas. Two very similar products, hydroprene and kinoprene, are used for controlling species such as cockroaches and white flies. Methoprene was registered with the EPA in 1975. Virtually no reports of resistance have been filed. A more recent type of IGR is the ecdysone agonist tebufenozide (MIMIC), which is used in forestry and other applications for control of caterpillars, which are far more sensitive to its hormonal effects than other insect orders.

Environmental Effects

Effects on Nontarget Species

Some insecticides kill or harm other creatures in addition to those they are intended to kill. For example, birds may be poisoned when they eat food that was recently sprayed with insecticides or when they mistake an insecticide granule on the ground for food and eat it.

Sprayed insecticide may drift from the area to which it is applied and into wildlife areas, especially when it is sprayed aerially.

DDT

The development of DDT was motivated by desire to replace more dangerous or less effective alternatives. DDT was introduced to replace lead and arsenic-based compounds, which were in widespread use in the early 1940s.

DDT was brought to public attention by Rachel Carson's book *Silent Spring*. One side-effect of DDT is to reduce the thickness of shells on the eggs of predatory birds. The shells sometimes become too thin to be viable, reducing bird populations. This occurs with DDT and related compounds due to the process of bioaccumulation, wherein the chemical, due to its stability and fat solubility, accumulates in organisms' fatty tissues. Also, DDT may biomagnify, which causes progressively higher concentrations in the body fat of animals farther up the food chain. The near-worldwide ban on agricultural use of DDT and related chemicals has allowed some of these birds, such as the peregrine falcon, to recover in recent years. A number of organochlorine pesticides have been banned from most uses worldwide. Globally they are controlled via the Stockholm Convention on persistent organic pollutants. These include: aldrin, chlordane, DDT, dieldrin, endrin, heptachlor, mirex and toxaphene.

Pollinator Decline

Insecticides can kill bees and may be a cause of pollinator decline, the loss of bees that pollinate plants, and colony collapse disorder (CCD), in which worker bees from a beehive or Western honey bee colony abruptly disappear. Loss of pollinators means a reduction in crop yields. Sublethal doses of insecticides (i.e. imidacloprid and other neonicotinoids) affect bee foraging behavior. However, research into the causes of CCD was inconclusive as of June 2007.

Herbicide

Weeds controlled with herbicide

Herbicide(s), also commonly known as weedkillers, are chemical substances used to control unwanted plants. Selective herbicides control specific weed species, while leaving the desired crop relatively unharmed, while non-selective herbicides (sometimes called "total weedkillers" in commercial products) can be used to clear waste ground, industrial and construction sites, railways and railway embankments as they kill all plant material with which they come into contact. Apart from selective/non-selective, other important distinctions include *persistence* (also known as *residual action*: how long the product stays in place and remains active), *means of uptake* (whether it is absorbed by above-ground foliage only, through the roots, or by other means), and *mechanism of action* (how it works). Historically, products such as common salt and other metal salts were used as herbicides, however these have gradually fallen out of favor and in some countries a number of these are banned due to their persistence in soil, and toxicity and groundwater contamination concerns. Herbicides have also been used in warfare and conflict.

Modern herbicides are often synthetic mimics of natural plant hormones which interfere with growth of the target plants. The term organic herbicide has come to mean herbicides intended for organic farming; these are often less efficient and more costly than synthetic herbicides and are based on natural materials. Some plants also produce their own natural herbicides, such as the genus *Juglans* (walnuts), or the tree of heaven; such action of natural herbicides, and other related chemical interactions, is called allelopathy. Due to herbicide resistance - a major concern in agriculture - a number of products also combine herbicides with different means of action.

In the US in 2007, about 83% of all herbicide usage, determined by weight applied, was in agriculture. In 2007, world pesticide expenditures totaled about $39.4 billion; herbicides were about 40% of those sales and constituted the biggest portion, followed by insecticides, fungicides, and other types. Smaller quantities are used in forestry, pasture systems, and management of areas set aside as wildlife habitat.

History

Prior to the widespread use of chemical herbicides, cultural controls, such as altering soil pH, salinity, or fertility levels, were used to control weeds. Mechanical control (including tillage) was also (and still is) used to control weeds.

First Herbicides

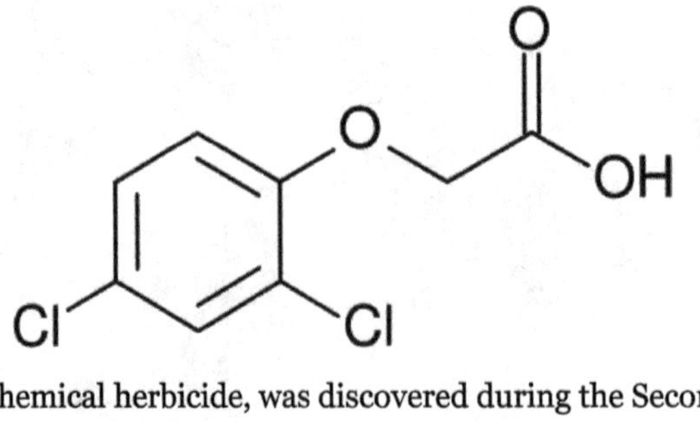

O
O
OH
Cl
Cl

2,4-D, the first chemical herbicide, was discovered during the Second World War.

Although research into chemical herbicides began in the early 20th century, the first major breakthrough was the result of research conducted in both the UK and the US during the Second World War into the potential use of agents as biological weapons. The first modern herbicide, 2,4-D, was first discovered and synthesized by W. G. Templeman at Imperial Chemical Industries. In 1940, he showed that "Growth substances applied appropriately would kill certain broad-leaved weeds in cereals without harming the crops." By 1941, his team succeeded in synthesizing the chemical. In the same year, Pokorny in the US achieved this as well.

Independently, a team under Juda Hirsch Quastel, working at the Rothamsted Experimental Station made the same discovery. Quastel was tasked by the Agricultural Research Council (ARC) to discover methods for improving crop yield. By analyzing soil as a dynamic system, rather than an inert substance, he was able to apply techniques such as perfusion. Quastel was able to quantify the influence of various plant hormones, inhibitors and other chemicals on the activity of microorganisms in the soil and assess their direct impact on plant growth. While the full work of the unit remained secret, certain discoveries were developed for commercial use after the war, including the 2,4-D compound.

When it was commercially released in 1946, it triggered a worldwide revolution in agricultural output and became the first successful selective herbicide. It allowed for greatly enhanced weed control in wheat, maize (corn), rice, and similar cereal grass crops, because it kills dicots (broadleaf plants), but not most monocots (grasses). The low cost of 2,4-D has led to continued usage today, and it remains one of the most commonly used herbicides in the world. Like other acid herbicides, current formulations use either an amine salt (often trimethylamine) or one of many esters of the parent compound. These are easier to handle than the acid.

Further Discoveries

The triazine family of herbicides, which includes atrazine, were introduced in the 1950s; they have the current distinction of being the herbicide family of greatest concern regarding groundwater contamination. Atrazine does not break down readily (within a few weeks) after being applied to soils of above neutral pH. Under alkaline soil conditions, atrazine may be carried into the soil profile as far as the water table by soil water following rainfall causing the aforementioned contamination. Atrazine is thus said to have “carryover”, a generally undesirable property for herbicides.

Glyphosate (Roundup) was introduced in 1974 for nonselective weed control. Following the development of glyphosate-resistant crop plants, it is now used very extensively for selective weed control in growing crops. The pairing of the herbicide with the resistant seed contributed to the consolidation of the seed and chemistry industry in the late 1990s.

Many modern chemical herbicides used in agriculture and gardening are specifically formulated to decompose within a short period after application. This is desirable, as it allows crops and plants to be planted afterwards, which could otherwise be affected by the herbicide. However, herbicides with low residual activity (i.e., that decompose quickly) often do not provide season-long weed control and do not ensure that weed roots are killed beneath construction and paving (and cannot emerge destructively in years to come), therefore there remains a role for weedkiller with high levels of persistence in the soil.

Terminology

Herbicides are classified/grouped in various ways e.g. according to the activity, timing of application, method of application, mechanism of action, chemical family. This gives rise to a considerable level of terminology related to herbicides and their use.

Intended Outcome

- Control is the destruction of unwanted weeds, or the damage of them to the point where they are no longer competitive with the crop.
- Suppression is incomplete control still providing some economic benefit, such as reduced competition with the crop.

- Crop safety, for selective herbicides, is the relative absence of damage or stress to the crop. Most selective herbicides cause some visible stress to crop plants.
- Defoliant, similar to herbicides, but designed to remove foliage (leaves) rather than kill the plant.

Selectivity (All Plants or Specific Plants)

- Selective herbicides: They control or suppress certain plants without affecting the growth of other plants species. Selectivity may be due to translocation, differential absorption, physical (morphological) or physiological differences between plant species. 2,4-D, mecoprop, dicamba control many broadleaf weeds but remain ineffective against turfgrasses.
- Non-selective herbicides: These herbicides are not specific in acting against certain plant species and control all plant material with which they come into contact. They are used to clear industrial sites, waste ground, railways and railway embankments. Paraquat, glufosinate, glyphosate are non-selective herbicides.

Timing of Application

- Preplant: Preplant herbicides are nonselective herbicides applied to soil before planting. Some preplant herbicides may be mechanically incorporated into the soil. The objective for incorporation is to prevent dissipation through photodecomposition and/or volatility. The herbicides kill weeds as they grow through the herbicide treated zone. Volatile herbicides have to be incorporated into the soil before planting the pasture. Agricultural crops grown in soil treated with a preplant herbicide include tomatoes, corn, soybeans and strawberries. Soil fumigants like metam-sodium and dazomet are in use as preplant herbicides.
- Preemergence: Preemergence herbicides are applied before the weed seedlings emerge through the soil surface. Herbicides do not prevent weeds from germinating but they kill weeds as they grow through the herbicide treated zone by affecting the cell division in the emerging seedling. Dithopyr and pendimethalin are preemergence herbicides. Weeds that have already emerged before application or activation are not affected by pre-herbicides as their primary growing point escapes the treatment.
- Postemergence: These herbicides are applied after weed seedlings have emerged through the soil surface. They can be foliar or root absorbed, selective or nonselective, contact or systemic. Application of these herbicides is avoided during rain because the problem of being washed off to the soil makes it ineffective. 2,4-D is a selective, systemic, foliar absorbed postemergence herbicide.

Method of Application

- Soil applied: Herbicides applied to the soil are usually taken up by the root or shoot of the emerging seedlings and are used as preplant or preemergence treatment. Several factors influence the effectiveness of soil-applied herbicides. Weeds absorb herbicides by both passive and active mechanism. Herbicide adsorption to soil colloids or organic matter often reduces its amount available for weed absorption. Positioning of herbicide in correct layer of soil is very important, which can be achieved mechanically and by rainfall. Herbicides on the soil surface are subjected to several processes that reduce their availability. Volatility and photolysis are two common processes that reduce the availability of herbicides. Many soil applied herbicides are absorbed through plant shoots while they are still underground leading to their death or injury. EPTC and trifluralin are soil applied herbicides.

- Foliar applied: These are applied to portion of the plant above the ground and are absorbed by exposed tissues. These are generally postemergence herbicides and can either be translocated (systemic) throughout the plant or remain at specific site (contact). External barriers of plants like cuticle, waxes, cell wall etc. affect herbicide absorption and action. Glyphosate, 2,4-D and dicamba are foliar applied herbicide.

Persistence

- Residual activity: A herbicide is described as having low residual activity if it is neutralized within a short time of application (within a few weeks or months) - typically this is due to rainfall, or by reactions in the soil. A herbicide described as having high residual activity will remains potent for a long term in the soil. For some compounds, the residual activity can leave the ground almost permanently barren.

Mechanism of Action

Herbicides are often classified according to their site of action, because as a general rule, herbicides within the same site of action class will produce similar symptoms on susceptible plants. Classification based on site of action of herbicide is comparatively better as herbicide resistance management can be handled more properly and effectively. Classification by mechanism of action (MOA) indicates the first enzyme, protein, or biochemical step affected in the plant following application.

List of Mechanisms Found in Modern Herbicides

- ACCase inhibitors compounds kill grasses. Acetyl coenzyme A carboxylase (ACCase) is part of the first step of lipid synthesis. Thus, ACCase inhibitors affect cell membrane production in the meristems of the grass plant. The ACCases of grasses are sensitive to these herbicides, whereas the ACCases of dicot plants are not.

- ALS inhibitors: the acetolactate synthase (ALS) enzyme (also known as acetohydroxyacid synthase, or AHAS) is the first step in the synthesis of the branched-chain amino acids (valine, leucine, and isoleucine). These herbicides slowly starve affected plants of these amino acids, which eventually leads to inhibition of DNA synthesis. They affect grasses and dicots alike. The ALS inhibitor family includes various sulfonylureas (such as Flazasulfuron and Metsulfuron-methyl), imidazolinones, triazolopyrimidines, pyrimidinyl oxybenzoates, and sulfonylamino carbonyl triazolinones. The ALS biological pathway exists only in plants and not animals, thus making the ALS-inhibitors among the safest herbicides.

- EPSPS inhibitors: The enolpyruvylshikimate 3-phosphate synthase enzyme EPSPS is used in the synthesis of the amino acids tryptophan, phenylalanine and tyrosine. They affect grasses and dicots alike. Glyphosate (Roundup) is a systemic EPSPS inhibitor inactivated by soil contact.

- Synthetic auxins inaugurated the era of organic herbicides. They were discovered in the 1940s after a long study of the plant growth regulator auxin. Synthetic auxins mimic this plant hormone. They have several points of action on the cell membrane, and are effective in the control of dicot plants. 2,4-D is a synthetic auxin herbicide.

- Photosystem II inhibitors reduce electron flow from water to NADPH2+ at the photochemical step in photosynthesis. They bind to the Qb site on the D1 protein, and prevent quinone from binding to this site. Therefore, this group of compounds causes electrons to accumulate on chlorophyll molecules. As a consequence, oxidation reactions in excess of those normally tolerated by the cell occur, and the plant dies. The triazine herbicides (including atrazine) and urea derivatives (diuron) are photosystem II inhibitors.

- Photosystem I inhibitors steal electrons from the normal pathway through FeS to Fdx to NADP leading to direct discharge of electrons on oxygen. As a result, reactive oxygen species are produced and oxidation reactions in excess of those normally tolerated by the cell occur, leading to plant death. Bipyridinium herbicides (such as diquat and paraquat) inhibit the Fe-S – Fdx step of that chain, while diphenyl ether herbicides (such as nitrofen, nitrofluorfen, and acifluorfen) inhibit the Fdx – NADP step.

- HPPD inhibitors inhibit 4-Hydroxyphenylpyruvate dioxygenase, which are involved in tyrosine breakdown. Tyrosine breakdown products are used by plants to make carotenoids, which protect chlorophyll in plants from being destroyed by sunlight. If this happens, the plants turn white due to complete loss of chlorophyll, and the plants die. Mesotrione and sulcotrione are herbicides in this class; a drug, nitisinone, was discovered in the course of developing this class of herbicides.

Herbicide Group (Labeling)

One of the most important methods for preventing, delaying, or managing resistance is to reduce the reliance on a single herbicide mode of action. To do this, farmers must know the mode of action for the herbicides they intend to use, but the relatively complex nature of plant biochemistry makes this difficult to determine. Attempts were made to simplify the understanding of herbicide mode of action by developing a classification system that grouped herbicides by mode of action. Eventually the Herbicide Resistance Action Committee (HRAC) and the Weed Science Society of America (WSSA) developed a classification system. The WSSA and HRAC systems differ in the group designation. Groups in the WSSA and the HRAC systems are designated by numbers and letters, respectively. The goal for adding the "Group" classification and mode of action to the herbicide product label is to provide a simple and practical approach to deliver the information to users. This information will make it easier to develop educational material that is consistent and effective. It should increase user's awareness of herbicide mode of action and provide more accurate recommendations for resistance management. Another goal is to make it easier for users to keep records on which herbicide mode of actions are being used on a particular field from year to year.

Chemical Family

Detailed investigations on chemical structure of the active ingredients of the registered herbicides showed that some moieties (moiety is a part of a molecule that may include either whole functional groups or parts of functional groups as substructures; a functional group has similar chemical properties whenever it occurs in different compounds) have the same mechanisms of action. According to Forouzesh *et al.* 2015, these moieties have been assigned to the names of chemical families and active ingredients are then classified within the chemical families accordingly. Knowing about herbicide chemical family grouping could serve as a short-term strategy for managing resistance to site of action.

Use and Application

Most herbicides are applied as water-based sprays using ground equipment. Ground equipment varies in design, but large areas can be sprayed using self-propelled sprayers equipped with long booms, of 60 to 120 feet (18 to 37 m) with spray nozzles spaced every 20–30 inches (510–760 mm) apart. Towed, handheld, and even horse-drawn sprayers are also used. On large areas, herbicides may also at times be applied aerially using helicopters or airplanes, or through irrigation systems (known as chemigation).

Herbicides being sprayed from the spray arms of a tractor in North Dakota.

A further method of herbicide application developed around 2010, involves ridding the soil of its active weed seed bank rather than just killing the weed. This can successfully treat annual plants but not perennials. Researchers at the Agricultural Research Service found that the application of herbicides to fields late in the weeds' growing season greatly reduces their seed production, and therefore fewer weeds will return the following season. Because most weeds are annuals, their seeds will only survive in soil for a year or two, so this method will be able to destroy such weeds after a few years of herbicide application.

Weed-wiping may also be used, where a wick wetted with herbicide is suspended from a boom and dragged or rolled across the tops of the taller weed plants. This allows treatment of taller grassland weeds by direct contact without affecting related but desirable shorter plants in the grassland sward beneath. The method has the benefit of avoiding spray drift. In Wales, a scheme offering free weed-wiper hire was launched in 2015 in an effort to reduce the levels of MCPA in water courses.

Misuse and Misapplication

Herbicide volatilisation or spray drift may result in herbicide affecting neighboring fields or plants, particularly in windy conditions. Sometimes, the wrong field or plants may be sprayed due to error.

Use Politically, Militarily, and in Conflict

Health and Environmental Effects

Herbicides have widely variable toxicity in addition to acute toxicity from occupational exposure levels.

Some herbicides cause a range of health effects ranging from skin rashes to death. The pathway of attack can arise from intentional or unintentional direct consumption, improper application resulting in the herbicide coming into direct contact with people or wildlife, inhalation of aerial sprays, or food consumption prior to the labeled preharvest interval. Under some conditions, certain herbicides can be transported via leaching or surface runoff to contaminate groundwater or distant surface water sources. Generally, the conditions that promote herbicide transport include intense storm events (particularly shortly after application) and soils with limited capacity to adsorb or retain the herbicides. Herbicide properties that increase likelihood of transport include persistence (resistance to degradation) and high water solubility.

Phenoxy herbicides are often contaminated with dioxins such as TCDD research has suggested such contamination results in a small rise in cancer risk after occupational exposure to these herbicides. Triazine exposure has been implicated in a likely relationship to increased risk of breast cancer, although a causal relationship remains unclear.

Herbicide manufacturers have at times made false or misleading claims about the safety of their products. Chemical manufacturer Monsanto Company agreed to change its advertising after pressure from New York attorney general Dennis Vacco; Vacco complained about misleading claims that its spray-on glyphosate-based herbicides, including Roundup, were safer than table salt and "practically non-toxic" to mammals, birds, and fish (though proof that this was ever said is hard to find). Roundup is toxic and has resulted in death after being ingested in quantities ranging from 85 to 200 ml, although it has also been ingested in quantities as large as 500 ml with only mild or moderate symptoms. The manufacturer of Tordon 101 (Dow AgroSciences, owned by the Dow Chemical Company) has claimed Tordon 101 has no effects on animals and insects, in spite of evidence of strong carcinogenic activity of the active ingredient Picloram in studies on rats.

The risk of Parkinson's disease has been shown to increase with occupational exposure to herbicides and pesticides. The herbicide paraquat is suspected to be one such factor.

All commercially sold, organic and nonorganic herbicides must be extensively tested prior to approval for sale and labeling by the Environmental Protection Agency. However, because of the large number of herbicides in use, concern regarding health effects is significant. In addition to health effects caused by herbicides themselves, commercial herbicide mixtures often contain other chemicals, including inactive ingredients, which have negative impacts on human health.

Ecological Effects

Commercial herbicide use generally has negative impacts on bird populations, although the impacts are highly variable and often require field studies to predict accurately. Laboratory studies have at times overestimated negative impacts on birds due to toxicity, predicting serious problems that were not observed in the field. Most observed effects are due not to toxicity, but to habitat changes and the decreases in abundance of species on which birds rely for food or shelter. Herbicide use in silviculture, used to favor certain types of growth following clearcutting, can cause significant drops in bird populations. Even when herbicides which have low toxicity to birds are used, they decrease the abundance of many types of vegetation on which the birds rely. Herbicide use in agriculture in Britain has been linked to a decline in seed-eating bird species which rely on the weeds killed by the herbicides. Heavy use of herbicides in neotropical agricultural areas has been one of many factors implicated in limiting the usefulness of such agricultural land for wintering migratory birds.

Frog populations may be affected negatively by the use of herbicides as well. While some studies have shown that atrazine may be a teratogen, causing demasculinization in male frogs, the U.S. Environmental Protection Agency (EPA) and its independent Scientific Advisory Panel (SAP) examined all available studies on this topic and concluded that "atrazine does not adversely affect amphibian gonadal development based on a review of laboratory and field studies."

Scientific Uncertainty of Full Extent of Herbicide Effects

The health and environmental effects of many herbicides is unknown, and even the scientific community often disagrees on the risk. For example, a 1995 panel of 13 scientists reviewing studies on the carcinogenicity of 2,4-D had divided opinions on the likelihood 2,4-D causes cancer in humans. As of 1992, studies on phenoxy herbicides were too few to accurately assess the risk of many types of cancer from these herbicides, even though evidence was stronger that exposure to these herbicides is associated with increased risk of soft tissue sarcoma and non-Hodgkin lymphoma. Furthermore, there is some suggestion that herbicides can play a role in sex reversal of certain organisms that experience temperature-dependent sex determination, which could theoretically alter sex ratios.

Resistance

Weed resistance to herbicides has become a major concern in crop production worldwide. Resistance to herbicides is often attributed to lack of rotational programmes of herbicides and to continuous applications of herbicides with the same sites of action. Thus, a true understanding of the sites of action of herbicides is essential for strategic planning of herbicide-based weed control.

Plants have developed resistance to atrazine and to ALS-inhibitors, and more recently, to glyphosate herbicides. Marestail is one weed that has developed glyphosate resistance. Glyphosate-resistant weeds are present in the vast majority of soybean, cotton and corn farms in some U.S. states. Weeds that can resist multiple other herbicides are spreading. Few new herbicides are near commercialization, and none with a molecular mode of action for which there is no resistance. Because most herbicides could not kill all weeds, farmers rotated crops and herbicides to stop resistant weeds. During its initial years, glyphosate was not subject to resistance and allowed farmers to reduce the use of rotation.

A family of weeds that includes waterhemp (Amaranthus rudis) is the largest concern. A 2008-9 survey of 144 populations of waterhemp in 41 Missouri counties revealed glyphosate resistance in 69%. Weeds from some 500 sites throughout Iowa in 2011 and 2012 revealed glyphosate resistance in approximately 64% of waterhemp samples. The use of other killers to target "residual" weeds has become common, and may be sufficient to have stopped the spread of resistance From 2005 through 2010 researchers discovered 13 different weed species that had developed resistance to glyphosate. But since then only two more have been discovered. Weeds resistant to multiple herbicides with completely different biological action modes are on the rise. In Missouri, 43% of samples were resistant to two different herbicides; 6% resisted three; and 0.5% resisted four. In Iowa 89% of waterhemp samples resist two or more herbicides, 25% resist three, and 10% resist five.

For southern cotton, herbicide costs has climbed from between $50 and $75 per hectare a few years ago to about $370 per hectare in 2013. Resistance is contributing to a massive shift away from growing cotton; over the past few years, the area planted with cotton has declined by 70% in Arkansas and by 60% in Tennessee. For soybeans in Illinois, costs have risen from about $25 to $160 per hectare.

Dow, Bayer CropScience, Syngenta and Monsanto are all developing seed varieties resistant to herbicides other than glyphosate, which will make it easier for farmers to use alternative weed killers. Even though weeds have already evolved some resistance to those herbicides, Powles says the new seed-and-herbicide combos should work well if used with proper rotation.

Biochemistry of Resistance

Resistance to herbicides can be based on one of the following biochemical mechanisms:

- Target-site resistance: This is due to a reduced (or even lost) ability of the herbicide to bind to its target protein. The effect usually relates to an enzyme with a crucial function in a metabolic pathway, or to a component of an electron-transport system. Target-site resistance may also be caused by an overexpression of the target enzyme (via gene amplification or changes in a gene promoter).
- Non-target-site resistance: This is caused by mechanisms that reduce the amount of herbicidal active compound reaching the target site. One important mechanism is an enhanced metabolic detoxification of the herbicide in the weed, which leads to insufficient amounts of the active substance reaching the target site. A reduced uptake and translocation, or sequestration of the herbicide, may also result in an insufficient herbicide transport to the target site.
- Cross-resistance: In this case, a single resistance mechanism causes resistance to several herbicides. The term target-site cross-resistance is used when the herbicides bind to the same target site, whereas non-target-site cross-resistance is due to a single non-target-site mechanism (e.g., enhanced metabolic detoxification) that entails resistance across herbicides with different sites of action.
- Multiple resistance: In this situation, two or more resistance mechanisms are present within individual plants, or within a plant population.

Resistance Management

Worldwide experience has been that farmers tend to do little to prevent herbicide resistance developing, and only take action when it is a problem on their own farm or neighbor's. Careful observation is important so that any reduction in herbicide efficacy can be detected. This may indicate evolving resistance. It is vital that resistance is detected at an early stage as if it becomes an acute, whole-farm problem, options are more limited and greater expense is almost inevitable. Table 1 lists factors which enable the risk of resistance to be assessed. An essential pre-requisite for confirmation of resistance is a good diagnostic test. Ideally this should be rapid, accurate, cheap and accessible. Many diagnostic tests have been developed, including glasshouse pot assays, petri dish assays and chlorophyll fluorescence. A key component of such tests is that the response of the suspect population to a herbicide can be compared with that of known susceptible and resistant standards under controlled conditions. Most cases of herbicide resistance are a consequence of the repeated use of herbicides, often in association with crop monoculture and reduced cultivation practices. It is necessary, therefore, to modify these practices in order to prevent or delay the onset of resistance or to control existing resistant populations. A key objective should be the reduction in selection pressure. An integrated weed management (IWM) approach is required, in which as many tactics as possible are used to combat weeds. In this way, less reliance is placed on herbicides and so selection pressure should be reduced.

Optimising herbicide input to the economic threshold level should avoid the unnecessary use of herbicides and reduce selection pressure. Herbicides should be used to their greatest potential by ensuring that the timing, dose, application method, soil and climatic conditions are optimal for good activity. In the UK, partially resistant grass weeds such as *Alopecurus myosuroides* (blackgrass) and *Avena* spp. (wild oat) can often be controlled adequately when herbicides are applied at the 2-3 leaf stage, whereas later applications at the 2-3 tiller stage can fail badly. Patch spraying, or applying herbicide to only the badly infested areas of fields, is another means of reducing total herbicide use.

Table 1. Agronomic factors influencing the risk of herbicide resistance development

Factor	Low risk	High risk
Cropping system	Good rotation	Crop monoculture
Cultivation system	Annual ploughing	Continuous minimum tillage
Weed control	Cultural only	Herbicide only
Herbicide use	Many modes of action	Single modes of action
Control in previous years	Excellent	Poor
Weed infestation	Low	High
Resistance in vicinity	Unknown	Common

Approaches to Treating Resistant Weeds

Alternative Herbicides

When resistance is first suspected or confirmed, the efficacy of alternatives is likely to be the first consideration. The use of alternative herbicides which remain effective on resistant populations can be a successful strategy, at least in the short term. The effectiveness of alternative herbicides will be highly dependent on the extent of cross-resistance. If there is resistance to a single group of herbicides, then the use of herbicides from other groups may provide a simple and effective solution, at least in the short term. For example, many triazine-resistant weeds have been readily controlled by the use of alternative herbicides such as dicamba or glyphosate. If resistance extends to more than one herbicide group, then choices are more limited. It should not be assumed that resistance will automatically extend to all herbicides with the same mode of action, although it is wise to assume this until proved otherwise. In many weeds the degree of cross-resistance between the five groups of ALS inhibitors varies considerably. Much will depend on the resistance mechanisms present, and it should not be assumed that these will necessarily be the same in different populations of the same species. These differences are due, at least in part, to the existence of different mutations conferring target site resistance. Consequently, selection for different mutations may result in different patterns of cross-resistance. Enhanced metabolism can affect even closely related herbicides to differing degrees. For example, populations of *Alopecurus myosuroides* (blackgrass) with an enhanced metabolism mechanism show resistance to pendimethalin but not to trifluralin, despite both being dinitroanilines. This is due to differences in the vulnerability of these two herbicides to oxidative metabolism. Consequently, care is needed when trying to predict the efficacy of alternative herbicides.

Mixtures and Sequences

The use of two or more herbicides which have differing modes of action can reduce the selection for resistant genotypes. Ideally, each component in a mixture should:

- Be active at different target sites
- Have a high level of efficacy
- Be detoxified by different biochemical pathways
- Have similar persistence in the soil (if it is a residual herbicide)
- Exert negative cross-resistance
- Synergise the activity of the other component

No mixture is likely to have all these attributes, but the first two listed are the most important. There is a risk that mixtures will select for resistance to both components in the longer term. One practical advantage of sequences of two herbicides compared with mixtures is that a better appraisal of the efficacy of each herbicide component is possible, provided that sufficient time elapses between each application. A disadvantage with sequences is that two separate applications have to be made and it is possible that the later application will be less effective on weeds surviving the first application. If these are resistant, then the second herbicide in the sequence may increase selection for resistant individuals by killing the susceptible plants which were damaged but not killed by the first application, but allowing the larger, less affected, resistant plants to survive. This has been cited as one reason why ALS-resistant *Stellaria media* has evolved in Scotland recently (2000), despite the regular use of a sequence incorporating mecoprop, a herbicide with a different mode of action.

Herbicide Rotations

Rotation of herbicides from different chemical groups in successive years should reduce selection for resistance. This is a key element in most resistance prevention programmes. The value of this approach depends on the extent of cross-resistance, and whether multiple resistance occurs owing to the presence of several different resistance mechanisms. A practical problem can be the lack of awareness by farmers of the different groups of herbicides that exist. In Australia a scheme has been introduced in which identifying letters are included on the product label as a means of enabling farmers to distinguish products with different modes of action.

Farming Practices and Resistance: A Case Study

Herbicide resistance became a critical problem in Australian agriculture, after many Australian sheep farmers began to exclusively grow wheat in their pastures in the 1970s. Introduced varieties of ryegrass, while good for grazing sheep, compete intensely with wheat. Ryegrasses produce so many seeds that, if left unchecked, they can completely choke a field. Herbicides provided excellent control, while reducing soil disrupting because of less need to plough. Within little more than a decade, ryegrass and other weeds began to develop resistance. In response Australian farmers changed methods. By 1983, patches of ryegrass had become immune to Hoegrass, a family of herbicides that inhibit an enzyme called acetyl coenzyme A carboxylase.

Ryegrass populations were large, and had substantial genetic diversity, because farmers had planted many varieties. Ryegrass is cross-pollinated by wind, so genes shuffle frequently. To control its distribution farmers sprayed inexpensive Hoegrass, creating selection pressure. In addition, farmers sometimes diluted the herbicide in order to save money, which allowed some plants to survive application. When resistance appeared farmers turned to a group of herbicides that block acetolactate synthase. Once again, ryegrass in Australia evolved a kind of "cross-resistance" that allowed it to rapidly break down a variety of herbicides. Four classes of herbicides become ineffective within a few years. In 2013 only two herbicide classes, called Photosystem II and long-chain fatty acid inhibitors, were effective against ryegrass.

List of Common Herbicides

Synthetic Herbicides

- 2,4-D is a broadleaf herbicide in the phenoxy group used in turf and no-till field crop production. Now, it is mainly used in a blend with other herbicides to allow lower rates of herbicides to be used; it is the most widely used herbicide in the world, and third most commonly used in the United States. It is an example of synthetic auxin (plant hormone).
- Aminopyralid is a broadleaf herbicide in the pyridine group, used to control weeds on grassland, such as docks, thistles and nettles. It is notorious for its ability to persist in compost.
- Atrazine, a triazine herbicide, is used in corn and sorghum for control of broadleaf weeds and grasses. Still used because of its low cost and because it works well on a broad spectrum of weeds common in the US corn belt, atrazine is commonly used with other herbicides to reduce the overall rate of atrazine and to lower the potential for groundwater contamination; it is a photosystem II inhibitor.
- Clopyralid is a broadleaf herbicide in the pyridine group, used mainly in turf, rangeland, and for control of noxious thistles. Notorious for its ability to persist in compost, it is an-

other example of synthetic auxin.

- Dicamba, a postemergent broadleaf herbicide with some soil activity, is used on turf and field corn. It is another example of a synthetic auxin.
- Glufosinate ammonium, a broad-spectrum contact herbicide, is used to control weeds after the crop emerges or for total vegetation control on land not used for cultivation.
- Fluazifop (Fuselade Forte), a post emergence, foliar absorbed, translocated grass-selective herbicide with little residual action. It is used on a very wide range of broad leaved crops for control of annual and perennial grasses.
- Fluroxypyr, a systemic, selective herbicide, is used for the control of broad-leaved weeds in small grain cereals, maize, pastures, rangeland and turf. It is a synthetic auxin. In cereal growing, fluroxypyr's key importance is control of cleavers, *Galium aparine*. Other key broadleaf weeds are also controlled.
- Glyphosate, a systemic nonselective herbicide, is used in no-till burndown and for weed control in crops genetically modified to resist its effects. It is an example of an EPSPs inhibitor.
- Imazapyr a nonselective herbicide, is used for the control of a broad range of weeds, including terrestrial annual and perennial grasses and broadleaf herbs, woody species, and riparian and emergent aquatic species.
- Imazapic, a selective herbicide for both the pre- and postemergent control of some annual and perennial grasses and some broadleaf weeds, kills plants by inhibiting the production of branched chain amino acids (valine, leucine, and isoleucine), which are necessary for protein synthesis and cell growth.
- Imazamox, an imidazolinone manufactured by BASF for postemergence application that is an acetolactate synthase (ALS) inhibitor. Sold under trade names Raptor, Beyond, and Clearcast.
- Linuron is a nonselective herbicide used in the control of grasses and broadleaf weeds. It works by inhibiting photosynthesis.
- MCPA (2-methyl-4-chlorophenoxyacetic acid) is a phenoxy herbicide selective for broadleaf plants and widely used in cereals and pasture.
- Metolachlor is a pre-emergent herbicide widely used for control of annual grasses in corn and sorghum; it has displaced some of the atrazine in these uses.
- Paraquat is a nonselective contact herbicide used for no-till burndown and in aerial destruction of marijuana and coca plantings. It is more acutely toxic to people than any other herbicide in widespread commercial use.
- Pendimethalin, a pre-emergent herbicide, is widely used to control annual grasses and some broad-leaf weeds in a wide range of crops, including corn, soybeans, wheat, cotton, many tree and vine crops, and many turfgrass species.

- Picloram, a pyridine herbicide, mainly is used to control unwanted trees in pastures and edges of fields. It is another synthetic auxin.
- Sodium chlorate *(disused/banned in some countries)*, a nonselective herbicide, is considered phytotoxic to all green plant parts. It can also kill through root absorption.
- Triclopyr, a systemic, foliar herbicide in the pyridine group, is used to control broadleaf weeds while leaving grasses and conifers unaffected.
- Several sulfonylureas, including Flazasulfuron and Metsulfuron-methyl, which act as ALS inhibitors and in some cases are taken up from the soil via the roots.

Organic Herbicides

Recently, the term "organic" has come to imply products used in organic farming. Under this definition, an organic herbicide is one that can be used in a farming enterprise that has been classified as organic. Commercially sold organic herbicides are expensive and may not be affordable for commercial farming. Depending on the application, they may be less effective than synthetic herbicides and are generally used along with cultural and mechanical weed control practices.

Homemade organic herbicides include:

- Corn gluten meal (CGM) is a natural pre-emergence weed control used in turfgrass, which reduces germination of many broadleaf and grass weeds.
- Vinegar is effective for 5–20% solutions of acetic acid, with higher concentrations most effective, but it mainly destroys surface growth, so respraying to treat regrowth is needed. Resistant plants generally succumb when weakened by respraying.
- Steam has been applied commercially, but is now considered uneconomical and inadequate. It controls surface growth but not underground growth and so respraying to treat regrowth of perennials is needed.
- Flame is considered more effective than steam, but suffers from the same difficulties.
- D-limonene (citrus oil) is a natural degreasing agent that strips the waxy skin or cuticle from weeds, causing dehydration and ultimately death.
- Saltwater or salt applied in appropriate strengths to the rootzone will kill most plants.
- Monocerin produced by certain fungi will kill certain weeds such as Johnson grass.

Of Historical Interest and Other

- 2,4,5-Trichlorophenoxyacetic acid (2,4,5-T) was a widely used broadleaf herbicide until being phased out starting in the late 1970s. While 2,4,5-T itself is of only moderate toxicity, the manufacturing process for 2,4,5-T contaminates this chemical with trace amounts of 2,3,7,8-tetrachlorodibenzo-p-dioxin (TCDD). TCDD is extremely toxic to humans. With proper temperature control during production of 2,4,5-T, TCDD levels can be held to about .005 ppm. Before the TCDD risk was well understood, early production facilities lacked proper temperature controls. Individual batches tested later were found to have as much as 60 ppm of TCDD. 2,4,5-T was withdrawn from use in the USA in 1983, at a time of heightened public sensitivity about chemical hazards in the environment. Public concern about dioxins was high, and production and use of other (non-herbicide) chemicals potentially containing TCDD contamination was also withdrawn. These included pentachlorophenol (a wood preservative) and PCBs (mainly used as stabilizing agents in transformer oil). Some feel that the 2,4,5-T withdrawal was not based on sound science. 2,4,5-T has since largely been replaced by dicamba and triclopyr.

- Agent Orange was a herbicide blend used by the British military during the Malayan Emergency and the U.S. military during the Vietnam War between January 1965 and April 1970 as a defoliant. It was a 50/50 mixture of the *n*-butyl esters of 2,4,5-T and 2,4-D. Because of TCDD contamination in the 2,4,5-T component, it has been blamed for serious illnesses in many people who were exposed to it. However, research on populations exposed to its dioxin contaminant have been inconsistent and inconclusive.
- Diesel, and other heavy oil derivatives, are known to be informally used at times, but are usually banned for this purpose.

Fungicide

Fungicides are biocidal chemical compounds or biological organisms used to kill fungi or fungal spores. A fungistatic inhibits their growth. Fungi can cause serious damage in agriculture, resulting in critical losses of yield, quality, and profit. Fungicides are used both in agriculture and to fight fungal infections in animals. Chemicals used to control oomycetes, which are not fungi, are also referred to as fungicides, as oomycetes use the same mechanisms as fungi to infect plants.

Fungicides can either be contact, translaminar or systemic. Contact fungicides are not taken up into the plant tissue and protect only the plant where the spray is deposited. Translaminar fungicides redistribute the fungicide from the upper, sprayed leaf surface to the lower, unsprayed surface. Systemic fungicides are taken up and redistributed through the xylem vessels. Few fungicides move to all parts of a plant. Some are locally systemic, and some move upwardly.

Most fungicides that can be bought retail are sold in a liquid form. A very common active ingredient is sulfur, present at 0.08% in weaker concentrates, and as high as 0.5% for more potent fungicides. Fungicides in powdered form are usually around 90% sulfur and are very toxic. Other active ingredients in fungicides include neem oil, rosemary oil, jojoba oil, the bacterium *Bacillus subtilis*, and the beneficial fungus *Ulocladium oudemansii.*

Fungicide residues have been found on food for human consumption, mostly from post-harvest treatments. Some fungicides are dangerous to human health, such as vinclozolin, which has now been removed from use. Ziram is also a fungicide that is thought to be toxic to humans if exposed to chronically. A number of fungicides are also used in human health care.

Natural Fungicides

Plants and other organisms have chemical defenses that give them an advantage against microorganisms such as fungi. Some of these compounds can be used as fungicides:

- Tea tree oil
- Cinnamaldehyde
- Citronella oil
- Jojoba oil
- Nimbin
- Oregano oil
- Rosemary oil
- Monocerin
- Milk

Whole live or dead organisms that are efficient at killing or inhibiting fungi can sometimes be used as fungicides:

- *Bacillus subtilis*
- *Ulocladium oudemansii*
- Kelp (powdered dried kelp is fed to cattle to help prevent fungal infection)
- *Ampelomyces quisqualis*

Resistance

Pathogens respond to the use of fungicides by evolving resistance. In the field several mechanisms of resistance have been identified. The evolution of fungicide resistance can be gradual or sudden. In qualitative or discrete resistance, a mutation (normally to a single gene) produces a race of a fungus with a high degree of resistance. Such resistant varieties also tend to show stability, persisting after the fungicide has been removed from the market. For example, sugar beet leaf blotch remains resistant to azoles years after they were no longer used for control of the disease. This is because such mutations often have a high selection pressure when the fungicide is used, but there is low selection pressure to remove them in the absence of the fungicide.

In instances where resistance occurs more gradually, a shift in sensitivity in the pathogen to the fungicide can be seen. Such resistance is polygenic – an accumulation of many mutations in different genes, each having a small additive effect. This type of resistance is known as quantitative or continuous resistance. In this kind of resistance, the pathogen population will revert to a sensitive state if the fungicide is no longer applied.

Little is known about how variations in fungicide treatment affect the selection pressure to evolve resistance to that fungicide. Evidence shows that the doses that provide the most control of the disease also provide the largest selection pressure to acquire resistance, and that lower doses decrease the selection pressure.

In some cases when a pathogen evolves resistance to one fungicide, it automatically obtains resistance to others – a phenomenon known as cross resistance. These additional fungicides are normally of the same chemical family or have the same mode of action, or can be detoxified by the same mechanism. Sometimes negative cross resistance occurs, where resistance to one chemical class of fungicides leads to an increase in sensitivity to a different chemical class of fungicides. This has been seen with carbendazim and diethofencarb.

There are also recorded incidences of the evolution of multiple drug resistance by pathogens – resistance to two chemically different fungicides by separate mutation events. For example, *Botrytis cinerea* is resistant to both azoles and dicarboximide fungicides.

There are several routes by which pathogens can evolve fungicide resistance. The most common mechanism appears to be alteration of the target site, in particular as a defence against single site of action fungicides. For example, Black Sigatoka, an economically important pathogen of banana, is resistant to the QoI fungicides, due to a single nucleotide change resulting in the replacement of one amino acid (glycine) by another (alanine) in the target protein of the QoI fungicides, cytochrome b. It is presumed that this disrupts the binding of the fungicide to the protein, rendering the fungicide ineffective. Upregulation of target genes can also render the fungicide ineffective. This is seen in DMI-resistant strains of *Venturia inaequalis.*

Resistance to fungicides can also be developed by efficient efflux of the fungicide out of the cell. *Septoria tritici* has developed multiple drug resistance using this mechanism. The pathogen had 5 ABC-type transporters with overlapping substrate specificities that together work to pump toxic chemicals out of the cell.

In addition to the mechanisms outlined above, fungi may also develop metabolic pathways that circumvent the target protein, or acquire enzymes that enable metabolism of the fungicide to a harmless substance.

Fungicide Resistance Management

The fungicide resistance action committee (FRAC) has several recommended practices to try to avoid the development of fungicide resistance, especially in at-risk fungicides including *Strobilurins* such as azoxystrobin.

Products should not be used in isolation, but rather as mixture, or alternate sprays, with another fungicide with a different mechanism of action. The likelihood of the pathogen's developing resistance is greatly decreased by the fact that any resistant isolates to one fungicide will be killed by the other; in other words, two mutations would be required rather than just one. The effectiveness of this technique can be demonstrated by Metalaxyl, a phenylamide fungicide. When used as the sole product in Ireland to control potato blight (*Phytophthora infestans*), resistance developed within one growing season. However, in countries like the UK where it was marketed only as a mixture, resistance problems developed more slowly.

Fungicides should be applied only when absolutely necessary, especially if they are in an at-risk group. Lowering the amount of fungicide in the environment lowers the selection pressure for resistance to develop.

Manufacturers' doses should always be followed.These doses are normally designed to give the right balance between controlling the disease and limiting the risk of resistance development. Higher doses increase the selection pressure for single-site mutations that confer resistance, as all strains but those that carry the mutation will be eliminated, and thus the resistant strain will propagate. Lower doses greatly increase the risk of polygenic resistance, as strains that are slightly less sensitive to the fungicide may survive.

It is also recommended that where possible fungicides are used only in a protective manner, rather than to try to cure already-infected crops. Far fewer fungicides have curative/eradicative ability than protectant. Thus, fungicide preparations advertised as having curative action may have only one active chemical; a single fungicide acting in isolation increases the risk of fungicide resistance.

It is better to use an integrative pest management approach to disease control rather than relying on fungicides alone.This involves the use of resistant varieties and hygienic practices, such as the removal of potato discard piles and stubble on which the pathogen can overwinter, greatly reducing the titre of the pathogen and thus the risk of fungicide resistance development.

Bactericide

A bactericide or bacteriocide, sometimes abbreviated Bcidal, is a substance that kills bacteria. Bactericides are disinfectants, antiseptics, or antibiotics.

Bactericidal Disinfectants

The most used disinfectants are those applying

- active chlorine (i.e., hypochlorites, chloramines, dichloroisocyanurate and trichloroisocyanurate, wet chlorine, chlorine dioxide, etc.),
- active oxygen (peroxides, such as peracetic acid, potassium persulfate, sodium perborate, sodium percarbonate, and urea perhydrate),
- iodine (povidone-iodine, Lugol's solution, iodine tincture, iodinated nonionic surfactants),
- concentrated alcohols (mainly ethanol, 1-propanol, called also n-propanol and 2-propanol, called isopropanol and mixtures thereof; further, 2-phenoxyethanol and 1- and 2-phenoxypropanols are used),
- phenolic substances (such as phenol (also called "carbolic acid"), cresols such as thymol, halogenated (chlorinated, brominated) phenols, such as hexachlorophene, triclosan, trichlorophenol, tribromophenol, pentachlorophenol, salts and isomers thereof),
- cationic surfactants, such as some quaternary ammonium cations (such as benzalko-

nium chloride, cetyl trimethylammonium bromide or chloride, didecyldimethylammonium chloride, cetylpyridinium chloride, benzethonium chloride) and others, non-quaternary compounds, such as chlorhexidine, glucoprotamine, octenidine dihydrochloride etc.),

- strong oxidizers, such as ozone and permanganate solutions;
- heavy metals and their salts, such as colloidal silver, silver nitrate, mercury chloride, phenylmercury salts, copper sulfate, copper oxide-chloride etc. Heavy metals and their salts are the most toxic and environment-hazardous bactericides and therefore their use is strongly discouraged or prohibited
- strong acids (phosphoric, nitric, sulfuric, amidosulfuric, toluenesulfonic acids), pH < 1, and
- alkalis (sodium, potassium, calcium hydroxides), such as of pH > 13, particularly under elevated temperature (above 60 °C), kills bacteria.

Bactericidal Antiseptics

As antiseptics (i.e., germicide agents that can be used on human or animal body, skin, mucoses, wounds and the like), few of the above-mentioned disinfectants can be used, under proper conditions (mainly concentration, pH, temperature and toxicity toward humans and animals). Among them, some important are

- properly diluted chlorine preparations (f.e. Dakin's solution, 0.5% sodium or potassium hypochlorite solution, pH-adjusted to pH 7 – 8, or 0.5 – 1% solution of sodium benzenesulfochloramide (chloramine B)), some
- iodine preparations, such as iodopovidone in various galenics (ointment, solutions, wound plasters), in the past also Lugol's solution,
- peroxides such as urea perhydrate solutions and pH-buffered 0.1 – 0.25% peracetic acid solutions,
- alcohols with or without antiseptic additives, used mainly for skin antisepsis,
- weak organic acids such as sorbic acid, benzoic acid, lactic acid and salicylic acid
- some phenolic compounds, such as hexachlorophene, triclosan and Dibromol, and
- cationic surfactants, such as 0.05 – 0.5% benzalkonium, 0.5 – 4% chlorhexidine, 0.1 – 2% octenidine solutions.

Others are generally not applicable as safe antiseptics, either because of their corrosive or toxic nature.

Bactericidal Antibiotics

Bactericidal antibiotics kill bacteria; bacteriostatic antibiotics slow their growth or reproduction.

Antibiotics that inhibit cell wall synthesis: the Beta-lactam antibiotics (penicillin derivatives (penams), cephalosporins (cephems), monobactams, and carbapenems) and vancomycin.

Also bactericidal are daptomycin, fluoroquinolones, metronidazole, nitrofurantoin, co-trimoxazole, telithromycin.

Aminoglycosidic antibiotics are usually considered bactericidal, although they may be bacteriostatic with some organisms

The distinction between bactericidal and bacteriostatic agents appears to be clear according to the basic/clinical definition, but this only applies under strict laboratory conditions and it is important to distinguish microbiological and clinical definitions. The distinction is more arbitrary when agents are categorized in clinical situations. The supposed superiority of bactericidal agents over bacteriostatic agents is of little relevance when treating the vast majority of infections with gram-positive bacteria, particularly in patients with uncomplicated infections and noncompromised immune systems. Bacteriostatic agents have been effectively used for treatment that are considered to require bactericidal activity. Furthermore, some broad classes of antibacterial agents considered bacteriostatic can exhibit bactericidal activity against some bacteria on the basis of in vitro determination of MBC/MIC values. At high concentrations, bacteriostatic agents are often bactericidal against some susceptible organisms. The ultimate guide to treatment of any infection must be clinical outcome.

References

- Francis Borgio J, Sahayaraj K and Alper Susurluk I (eds) . Microbial Insecticides: Principles and Applications, Nova Publishers, USA. 492pp. ISBN 978-1-61209-223-2
- Miller, GT (2002). Living in the Environment (12th Ed.). Belmont: Wadsworth/Thomson Learning. ISBN 0-534-37697-5
- Metcalf, Robert L. (2002). "Ullmann's Encyclopedia of Industrial Chemistry". Ullmann's Encyclopedia of Industrial Chemistry. Wiley-VCH. doi:10.1002/14356007.a14_263. ISBN 3527306730. |chapter= ignored (help)
- Quastel, J. H. (1950). "2,4-Dichlorophenoxyacetic Acid (2,4-D) as a Selective Herbicide". Agricultural Control Chemicals. Advances in Chemistry. 1. p. 244. doi:10.1021/ba-1950-0001.ch045. ISBN 0-8412-2442-0.
- Modern crop protection compounds (2., rev. and enl. ed. ed.). Weinheim: Wiley-VCH-Verl. 2012. pp. 197–276. ISBN 978-3-527-32965-6. |first1= missing |last1= in Authors list (help)
- Smith (18 July 1995). "8: Fate of herbicides in the environment". Handbook of Weed Management Systems. CRC Press. pp. 245–278. ISBN 978-0-8247-9547-4.
- Powles, S. B.; Shaner, D. L., eds. (2001). Herbicide Resistance and World Grains. CRC Press, Boca Raton, FL. p. 328. ISBN 9781420039085.
- Moss, S. R. (2002). "Herbicide-Resistant Weeds". In Naylor,, R. E. L. Weed management handbook (9th ed.). Blackwell Science Ltd. pp. 225–252. ISBN 0-632-05732-7.
- Vats, S. (2015). "Herbicides: history, classification and genetic manipulation of plants for herbicide resistance". In Lichtfouse, E. Sustainable Agriculture Reviews 15. Springer International Publishing. pp. 153–192.
- "CDC - Pesticide Illness & Injury Surveillance - NIOSH Workplace Safety and Health Topic". Cdc.gov. 2013-09-11. Retrieved 2014-01-28.
- "Pesticides 101 - A primer on pesticides, their use in agriculture and the exposure we face | Pesticide Action Network". Panna.org. Retrieved 2014-01-28.
- IUPAC (2006). "Glossary of terms relating to Pesticides" (PDF). IUPAC. p. 2123. Retrieved Jan-uary 2014. Check date values in: |access-date= (help)

Permissions

All chapters in this book are published with permission under the Creative Commons Attribution Share Alike License or equivalent. Every chapter published in this book has been scrutinized by our experts. Their significance has been extensively debated. The topics covered herein carry significant information for a comprehensive understanding. They may even be implemented as practical applications or may be referred to as a beginning point for further studies.

We would like to thank the editorial team for lending their expertise to make the book truly unique. They have played a crucial role in the development of this book. Without their invaluable contributions this book wouldn't have been possible. They have made vital efforts to compile up to date information on the varied aspects of this subject to make this book a valuable addition to the collection of many professionals and students.

This book was conceptualized with the vision of imparting up-to-date and integrated information in this field. To ensure the same, a matchless editorial board was set up. Every individual on the board went through rigorous rounds of assessment to prove their worth. After which they invested a large part of their time researching and compiling the most relevant data for our readers.

The editorial board has been involved in producing this book since its inception. They have spent rigorous hours researching and exploring the diverse topics which have resulted in the successful publishing of this book. They have passed on their knowledge of decades through this book. To expedite this challenging task, the publisher supported the team at every step. A small team of assistant editors was also appointed to further simplify the editing procedure and attain best results for the readers.

Apart from the editorial board, the designing team has also invested a significant amount of their time in understanding the subject and creating the most relevant covers. They scrutinized every image to scout for the most suitable representation of the subject and create an appropriate cover for the book.

The publishing team has been an ardent support to the editorial, designing and production team. Their endless efforts to recruit the best for this project, has resulted in the accomplishment of this book. They are a veteran in the field of academics and their pool of knowledge is as vast as their experience in printing. Their expertise and guidance has proved useful at every step. Their uncompromising quality standards have made this book an exceptional effort. Their encouragement from time to time has been an inspiration for everyone.

The publisher and the editorial board hope that this book will prove to be a valuable piece of knowledge for students, practitioners and scholars across the globe.

Index

www.ingramcontent.com/pod-product-compliance
Lightning Source LLC
Jackson TN
JSHW052205130125
77033JS00004B/213

9781635490510